STUDIES IN
PLANT SURVIVAL

*Ecological case histories of
plant adaptation to adversity*

STUDIES IN ECOLOGY

GENERAL EDITORS

D.J. ANDERSON BSc, PhD
Department of Botany
University of New South Wales
Sydney

P. GREIG-SMITH MA, ScD
School of Plant Biology
University College of New South Wales
Bangor

and

FRANK A. PITELKA PhD
Department of Zoology
University of California
Berkeley

STUDIES IN ECOLOGY VOLUME 11

STUDIES IN PLANT SURVIVAL

*Ecological case histories of
plant adaptation to adversity*

RMM CRAWFORD

BSc DNatSci (Liège) FRSE
Professor of Plant Ecology
University of St Andrews

OXFORD

BLACKWELL SCIENTIFIC PUBLICATIONS

LONDON EDINBURGH BOSTON

MELBOURNE PARIS BERLIN VIENNA

© 1989 by
Blackwell Scientific Publications
Editorial Offices:
Osney Mead, Oxford OX2 0EL
25 John Street, London WC1N 2BL
23 Ainslie Place, Edinburgh EH3 6AJ
238 Main Street, Cambridge
    Massachusetts 02142, USA
54 University Street, Carlton
    Victoria 3053, Australia

Other Editorial Offices:
Librairie Arnette SA
1, rue de Lille
75007 Paris
France

Blackwell Wissenschafts-Verlag GmbH
Kurfürstendamm 57
10707 Berlin
Germany

Blackwell MZV
Feldgasse 13
1238 Wien
Austria

First published 1989
Reprinted 1990, 1994

Set by Setrite Typesetters, Hong Kong
Printed and bound in Great Britain
at the Alden Press Limited
Oxford and Northampton

DISTRIBUTORS

Marston Book Services Ltd
PO Box 87
Oxford OX2 0DT
(*Orders*: Tel: 0865 791155
         Fax: 0865 791927
         Telex: 837515)

USA
Blackwell Scientific Publications, Inc.
238 Main Street,
Cambridge, MA 02142
(*Orders*: Tel: 800 759-6102
         617 876-7000)

Canada
Times Mirror Professional Publishing, Ltd
130 Flaska Drive
Markham, Ontario L6G 1B8
(*Orders*: Tel: 800 268-4178
         416 470-6739)

Australia
Blackwell Scientific Publications Pty Ltd
54 University Street
Carlton, Victoria 3053
(*Orders*: Tel: 03 347-5552)

A catalogue record for this title is available
from the British Library

ISBN 0-632-01475-X
ISBN 0-632-01477-6 Pbk

Library of Congress
Cataloging-in-Publication Data

Crawford, R. M. M.
    Studies in plant survival
    1. Environment, Adaption of plants
    I. Title II. Series
    581.5

Contents

Preface

The purpose of the book is to examine how plants function and survive under different types of adversity. These case studies attempt to present within an ecological framework a comparison of the adaptations of plants to environmental limitations, ranging from the Arctic to the anaerobic muds of lake margins and shaded habitats of the forest floor. The polluted regions of our modern industrial countryside are also included.

Survival is not dependent on physiology alone and as full a discussion as possible is given of other relevant features which aid survival, such as adaptations in reproduction, morphology and timing of various biological activities. The study of Plant Biology and Ecology is a constant alternation between field observation and laboratory experimentation. Numerous photographs have therefore been included to remind us that whatever aspect of the biology of a species may be under study in the laboratory, it is but one piece in the total sum of adaptations that are necessary to ensure survival in the ever-changing conditions in the field.

Much of the stimulus for attempting this comparative study of plant survival has come from opportunities to observe plants in many different parts of the world from Arctic-Spitsbergen to the flooded forests on the banks of the Amazon and the great Southern Beech forests of Patagonia. I am therefore particularly grateful to the Carnegie Trust for the Universities of Scotland, the Leverhulme Foundation and the Royal Society for support which made these visits possible. My botanical colleagues in St Andrews and Drs Jean Balfour and John Grace have individually read different chapters and have provided much valuable guidance. Time spent in laboratories overseas has brought me in contact with stimulating ecological situations and my special thanks are due therefore to Dr R. Braendle (Bern), Professor D.D. Hook (Charleston), Professor O. Kandler (Munich), Dr I. Mendelssohn (Baton Rouge), Dr A. Pradet (Bordeaux), and Professor B.B. Vartapetian (Moscow). My wife and son have been my constant companions on many field-outings and their support and enthusiasm for exploration has enormously increased my awareness of the natural world and provided the incentive to finish this book.

R.M.M. Crawford
St Andrews

Fig. 1.1 Plant survival in an exposed site in the Norwegian Jotunheimen Mountains. The mountain cudweed (*Gnaphalium norvegicum*) flowering in late summer snow at 1300 m.

1
Investigating survival

1.1 Introduction

There is scarcely any habitat in the world that is too hostile to support plant life provided some moisture and light is available. The capacity for adaptation in green plants is so great that there are species capable of surviving on the summits of mountains in the far north (Fig. 1.1), while others can live in the deep shade of the forest floor or perpetually submerged several metres below the surface of freshwater lakes. Some plants even grow where it never rains as they can extract enough moisture to complete their life cycles from fog and dew. This adaptability of plants has been exploited by man for his own ends and crops are now grown from the equator to the Arctic circle. Forestry spreads over an even greater latitudinal range with trees being extracted from the natural forests of the tropics as well as in regions well within the Arctic. The evolutionary adaptability of plants is so great that there is nearly always a species or ecotype that is capable of surviving in every possible ecological niche. The survival capacity of plants also compels attention for its enormous practical importance. The greater the tolerance of plants to climatic extremes the less likely the occurrence of crop failures and famine. For this reason the impact of natural hazards on plant survival has probably been a subject of interest to man ever since the beginnings of agriculture.

Crop varieties and the need for selection of suitable strains was already a well-developed subject in the classical writings of the Roman agriculturists Cato, Varro, and Columella. Even earlier in Greece, Theophrastus (late fourth century BC) in his *Enquiry into Plants* noted the need for the reservation of plants of prime quality for seed selection. More precisely he recorded that varieties from lands with severe winters were late in coming into ear and destroyed by drought unless rain late in the season saved them. Roman farmers also experimented with imported seed and were aware that spring wheat from Sicily was a failure when used further north (White, 1970).

In more modern times the study of survival in species and populations has provided much of the subject matter of ecology. The development of ecology as a separate biological discipline took place in the wake of Darwin's theory of natural selection and to some extent reflected the need for a scientific quantification of survival as a measure of relative fitness. Ecological field methods employ both the study of numbers and distribution in living organisms. The ability to survive can be estimated quantitatively in any population by recording changes in abundance with the lapse of time. Recently the numerical study of populations has stimulated much research, particularly with animals where discrete individuals lend themselves to ready counting (Begon and Mortimer, 1981). With plants, numbers are not always readily obtainable and demographic studies often have to be limited either to seed production or else to species which do not reproduce vegetatively.

Plants however lend themselves more readily than animals to the study of distribution. The static nature of most plants facilitates the process of recording as only a few species have the ability to change their location while growing. The water soldier (*Stratiotes aloides*) which migrates from the bottom of ponds where it spends the winter to the surface in order to flower and the South American desert bromeliad *Tillandsia latifolia* which passes its life being blown about the Atacama Desert (Fig.1.2) both enjoy a degree of mobility that is rare in higher plants. Most species remain rooted to the spot from germination until death. Comparisons of differences in local distribution between plants, whether it be in terms of species or phenotypes can be matched against environmental variables. Some degree of caution is needed in interpreting the causes of distribution patterns as limits to distribution may not always reflect the survival capacity of the plants. This is particularly evident on the wider biogeographical scale where distribution is frequently limited by geographical barriers to dispersal. The development of vicarious species, (for definition *see* Table 1.4) which replace one another in disjunct geographical areas, can give a false impression of the range of habitat that would be potentially open to any one species if its close relative had not evolved to fill the same niche (Table 1.1). Continental drift can move whole floras so that the boundary between contrasting types is not related in any way to modern environ-

Fig. 1.2 *Tillandsia latifolia* (Bromeliaceae) which lives without rooting in the soil in the Atacama Desert of Peru and northern Chile and moves about the desert as blown by the wind.

mental conditions but is due entirely to historical events. Such a case can be seen in the Wallace line (*see* Chapter 2). The success with which plant introductions have been made clearly demonstrates that for many species it is dispersal limitations rather than survival capacity that restrict their distribution. However for most successful introductions it is usual for there to be a reasonable degree of similarity in environmental conditions between the place of origin and the newly colonized habitat.

The distribution of plants is affected therefore by the sum total of historical, geographical, and environmental factors. Whether or not a species is capable of maintaining its distribution over a certain area is just as fundamental an ecological question as the study of abundance and the control of populations.

An essential element of this survival is the ability to tolerate the local environmental conditions. A direct consequence of the stationary behaviour of plant populations is the selection of locally adapted forms to a greater extent than is found in animals. Differences in micrometeorology and soil conditions can result in population differentiation in particular habitats. Habitat specialization arises also as a result of the biological characteristics of the habitat. As evolution proceeds varying genotypes will arise which will compete with each other either inter- or intra-specifically for survival. As will become apparent in the following chapters, much of this habitat specialization is achieved by optimizing the biotype in the face of competition under one particular set of conditions.

North America	Europe−Western Asia	Eastern Asia
Acer nigrum	*A. pseudoplatanus*	*A. cappadocicum*
Aesculus flava	*A. hippocastanum*	*A. indica*
Betula nigra	*B. pubescens*	*B. utilis*
Fagus grandifolia	*F. sylvatica*	*F. longipetiolata*
Juglans nigra	*J. regia*	*J. mandshurica*
Pinus contorta	*P. sylvestris*	*P. armandii*
Populus tremuloides	*P. tremula*	*P. yunnanensis*
Quercus phellos	*Q. petraea*	*Q. myrsinifolia*
Salix amygdaloides	*S. alba*	*S. babylonica*
Sorbus americana	*S. aucuparia*	*S. pohuashanensis*
Taxus canadensis	*T. baccata*	*T. cuspidata*
Ulmus rubra	*U. glabra*	*U. pumila*

Table 1.1 Examples of closely-related and ecologically-equivalent species (i.e. *vicarious* species) of trees of North American, European−West Asiatic and East Asiatic distribution. For each genus only three species are listed and a more exhaustive study could match up many more species.

1.2 **Adaptation**

Adaptation has been a concept that has run through biology ever since Jean Baptiste de Lamarck (1744–1829) made his first attempts to throw doubt on the theory of special creation. In simple terms, adaptation can be defined as the possession of properties which increase the probability of survival of a genotype in a particular habitat. With this definition we are not concerned therefore with any characteristic that an individual may have developed during its lifetime. Adaptation has to be heritable in order to have any lasting and predictable effect in increasing the survival capacity of a biotype. Sometimes the use of the term adaptation can lead to confusion when it is not clear whether or not it relates to the fitness of the genotype. The term has been used to describe the process of adjustment of an individual to an environmental stress. For this process it is better to use the term acclimatization in which it is only phenotypic adjustment that takes place in response to change in environmental conditions (*see* Table 1.4). Animal ecologists also consider that certain aspects of behaviour which are learnt and therefore not genetic can have a survival value (Clutton-Brock and Harvey, 1979).

Two different outlooks have persisted side by side in most studies concerned with the role of adaptation in increasing the survival capacity of organisms. Darwin in his *Origin of Species* (1859) strongly emphasized the competitive interaction between members of the same species which resulted from the effects of population pressure. His contemporary, Wallace, regarded the struggle for existence as taking place mainly against the elements. This latter approach is still a motive force in many ecophysiological studies where the relative tolerance of species or populations is compared in relation to minor alterations in soil and climatic conditions. A common feature in many older ecological studies was the description of species in terms of their habitat preferences as defined in purely physical terms. Table 1.2 lists a number of ecological terms used in describing the habitat preferences of plants. As well as classifying the plants on a positive or negative basis in relation to factors such as salt, calcium, temperature, and light there is a supplementary terminology which defines the degree of tolerance as being either narrow or broad using the Greek prefixes 'steno' and 'eury' to define these respective properties.

The limitations of these definitions of adaptation based purely on tolerance of individual environmental stresses neglects the multitude of factors that produce

Table 1.2 A selection of ecological terms used in describing habitat preferences in plants.

Aerophyte	An epiphyte growing on a terrestrial plant and lacking direct contact with air or water.
Anemophile	A plant occurring in wind-swept situations.
Calcicole	A plant occurring in soil rich in calcium.
Calcifuge	A plant which is intolerant of soils rich in calcium.
Glycophyte	A plant able to thrive in soils of low salt concentration, usually less than 0.5% NaCl.
Halophyte	A plant which tolerates or thrives in saline or alkaline soils.
Nitrophilous	The ability of plants to exploit soils rich in nitrogenous compounds.
Photophilic	The ability to live in conditions of full light.
Phreatophyte	A plant that has roots deep enough to extract water from the permanent water-table.
Rheophyte	A plant inhabiting running water.
Saprophyte	A plant which obtains nutrients from dead or decaying living matter.
Xerophyte	A plant which survives in dry habitats.

The terms may be modified by addition of the Greek prefixes steno (narrow) and eury (broad) to describe the range of ecological tolerance of a species, for example euryhaline and stenohaline describe plants that are respectively tolerant of broad and narrow ranges of saline conditions (*see* Chapter 6).

the many ecotypes (local races) that are such a characteristic feature of higher plants. The sea plantain (*Plantago maritima*, Fig. 1.3) is a particularly good example as it is well known for its ability to produce a wide range of varying biotypes which coexist within a few hundred metres of each other. Fig. 1.4 shows the distribution of hairy and less hairy ecotypes of *Plantago maritima* on a cliff top in the Orkney Islands off the north coast of Scotland. These cliff tops are drenched regularly with salt spray by the frequent gales that occur in this region. The sea plantain (*Plantago maritima*) is one of the most characteristic species of such exposed cliff top habitats. At the seaward edge of the transect, at the cliff edge the extremely hairy form (Fig. 1.4) is found with particular abundance relative to the other forms. Due to the frequency of salt drenching from north Atlantic gales all the plants that occur anywhere

Fig. 1.3 Close up view (scale divisions = 10 mm) of the scape and leaves of the extreme hairy form of sea plantain (*Plantago maritima*) as found in cliff top vegetation at Birsay, Orkney.

near the cliff top have to be salt tolerant. The hairy form of the sea plantain occurs only with any regularity near the cliff top where it shares the habitat with the other forms. Both forms survive side by side in this exposed site. The hairy form however does not persist with any appreciable frequency in sites away from the cliff top. The reduced level of competition allows the sea plantain to maintain this particular variant at the cliff top but greater uniformity is imposed on the more stable and competitive inland sites.

Another pioneering species which maintains different forms within distances that should permit interbreeding is the mountain avens (*Dryas octopetala*). This species is circumpolar and in its Alaskan sites two distinct subspecies are found namely *D.*

octopetala ssp. *octopetala* which is a circumpolar form with small deciduous leaves and commonly lives for more than 100 years. It is found on exposed ridges where it often forms almost pure stands. The second type *D. octopetala* ssp. *alaskensis* (Fig. 1.5) probably evolved from ssp. *octopetala* during the pleistocene period in Alaska (Hulten, 1959) and has large evergreen leaves. This form can be found in several vegetation types but is most frequent in snowbeds. In an experimental study of the factors which maintain genetic differentiation between these forms McGraw and Antonovics (1983) were able to show that although there were no breeding barriers to hybridization the difference between them in flowering time reduced the chance of inter-pollination. More than 99 per cent of pollen movement was

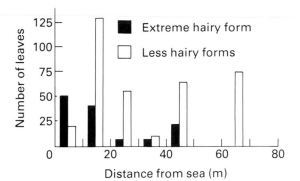

Fig. 1.4 Distribution of the extreme hairy form of the sea plantain (*Plantago maritima*) in relation to other forms as found on an exposed transect of cliff top vegetation at Birsay, Orkney. The proportions of the different ecotypes were scored by sampling along the transect with a point quadrat.

between flowers of the same ecotype. By using reciprocal transplant experiments and assessing survival at different stages in the life cycle it was possible to show which stages differentiated the relative success of the two ecotypes in their respective habitats. Strong selection was observed against fellfield ecotype seedlings in the snow-bed and against adult snow-bed plants in the fellfield. However seed germination was higher and seedling survival lower at the fellfield sites than the snow-bed for both ecotypes. Therefore the 'at home' site was not always the better. However the characteristics of the snow-bed site with better shelter from winter frost desiccation and a superior water supply during the growing

season supports adult plants that possess larger shoots which produce more flowers and therefore more seed and thus ensures the separate existence of the two subspecies in the same geographical area (Fig. 1.6).

From these two examples it can be seen that the existence of different subspecies or ecotypes can arise either on the basis of site selection as in the case of *Dryas octopetala* or else through competition as in the case of *Plantago maritima*. There can be therefore both biotic and environmental factors in the nature of adaptation. There is a clear parallel between the phenomenon of genetic polymorphism within species and the structure of communities in that the existence of genetic polymorphism represents the stable coexistence of genotypes while the persistence of communities is due to the stable coexistence of species (Clarke, 1979). Once any two genotypes become sufficiently distinct to be regulated independently they could be treated ecologically as two non-competing species.

1.3 Survival in a changing environment

In an unchanging environment it is conceivable that evolution could reach a stable end-point with the production of populations that were fully compatible with the physical elements of the environment. Such adaptation has been described as immediate fitness. However as soon as a population has to face competition, parasitism, or predation, survival depends on

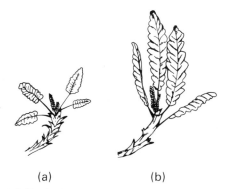

Fig. 1.5 Single shoots from the two ecotypes of *Dryas octopetala* which inhabit (a) the fellfield sites (ssp. *octopetala*) and (b) the snow-bed sites (ssp. *alaskensis*). (Reproduced with permission from McGraw and Antonovics, 1983.)

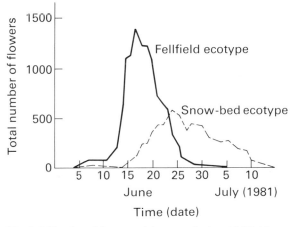

Fig. 1.6 Number of flowers of the snow-bed and fellfield ecotypes of *Dryas octopetala* produced in a total of 357 2 × 2 m quadrats in 1981 near Eagle Summit, Alaska. (Reproduced with permission from McGraw and Antonovics, 1983.)

continual adaptation. Similarly fluctuating environments will keep populations in a never-ending state of flux as different biotypes are favoured in turn by changing climatic differences. The frequency of different alleles for certain alcohol dehydrogenase isozymes was found to differ in wild sunflower populations depending on whether they had experienced a run of wet or dry years (Fig. 1.7) (Torres and Diedenhofen, 1979). This continual process of evolutionary change necessary to maintain a presence in a habitat has been described as the 'Red Queen' behaviour where constant running is needed just to stay in the same place (Maynard Smith, 1976).

Evolutionary forces, although varied, have one property in common, they tend to constrain and narrow the range of activities of organisms even

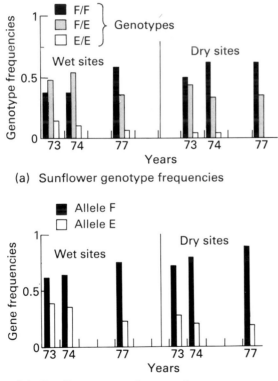

(a) Sunflower genotype frequencies

(b) Sunflower gene frequencies

Fig. 1.7 (a) Genotype frequencies and (b) allele frequencies at the Adh-1 locus in populations of sunflower as recorded from 1973 to 1977 by Torres and Diedenhofen (1979). Allele frequencies were calculated from the corresponding genotype frequencies. Note the lower frequencies of allele E in the dry sites and its decline in the wet site after a number of years of reduced flooding stress from 1974 to 1977.

although they may increase the variation between them (Harper, 1982). With time, therefore, in any habitat specialization should increase with the progressive reduction in the tolerance of species or biotypes to unaccustomed environmental change.

An example of this progressive habitat specialization with time has been very clearly demonstrated in the lesser spearwort (*Ranunculus flammula*). This widespread temperate species occurs both in Europe and North America at the edges of ponds and lakes where it exploits an amphibious habitat being able to produce both terrestrial and aquatic leaves. The species is described as heterophyllous as the leaves produced in water are linear and differ in form from the lanceolate leaves that are produced when the plants are not submerged. By examining the degree of morphological difference between aerial and aquatic leaves in populations from different habitats, Cook and Johnson (1968) assessed the degree of heterophylly in relation to habitat stability and age. They found that there was no predominant successful 'jack-of-all-trades' genotype. Instead there was a continuum of genotypes from those that were highly adaptable and which occurred over a broad ecogeographic spectrum to others which were more specialized to local conditions. The adaptable phenotypes with the greatest degree of heterophylly (Fig. 1.8) were found in transient habitats such as dune slacks and lakes susceptible to unpredictable fluctuations in water-table. The soils were young and poor, the ground open and the competing vegetation sparse. The specialized plants grew under more predictable conditions either in permanently aquatic or terrestrial environments. The soils were richer and the plants survived in competition in close communities with diverse species. These specialized populations which had the minimum development of heterophylly also had the least capacity to survive changes in their accustomed habitat as tested by submerging the terrestrial populations and exposing the aquatic ones to prolonged desiccation. There also appeared to be a suggestion that the specialized populations were demonstrably older. One population of an adaptable type came from a lake which was no more than 57 years old while the most specialized populations were collected from the margins of glacial lakes where the ice had retreated about 10 000 years ago.

The *Ranunculus flammula* study provides a simple illustration of how time and natural selection can bring about the evolution of genotypes with ever decreasing tolerance of environmental variation.

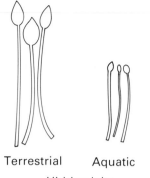

Terrestrial   Aquatic

Hidden lake

Terrestrial   Aquatic

Cleawox lake

Fig. 1.8 Leaf silhouettes from plants of *Ranunculus flammula* showing variation in blade width in terrestrial and aquatic leaves. Hidden lake had moderate to large temperature changes (altitude 1009 m) and had a fluctuating water-level. Both aquatic and terrestrial leaves in this lake develop a lamina. Cleawox lake was a coastal lake with small temperature changes (altitude 24 m) and had a stable water-level. Here all plants are uniformly narrow leaved. (Adapted from Cook and Johnson, 1968.)

This progressive process of moving towards a greater degree of habitat specialization suggests one reason why some species become extinct. Normally it is not a decrease in viability which causes species to die out but a disappearance of the accustomed habitat. It might even be said that adaptation is the first step on the road to extinction.

1.4 Environmental and physiological tolerance

Terms such as adaptation and tolerance have to be used in ecology with great care. An adapted species, as described above, means one which by natural selection has gained a high chance of survival in a particular habitat. Tolerance will describe more accurately the range of environmental factors that an individual organism or population may endure and

still survive. In assaying tolerance it would be misleading to measure ecological success in terms which reflect human exploitation of plants such as seed yield, or harvestable biomass. Survival is the only true measure of success. For individuals this can be quantified in terms of fitness which can be measured in their genetic contribution to future generations. Whatever weight or number of seeds a plant requires to ensure its successful reproduction is dependent on its own particular ecological circumstances. Thus comparisons of productivity between species or ecotypes may give no indication of relative survival chances and its ability to perpetuate itself in a given habitat.

A quantitative means of measuring tolerance of a biotype can be obtained by means of the 'environmental discrimination coefficient' used by McGraw and Antonovics (1983) in their study of *Dryas octopetala* ecotypes in Alaska, defined as:

$$a = 1 - \frac{b}{c},$$

where a = environmental discrimination coefficient
b = survival of a given ecotype in a given environment
c = survival of that ecotype in the environment giving maximum survival.

A coefficient value of one represents the maximum degree of discrimination while a figure of zero indicates the greatest degree of survival possible. When this data is presented in table form (Table 1.3) it allows some measure of the performance of the plant in a habitat that is suspected to be environmentally sub-optimal with one where the biotype is known to have the greatest chance of survival. This greatest chance of survival does not however mean that all factors are optimal for the species. Thus in the fellfield ecotype, seed germination was favoured in the native habitat, the fellfield. Seedling survival however was greater in the snow-bed habitat. The overall effect of the sum total of factors acting at different stages in the life cycle determines the end result and in this case perpetuates the continued existence of two separate ecotypes.

In natural habitats where plants are exposed to environmental fluctuation together with competition for light, space, water, and nutrients, any or all of the factors are liable to be sub-optimal. Walter (1962) has defined therefore two types of environmental tolerance namely: physiological tolerance where the endurance of the plants is measured in

Table 1.3 Environmental discrimination coefficients (*see* text) for *Dryas octopetala* ecotypes reciprocally transplanted in home and foreign sites in Alaska. Symbols: *P<0.05; **P<0.01; ***P<0.001; NS = not significant. Significance levels refer to differences from a value of zero which represents no environmental discrimination. (Reproduced from McGraw and Antonovics, 1983.)

Ecotype	Site	Pollination success	Seed germination	Seedling survival Winter	Summer	1 year	Adult survival
Fellfield	Fellfield	0	0	0.89**	0	0.85*	0
	Snow-bed	0.55^{NS}	0.85**	0	0.25^{NS}	0	0
Snow-bed	Fellfield	0.32^{NS}	0	0.89***	0.74***	0.97***	0.50**
	Snow-bed	0	0.63^{NS}	0	0	0	0

culture with no competition and only one variable being altered at a time; and ecological tolerance where the response of the plant is assayed under field conditions. The difference between these two types of tolerance is brought out in the distribution of the common heath species *Erica tetralix*, *E. cinerea* and *Calluna vulgaris* to soil moisture (Bannister, 1964). All three species grow best in mesic conditions. However their varying tolerance to flooding (Fig. 1.9) and drought causes the less competitive *Erica* species to be excluded from the more mesic sites where the common heather *Calluna vulgaris* grows with greatest vigour. Thus the wet sites are preferred habitat for the flood-tolerant *E. tetralix* and the drier sites for the drought resistant *E. cinerea*.

The relative tolerance of salt effects controls the

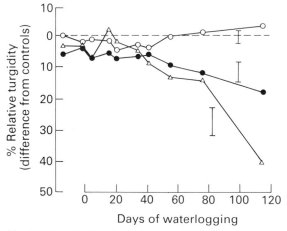

Fig. 1.9 Relative flooding tolerance of British heaths as measured by loss in relative turgidity of shoots of waterlogged plants of *Calluna vulgaris* (●), *Erica cinerea* (△) and *E. tetralix* (○). (Reproduced with permission from Bannister, 1964.)

distribution of *Armeria maritima*) and (*Festuca rubra* in many coastal habitats. Sea thrift (*Armeria maritima*) and red fescue (*Festuca rubra*) are common coastal species which can survive both salt spray and actual flooding with sea water. Neither species will completely exclude the other but their relative dominance in coastal habitats is determined by the amount of salt to which the vegetation is exposed. In the sites with the greatest input of salt *Armeria maritima* is dominant while in the less saline sites *Festuca rubra* becomes the major component of the vegetation. The relative competitive effects of one species on the other are clearly illustrated in the replacement experiments shown in Fig. 1.10 (Goldsmith, 1973). When the plants are watered with sea water the density of *Armeria maritima* increases and suppresses *Festuca rubra*. However when the plants are watered with tap water the reverse situation results. The equilibrium point producing a stable mixture between the two species is controlled directly by the amount of salt present.

1.5 **Distribution and survival**

In assessing the success or fitness of a species the implicit question is a demographic one of whether or not it can be expected to survive on a long-term basis. Demography alone, however, cannot give a true impression of the establishment of a species. Whole populations can run the gamut from superabundance to near extinction in very short periods. Species can be rare or common according to the likelihood of finding them in any particular region. Once found, a rare species may have considerable populations within a limited area, yet still be 'at risk' from a point of view of survival. The lesser bog rush *Schoenus ferrugineus* is extremely rare in Scotland and the raising of a loch level for a hydroelectric

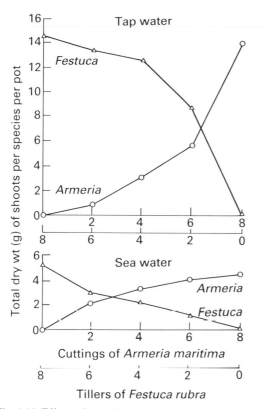

Fig. 1.10 Effects of experimental competition between *Festuca rubra* and *Armeria maritima* when watered with tap water and sea water. Each point is the mean of two replicates. (Adapted from Goldsmith, 1973.)

than any indication of numbers making up a population. In addition a wide-ranging distribution is likely to be coupled with a greater degree of ecological and physiological tolerance. This in turn will ensure a greater potential capacity to respond to changing environmental conditions and lessen the likelihood of extinction.

1.6 Conclusions

A study of plant ecology from the point of view of survival, although it can use natural distribution as a guideline to understanding the relative degree of adaptation and tolerance in plants, must take care not to assume direct causal relationships between environmental parameters and the presence or absence of species. The fact that the north-eastern distribution of the oak (*Quercus robur*) in Europe follows the course of the January isotherm for $-2°C$ and that the northern limit of the spruce boundary coincides with a July isotherm of $10°C$ does not necessarily indicate a causal relationship. At the most we can conclude only that as these temperature boundaries are reached the competitive power of these species is radically reduced. As there are many types of plant distributions these have to be discussed before seeking physiological explanations for limits to species dispersal. In many cases limits to distribution are due to dispersal barriers caused by past or present geographical boundaries. Such is the complexity of plant and animal communities that once established they may retain their identity and distribution long after the original limits to their dispersal have been removed. If this were not the case then it would not be possible to identify the Wallace line in terms of living species (*see* p.15).

The next chapter therefore begins by examining the limits to plant distribution both geographically and ecologically. Subsequent chapters take a number of case histories and examine what is known about survival adaptations as well as the constraints which limit species tolerance and dispersal into other areas. In current physiological studies it is common practice to speak of adaptation when discussing the suitability of plants to any one particular area. Adaptation however in the manner in which it has arisen by natural selection also implies a form of limitation. It is the purpose of the case histories discussed in the following chapters to explore as far as possible both positive and negative aspects of adaptation in plants and their consequences for survival. Table 1.4 summarizes some of the terms used in this discussion and clarifies their use in the following chapters.

scheme was thought to have made it extinct (Perring and Farrell, 1977). Recently, however, another site has been found where the species is abundant. Nevertheless it must still be considered 'at risk' due to the very few locations in which it occurs in the British Isles. Similar examples occur in every region of the world and Fig. 1.11 shows the British and Scandinavian distribution of *Artemisia norvegica* which although locally abundant is still a very rare species confined in Norway to the Dovrefjell, an upland refuge that is thought to have been free of pleistocene ice.

For these reasons the British *Red Data Book* lists species whose survival is 'at risk' if they occur in 15 or less of the 10 km squares used for recording the British flora (Perring and Farrell, 1977). The list includes 321 species or approximately 18 per cent of the British flora. Survival predictions based on the amount of terrain over which a species is found can give a better indication of successful establishment

(a)

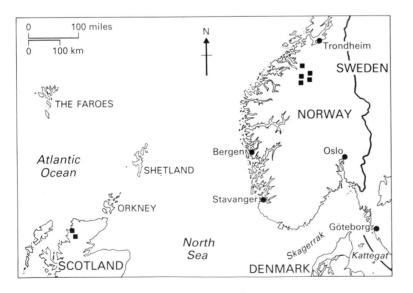

(b)

Fig. 1.11 *Artemisia norvegica*. (a) Growing on the Dovrefjell, Norway. (b) Distribution of *Artemisia norvegica* in Scotland and Scandinavia. (Data from Perring and Walter, 1976 and Hulten, 1971, reproduced with permission from Esselle Map Service, Stockholm.) The isolated and rare occurrences of this species are thought to be due to its survival during the pleistocene glaciation in Norway and possibly also in Scotland in areas that remained relatively free of ice.

Table 1.4 Summary and definitions of a number of ecological terms used in this chapter. (Adapted in part from Lincoln *et al.*, 1982.)

Acclimatization. Phenotypic adaptation to environmental fluctuation; the gradual and reversible adjustment of physiology and morphology to changes in environmental conditions. Sometimes used in another sense to represent changes observed in a species over a number of generations. Botanic gardens used for the introduction of species have been called *Jardins d'acclimatization.*

Adaptation. Process of evolutionary modification which results in improved survival and reproductive efficiency; any heritable, morphological, physiological, or developmental character that enhances survival or reproductive success. For non-heritable characters *see* Acclimatization.

Competition. The simultaneous demand by two or more organisms or species for an essential resource that is actually or potentially in limited supply (exploitation competition), or the detrimental interaction between two or more organisms or species seeking a common resource that is not limiting (interference competition).

Ecological niche. The concept of the space occupied by a species, which includes the physical space as well as the functional role of the species in terms of a multidimensional hypervolume occupied by a species which is unique to the species; the dimensions of the space are the parameters of the niche. The niche therefore defines the location of the species in the habitat in both space and time and the demands made on the habitat by its existence.

Ecophysiology. The study of the physiological adaptations of organisms to habitat or environment.

Ecotype. A locally-adapted population; a race or intraspecific group having distinctive characters which result from the selective pressures of the local environment; ecological race.

Fellfield. A type of tundra ecosystem having sparse, dwarfed vegetation and flat, very stony soil.

Fitness. The competitive ability of a genotype as measured by its capacity to contribute genetically to future generations; the average number of surviving progeny of one genotype as compared with the average number of surviving progeny of competing genotypes.

Habitat. The locality, and particular type of local environment occupied by an organism.

Red Queen hypothesis. That evolutionary advance by any one species represents a deterioration of the environment for other species so that each species must evolve as fast as it can merely to survive.

Tolerance. The range of an environmental factor that an organism or population can survive.

Stress. Any environmental factor which restricts growth and reproduction of an organism or population.

Survival of the fittest. The differential and greater success of the best-adapted genotypes.

Vicarious species. Closely related and ecologically equivalent species that tend to be mutually exclusive occupying disjunct geographical areas.

2
Limits to plant distribution

2.1 Biogeographical boundaries

In many species the limits to distribution are not the result of an ecological inability to survive outside their area of natural occurrence. The dangers of population explosions due to the accidental introduction of exotic species is proof enough that the limitations to distribution can be dependent solely on the existence of geographical barriers. The successful introduction of many plants from one part of the world to another without the need to modify their ecological tolerances by an extensive breeding programme also shows that many species have a greater potential distribution than that which dispersal limitations permit. The high growth rate of the American Monterey pine (*Pinus radiata*) in the New Zealand is one striking example of where an introduced species performs better in its new location than in the one where it originally evolved.

Before we can examine the physiological and ecological limits to species survival it is necessary to consider how various populations have diverged from one another and what the mechanisms are that have given rise to distinct breeding units. As species will vary in their evolutionary history there will also be differences in the nature of the limits to their natural distribution. The occurrence of whole biotas having a common range in the Gondwana continents and including such diverse groups as freshwater fish, crustacea, earthworms, and plant families such as the Proteaceae, Winteraceae, and Philesiaceae, and genera such as *Drimys*, *Nothofagus*, and *Gunnera* (Fig. 2.2) among many others, would be surprising if speciation occurred first and dispersal afterwards (Croizat *et al.*, 1974). The existence of sister biotas with coincident ranges is due to separation preceding speciation and results in many plant and animal species occurring together within a defined geographical zone. These limits to distribution are not in any way due to an ecological inability to grow over a wider area (Fig. 2.3).

The varying kinds of coexistence that are found between species can be usefully categorized by the terms allopatry, parapatry, and sympatry (Mayr, 1942) which describe species in terms of their evolutionary history.

1 *Allopatric species* are species which have evolved in different and disjunct areas.

2 *Sympatric species* are related species which occur in the same geographical area.

3 *Parapatric species* are species having contiguous geographical ranges which do not overlap and where some gene flow is possible between populations.

Thus speciation that has resulted as a consequence of geographical isolation is described as allopatric. For animals this geographical separation is probably the principal process of speciation as reproductive isolation is much harder in species that are 'within cruising range of each other during the breeding season' (Mayr, 1970). Distribution maps are frequently used to show the restriction of species within regions delimited by major geographical discontinuities. Sometimes these discontinuities are due to previous positions of the major land masses. The movements of continents, the erection of mountain chains, and the extension of deserts can come and go with geological time and the only mark that outlives them is a break in the distribution of plant and animal species. The most famous of all such breaks in plant and animal dispersal is the Wallace line, dividing the biotas of Australasia and south-east Asia (Fig. 2.4). For the plants and animals on either side of the Wallace line their distribution limits are not related to any known physiological restrictions. It is more likely that the long-established inter-relationships that will have evolved among the species making up the communities on either side of the line will have erected a biological barrier to the invasion of new species. Mutual interactions in shelter requirements, resource partitioning, and the precise timing of phenological events will inevitably fully exploit a habitat so that invasion by other species into the established community will not be easy. This community stability has probably played an essential role for the continuing existence of the Wallace line in the absence of any modern geographical boundary.

Fig. 2.1 (*opposite*) *Nothofagus* forest on the frontier between Chile and Argentina; five species grow in this region alone. The genus contains some 35—40 species all of which are restricted to regions that formerly belonged to Gondwanaland.

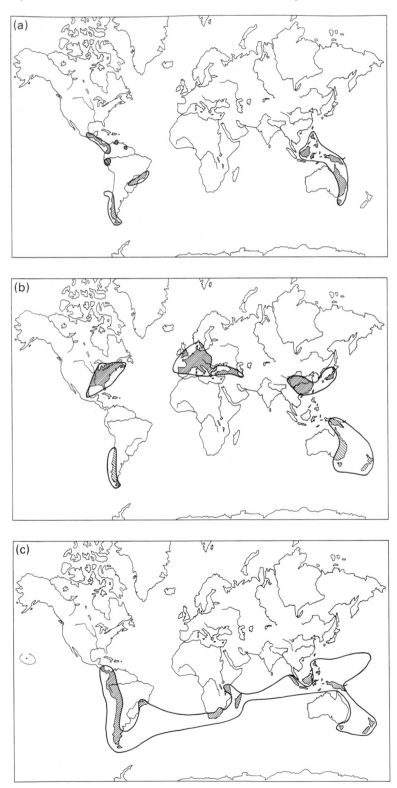

Fig. 2.2 The southern hemisphere distribution of genera that evolved within the former frontiers of Gondwanaland. (a) *Drimys*. (b) *Nothofagus* (cf. N. hemisphere *Fagus*). (c) *Gunnera*. (Redrawn from Walter and Straka 1970.)

Fig 2.3 *Senecio smithii* a long-range introduction from southern Patagonia to the Orkney and Shetland islands (off the north coast of Scotland). The species was introduced to these Islands by men returning from whaling in the Antarctic and illustrates the ability of a species to survive well outside its natural distributional range.

2.1.1 ALLOPATRIC SPECIATION

As allopatric species have evolved in disjunct areas there is no need for any genetical or ecological barrier to prevent hybridization. The chance introduction of one species across a geographical barrier as in the case of the arrival of the American cord grass *Spartina alterniflora* (Fig. 2.5) to Southampton water allowed a hybridization which would not other-wise have taken place with the European *S. maritima*. The resulting amphidiploid *S. anglica* possessed an ability not seen before in any species of higher plant in its capacity to colonize the salt-inundated, anaerobic mud-flats of river estuaries. Consequently large areas of mud-flats that had previously been bare of higher plants were rapidly invaded by *S. anglica* and subsequently developed to highly productive coastal grasslands.

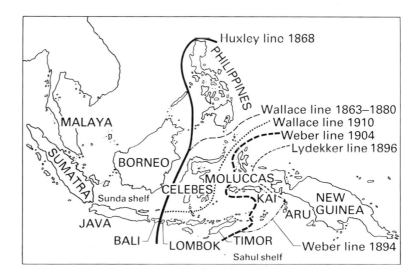

Fig. 2.4 The Wallace line together with various modifications suggested between 1868 and 1910. (Adapted from George, 1964.)

Fig. 2.5 The Mississippi delta showing a large stand of the American cord grass *Spartina alterniflora* in one of its most successful habitats.

2.1.2 *Sympatric speciation*

Sympatry is used to describe species that live in the same territory within normal dispersal range of each other. Fig. 2.6 shows an example of two closely related sedge species (*Carex panicea* and *C. nigra*) which share the same geographical area but are segregated by distinctive preferences for specific soil types. The effect of slope and aspect are particularly noticeable in their powerful effect in segregating the ecological preferences of these morphologically similar species. *C. nigra* occupies the level and presumably therefore wetter sites, while *C. panicea* is most frequently found on slopes of northern and eastern aspect. Similarly the common field buttercup (*Ranunculus acris*) and the closely related *R. bulbosus* (Fig. 2.7) can exist in close proximity with each

other with *R. acris* inhabiting principally the damper hollows and *R. bulbosus* the drier hummocks. The ecological distinction between these two species has been shown to be affected by the degree of soil waterlogging during the early stages of establishment. Seedlings of *R. bulbosus* show a reduced probability of surviving in the wetter sites than seedlings of either *R. acris* or *R. repens* (Harper and Sagar, 1953).

2.1.3 PARAPATRIC SPECIATION

In plants the static nature of populations results in many species being closely bound to one particular habitat and therefore adaptation to local conditions will be ecologically advantageous. The resulting divergent evolution between adjacent populations which

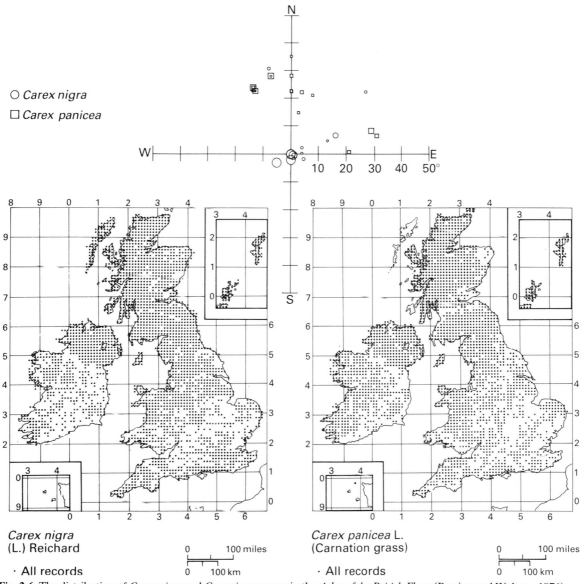

Fig. 2.6 The distribution of *Carex nigra* and *C. panicea* as seen in the *Atlas of the British Flora* (Perring and Walters, 1976) and in an ecological atlas relating distribution to slope and aspect (Grime and Lloyd, 1973).

still possess some degree of interfertility comes under the classification of parapatric speciation as defined above. The difference between allopatric and parapatric speciation is that while the former requires isolation to take place, parapatric differentiation persists in spite of a certain amount of gene flow between populations. Thus the selective forces favouring a biotype (a group of individuals with measureable common characteristics) in its own region

must be sufficiently great to prevent it being swamped by neighbouring biotypes (Pielou, 1979). Variations in soil pH, moisture, and exposure can be sufficient to induce such speciation with the evolution of closely-related species in proximity to one another. In the Mississippi delta *Iris fulva* occurs parapatrically with *I. giganticaerulea*, and leads to the production of hybrids between the two species. *I. giganticaerulea* occurs in the waterlogged soils of marshes and *I. fulva*

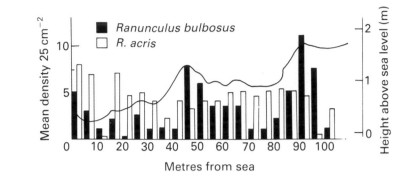

Fig. 2.7 The occurrence of *Ranunculus bulbosus* and *R. acris* in the same coastal pasture in Orkney. The figure illustrates the greater density of *R. bulbosus* in the drier hummocks. On this scale the almost complete separation of the species microtopographically in rises and hollows is not visible.

on the drier soils of river margins (Riley, 1938). The disturbance of the natural vegetation by man with drainage schemes and the establishment of periodically flooded pastures has created a habitat where the hybrid between these two species predominates. Similarly in the Orkney Islands to the north of Scotland the partially drained pasture lands are the preferred habitat of the hybrid between *Senecio jacobea* and *S. aquaticus*. Thus in Fig. 2.8 the tendency for *Senecio*

jacobea to avoid level ground, the preferred habitat of its relative *S. aquaticus* is clearly evident. Similarly *C. caryophyllea* differs in its habitat preferences from many sedges by occurring on sloping ground with better drainage (Fig. 2.9).

Barriers to hybrid swarm formation are not due merely to lack of initial production. In those species that do have a substantial degree of interfertility it can be the difficulty of establishing appropriate hybrids

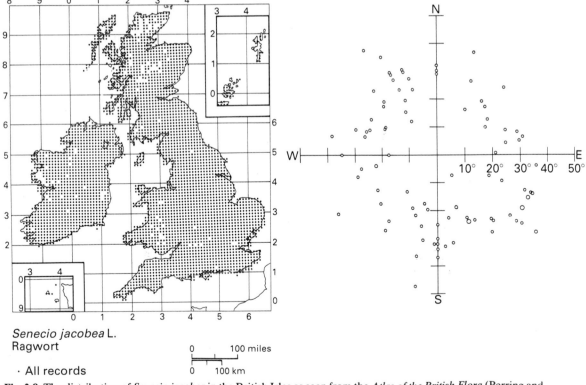

Senecio jacobea L.
Ragwort

· All records

Fig. 2.8 The distribution of *Senecio jacobea* in the British Isles as seen from the *Atlas of the British Flora* (Perring and Walters, 1976) and from an ecological atlas relating distribution to slope and aspect (Grime and Lloyd, 1973).

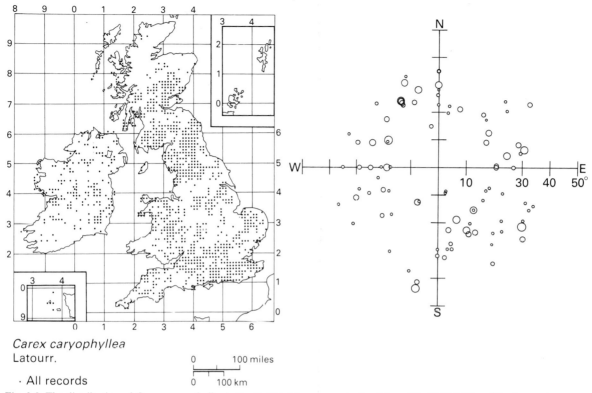

Carex caryophyllea
Latourr.

· All records

0   100 miles

0   100 km

Fig. 2.9 The distribution of *Carex caryophyllea* in the British Isles as seen in the *Atlas of the British Flora* (Perring and Walters, 1976) and from an ecological atlas relating distribution to slope and aspect (Grime and Lloyd, 1973).

in intermediate habitats between those occupied by the parents that limits the success of hybrid swarms. The hybrids between *Senecio jacobea* and *S. aquaticus* commonly have a 50 per cent fertility failure (Ingram, personal communication). Where conditions are not conducive to the production of hybrid swarms introgressive hybridization can still occur. In this way back-crossing of the hybrid to one parent will infiltrate germplasm of one species into that of another. Introgressive hybridization is also possible between species that do not occupy the same area. This requires the transport of pollen over considerable distances and is a form of hybridization to which wind-pollinated plants are well suited (Stace, 1975). Introgressive hybridization is thus a method whereby the ecological tolerance of one species can be extended and in which there may be only a territorial conflict with one parent. Examples of these different types of hybrids are given in Table 2.1.

In many species where geographical and ecological boundaries do not prevent interchange of pollen it is

not necessary to have complete absence of hybridization in order that distinct populations may evolve. The degree of genetic barrier necessary according to Lewis (1966) will depend on the precise ecological situation. If hybridization is reduced to a level where loss in reproductive potential is no greater than that due to vagaries of the environment then it will exist at a level where selection is ineffective and the distinctive feature of the populations will not be in danger of merging.

In sympatric species it is not always necessary to have complete barriers to hybridization. Lewis (1966) examined two diploid populations of *Clarkia* which occupy comparable ecological sites less than a mile apart and where the hybrids are sterile even although the parents show no obvious morphological differentiation. The genetic differentiation of these populations by saltation (an abrupt mutational or other evolutionary change) was considered the method which allowed these populations to become distinct from one another without either geographical or

1 Ecologically maintained species producing hybrid swarms.	
Geum rivale	*G. urbanum*
Primula elatior	*P. veris*
Linaria repens	*L. vulgaris*
Calystegia sepium	*C. silvatica*
Senecio aquaticus	*S. jacobea*
Tragopogon porrifolius	*T. pratensis*
Prunella laciniata	*P. officinarum*

Table 2.1 Hybridization in the British flora. (Adapted from Stace, 1975.)

2 Ecologically maintained species showing introgressive hybridization.	
Betula pendula	*B. pubescens*
B. pubescens	*B. nana*
Gentianella amarella	*G. uliginosa*
Silene alba	*S. dioica*
Euphrasia anglica	*E. confusa*
Senecio squalidus	*S. vulgaris*
Viola lactea	*V. riviniana*
Erica mackaiana	*E. tetralix*

3 Allopatric species forming introgressive hybrids in the UK.	
Parent sharing territory	Parent absent or rare in territory
Quercus petrea	*Q. robur*
Gentianella amarella	*G. germanica*
Saxifraga spathularis	*S. hirsuta*
Nuphar lutea	*N. pumila*
Ranunculus flammula	*R. reptans*
Carex aquatilis	*C. recta*

N.B. The division between allopatric and parapatric speciation although clear in theory demands in practice a complete knowledge of past distribution and evolutionary history that is seldom available. Similarly, the production of hybrid swarms as opposed to introgressive hybridization is not absolute. In *Geum* spp. the hybrid can back cross with *G. rivale* but not with *G. urbanum* (Stace, 1975).

ecological segregation. In northern habitats plants appear to make particular use of genetic isolation mechanisms in species production. Allopolyploids (amphidiploids) are frequent in Arctic and sub-Arctic floras. When apomixis (seed production without gamete formation) is combined with polyploidy then the ability to preserve advantageous microspecies free from contamination with closely related neighbours is enormously increased. Apomixis like polyploidy is frequent in northern habitats and is examined in greater detail in the discussion of Arctic vegetation in Chapter 3.

2.2 Ecological boundaries

Species which have evolved parapatrically present the greatest ecological interest as they persist under the constant threat of gene flow from neighbouring populations. Many such species (Grant, 1981) hybridize with greater difficulty than those that have evolved allopatrically. The strength of the breeding barrier is a function of the mode of speciation and the ecological distinctness of their habitats. Habitat specialization may present a further barrier to interfertility in that the hybrids may be produced but do not possess the required degree of physiological tolerance to survive in the available habitats. Each species in such a group must have properties that fit it to its own preferred habitat and what is even more important, place it at a disadvantage in other habitats. The restriction in adaptability provides one of the most intriguing problems of physiological ecology.

Such are the forces of selection in plant populations that even within a small geographical area different types may evolve in response to varying ecological pressures. Such ecotypes (locally adapted populations of a widespread species) can be readily recognized even by eye in certain species. As discussed in Chapter 1 (p.5) cliff top populations of *Plantago maritima* frequently have different genotypes that can be recognized by the hairiness of their leaves. Usually the occurrence of the hairy form is very localized as on moving out of the immediate region of salt spray, pubescent types give way to more glabrous ecotypes. The genetic basis of ecotypic differentiation was first recognized by the Swedish botanist Turesson (1922) using transplants from many regions and growing them together in the same experimental garden.

The degree of habitat specialization or ecotype development found in natural populations is demonstrated by the great variation in growth and phenology that can be observed in transplant experiments. Fig. 2.10 shows the relative performance of a number of transplants of different populations of *Plantago lanceolata* made between different sites in North Carolina (Antonovics and Primack, 1982). In the wooded site which had a high diversity of species growing under a cover of loblolly pine (*Pinus taeda*) there is very little difference in the performance of the different populations when they are transferred to this habitat. However in an adjoining site on an unshaded lawn where the natural population of plantains consisted of large clumps forming a conspicuous feature of the vegetation there were observed marked differences in the performance of the various populations. In such situations habitat specialization has evolved a specific population which is most likely to survive in one particular ecological situation and be at a disadvantage in others. Only in introgressive hybrids is specialization sometimes modified by the infusion of increased tolerance and enhanced ecological range as illustrated in the examples cited above with *Senecio aquaticus* and the *Iris* species on the banks of the Mississippi.

2.3 Climatic boundaries

The existence of past and present geographical boundaries to plant distribution should not obscure the fact that many species reach a dispersal limit that cannot be related to any contemporary or ancient geographical barrier. Non-geographical boundaries are frequently climatic in origin. However as pointed out in Chapter 1 the relationship may not be direct but may instead reflect the relative tolerances of different species competing for the same ecological

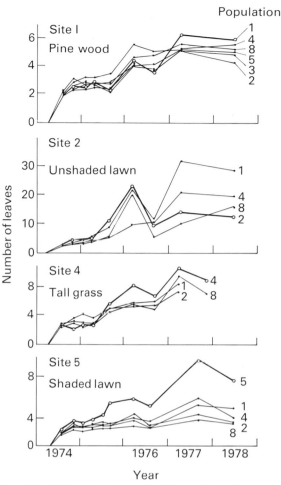

Fig. 2.10 Differences in growth in *Plantago lanceolata* as measured in number of leaves produced in different populations transplanted into varied sites in North Carolina. The heavy black line (o) indicates the population transplanted back into its own site of origin. (Adapted from Antonovics and Primack, 1982.)

niche. Thus the northern limits of the deciduous broad-leaved forest in Europe and North America match the upper limit of approximately 2 months of winter frost. This deciduous forest has too dense a shade for the regeneration of conifers and thus the southern limit of the boreal conifers is set indirectly by frost, not acting on them but on the survival capacity of their competitors. This boundary in forest types also marks the southern extension of podzol soils which once developed, reinforce the ecological division between the two forests. Certain temperature−threshold values have very marked effects on the existence of entire vegetation types. In particular frost and chilling temperatures produce ecological boundaries that affect large numbers of species in a similar manner. The following two sections therefore discuss these ecophysiological boundaries of chilling and frost injury before we examine the more gradual effects of temperature gradients and plant distribution.

2.3.1 CHILLING INJURY

Probably the greatest disjunction in the whole study of plant distribution is the division of the world's flora between tropical and non-tropical species. The division between tropical and temperate species

occurs in both the Old and New World and is not linked to past or present geographical barriers. Attempts to grow tropical species in the more temperate regions of the earth are dependent on maintaining the ambient temperature well above freezing. Most tropical species suffer injury leading to death when their leaves are subjected to temperatures in the 6–10°C range (*see* Table 2.2). As these lethal temperatures are above freezing point and therefore distinct from frost killing, this damage is referred to as chilling injury. It is not unique to plants. In the Amazon basin on the rare occasions when cold weather moves up from the south and the temperature drops below 20°C the rivers are filled with dead fish.

Table 2.2 Some commercially important crops which are known to be chill sensitive. Latin name, classification (family) and common name(s) are indicated (Guye, 1986).

1 Grain, vegetable and fibre crops		
Arachis hypogaea L.	Leguminosae	Peanut, groundnut
Andropogon sudanense L.	Gramineae	Sudan grass
Boehmeria argentia Linden.	Urticaceae	Ramie
Coffea arabica L.	Rubiaceae	Coffee
Eragrostis abyssinica Jacq. Lin.	Gramineae	Teff grass
Fagopyrum esculentum Moench.	Polygonaceae	Buckwheat
Glycine max (L.) Merr.	Leguminosae	Soybean
Gossypium hirsutum L.	Malvaceae	Cotton
Ipomea batatus (L.) Lam.	Convolvulaceae	Sweet potato
Oryza sativa L.	Gramineae	Rice
Phaseolus spp. L.	Leguminosae	French, runner beans
Saccharum officinarum L.	Gramineae	Sugar cane
Sorghum spp. Moench.	Gramineae	Sorghum
Stizolobium deerinianum Bort.	Leguminosae	Florida velvet beans
Vigna spp. Savi.	Leguminosae	Cowpea
Zea mays L.	Gramineae	Maize
2 Fruit crops		
Ananas comosus (L.) Merr.	Bromeliaceae	Pineapple
Capsicum annuum L.	Solanaceae	Peppers
Citrullus lanatus (Thunb.) Matsum & Nakai	Cucurbitaceae	Water melon
Citrus spp. L.	Rutaceae	Citrus
Cucumis spp. L.	Cucurbitaceae	Gourds
Lycopersicon spp. Mill.	Solanaceae	Tomato
Malus domestica L.	Rosaceae	Apple
Mangifera indica L.	Anacardaceae	Mango
Musa spp. L.	Musaceae	Banana
Nephelium luppaceum L.	Sapindaceae	Rambutan
Passiflora edulis Sims.	Passifloraceae	Passion fruit
Persea americana Mill.	Lauraceae	Avocado
3 *Ornamental crops*		
Coleus blumei Benth.	Labiatae	Coleus, flame nettle
Episcia spp. Mart.	Gesneriaceae	Episcia
Nautilocalyx lynchii Sprague.	Gesneriaceae	
Passiflora spp. L.	Passifloraceae	Passionflower
Saintpaulia grandiflora B.L. Burt.	Gesneriaceae	African violet

The difference between tropical and temperate vegetation is not a quantitative shift of the entire temperature-response range of the plants to lower temperatures. Many temperate species can withstand tissue temperatures as high as those tolerated by tropical species. Tropical and warm temperate species however do differ markedly from plants of cooler climates in their inability to withstand low temperatures above the freezing point. Attempts to determine a prime cause of chilling injury are complicated by the number of cellular processes that are disrupted as soon as the lower critical threshold temperature is reached. Cytoplasmic streaming is reduced, respiration increases, due to the uncoupling of oxidative phosphorylation, and there is a decline in the respiratory control of isolated mitochondria. Photoreductive capacity is also reduced (Fig. 2.12). These cellular dysfunctions are followed rapidly by an increase in membrane permeability causing a leakage of cellular contents (Simon, 1974). Degeneration of cell ultrastructure can follow either in less than 2 hours as with tomato cotyledons, or after a considerably longer exposure to chilling. Some argument of exists as to whether physiological dysfunction or a degeneration of cellular architecture is the prime cause of tissue collapse as a result of chilling injury.

In both plants and animals considerable use has been made of the Arrhenius plot in investigating the effects of decreasing temperature on physiological functions. This plot of the log of reaction rate against the inverse of temperature on the absolute scale will give a linear plot for purely chemical reactions where the rate, or change in rate as measured by activation energy, remains constant irrespective of temperature. In biological systems with reactions catalysed by enzymes there are upper and lower limits to the linear Arrhenius plot which are determined by the temperature denaturation of enzymes. Within these limits however the

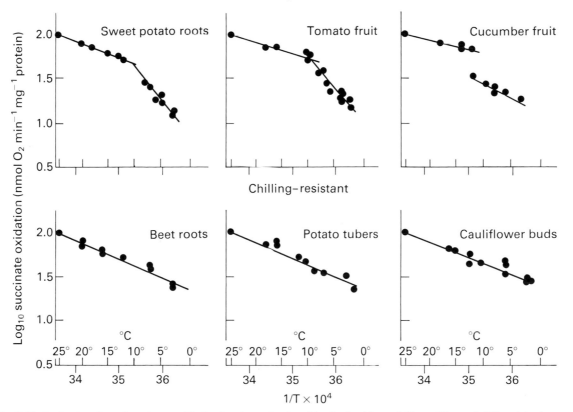

Fig. 2.11 Arrhenius plots of succinate oxidation by plant mitochondria obtained from chill-sensitive and chill-resistant tissues. The log values were adjusted to a common factor of 25°C in order to compensate for differences in rates between the different preparations. (Redrawn and adapted from Lyons and Raison, 1970.)

Arrhenius plot for biological reactions frequently departs from the expected straight line. In studies in chilling injury many experiments have been carried out to determine what aspects of cell metabolism show a coincidence in the departure of the Arrhenius plot from linearity with the temperature that causes chilling injury (Raison, 1973).

In experiments on the effects of chilling on higher plants and the temperature response of enzymes it is possible to see an apparent break in the linear Arrhenius plot for succinate oxidation by plant mitochondria from chill-sensitive species while straight lines are obtained with chill-resistant plants (Lyons and Raison, 1970) (Fig. 2.11). Similar breaks can also be seen in whole physiological processes as in the photosynthetic reduction of $NADP^+$ (Fig. 2.12) as well as in non-metabolic processes such as water uptake. These breaks in temperature response above freezing and well away from the temperatures causing a cessation of enzymatic activity have been linked with a hypothesis which suggests that phase changes in lipids are the initial cause. In pure lipids it is possible to demonstrate such phase changes which can be related to definite breaks in the Arrhenius plot. However membranes and the enzymes that are associated with them are not made of pure lipid. Mixtures of lipids and proteins together with other minor constituents of membranes such as cholesterol do not give breaks in the Arrhenius plot which can be statistically justified. An examination of many plots of temperature responses has lead some reviewers on this subject (Graham and Patterson, 1981) to suggest that no injustice is done if instead of straight lines with breaks, curves are fitted to some Arrhenius plots.

The hypothesis that chilling injury is related to lipid phase changes should be expected to lead to the observation of differing lipid composition between chill-sensitive and chill-resistant species. Keeping plants at low temperatures above freezing will bring about a general increase in lipids, especially phospholipids as well as increasing the degree of unsaturation of fatty acids (Clarkson *et al.*, 1980). However there are no apparent compositional differences between chill-resitant and chill-sensitive species (Patterson *et al.*, 1978).

In beans (*Phaseolus vulgaris*) it is possible to harden plants against chilling injury merely by exposing them for a number of days to a drought regime. Under these conditions there is no change in lipid composition associated with chilling injury. Chilling in this species induces a locking open of the stomata which, together with the increased resistance to water uptake at lower temperatures, induces wilting injury to the leaves at temperatures below 6°C followed by a sharp drop in the ATP levels in the leaves (Fig.2.13) (Wilson, 1976).

Experiments such as these have led to controversy (Graham and Patterson, 1982; Berry and Raison, 1981) as to whether alterations to cellular architecture, as typified by the supposed changes in membrane condition, are the principal cause of chilling injury or whether it is a cellular dysfunction due to an imbalance of metabolism below a critical temperature. Evidence for the metabolic imbalance view comes in part from the observation that when plants are kept at low temperatures covered with a polythene bag there is an increase in the level of unsaturated fatty acids without any parallel increase in

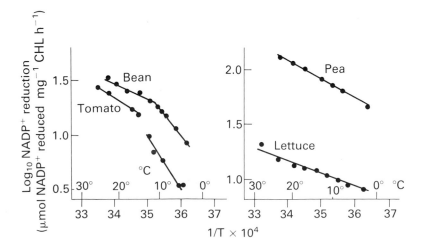

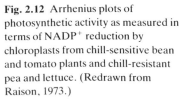

Fig. 2.12 Arrhenius plots of photosynthetic activity as measured in terms of $NADP^+$ reduction by chloroplasts from chill-sensitive bean and tomato plants and chill-resistant pea and lettuce. (Redrawn from Raison, 1973.)

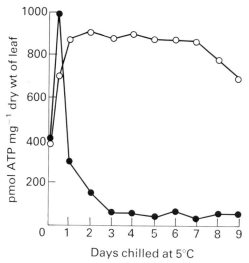

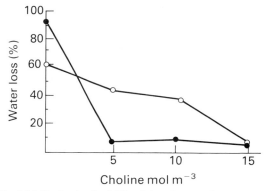

Fig. 2.13 Changes in level of ATP in *Phaseolus vulgaris* leaves during chilling at 5°C, 85 per cent rh (●) and 5°C, 100 per cent rh (○). (Reproduced with permission from Wilson, 1978.)

Fig. 2.14 Reduction in chilling injury in mung bean seedlings (*Phaseolus aureus*) when grown hydroponically in dilute solutions of choline. Chilling injury is measured by percentage water loss (○) during the chilling treatment and (●) in the subsequent warm period. (Reproduced with permission from Guye, 1986.)

resistance to chilling temperatures (Wilson, 1976). This, together with a growing realization that Arrhenius plots when critically examined show no clear point of membrane damage, has resulted in the cellular dysfunction view being currently favoured. Nevertheless it is probable that this is initiated by alterations in membrane-controlled processes which therefore implies a type of architectural damage to cell ultrastructure. However this situation leaves the problem of chilling injury still unsolved as it is not clear how the chill-resistant plants avoid cellular dysfunction at low temperatures.

It is more difficult to harden plants against chilling injury than frost damage. In some species damage can be postponed by preventing water loss from the leaves. Simply placing a polythene bag over the plant during the period of chilling can obviate injury in *Phaseolus vulgaris* (Wright and Simon, 1973). Drought hardening by depriving the plants of water for a period before the imposition of chilling can also reduce injury by closing the stomata and thus reducing water loss (Wilson, 1976). These preventive measures result only in a temporary postponement of injury. In the laboratory it is possible by growing mung bean seedlings (*Phaseolus aureus*) hydroponically in dilute solutions of choline to reduce the damage caused by chilling both during the cold stress and in a subsequent warm period (Guye, 1986) (Fig. 2.14). Thus susceptibility to chilling injury

is still not satisfactorily explained physiologically and in practical terms remains one of the greatest barriers to plant distribution that exists.

Chilling injury can also act on fruits. Damage here takes longer to appear and is apparently related to mitochondrial injury. From an economic standpoint fruit injury by chilling is of enormous importance as it places many difficulties in the way of transporting the abundant fruits and vegetables of the tropics to temperate markets. There is therefore much research

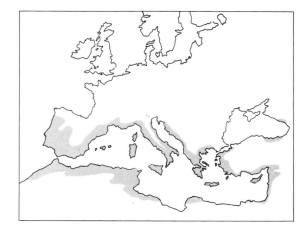

Fig. 2.15 The distribution of the olive in the Mediterranean basis showing its narrow limitation to the coastal region with the exception of Spain where there is a marked inland extension. (Redrawn from Polunin and Huxley, 1965.)

(Graham and Patterson, 1982) in progress to determine the cause of this injury and whether it can be prevented. Ecologically, chilling injury to fruits is of little importance to plant survival. If the leaves are sensitive to chilling this is reason enough for survival to be limited before the plant has reached the point of fruiting.

2.3.2 FREEZING INJURY AND CRYOPROTECTION

Frost, like chilling, can produce an environmental barrier which segregates distinct floras. One of the most dramatic frost boundaries occurs on moving inland from the Mediterranean littoral. Turrill (1929) lists 700 species that are endemic to the Mediterranean region. This frost-free zone is usually defined botanically by the distribution of the olive, which (with the exception of the Iberian Peninsula where there is a considerable inland distribution) is confined to the frost-free zone of the Mediterranean littoral (Figs. 2.15 and 2.16).

Different regional floras vary in their tolerance of frost. Tropical and sub-tropical trees as well as the leaves of some evergreen trees of warm temperate coastal regions have a life-long susceptibility to frost injury as do all chill-sensitive plants and some C_4 grass species (Fig. 2.17) (for general review, *see* Larcher and Bauer, 1981). With the onset of freezing, temperate species differ in the degree to which they can become hardened against frost damage. The specific ability to harden against frost injury is primarily genetically determined. There is consequently great variation not only between species but also between individuals, ecotypes, and varieties. In some

Fig. 2.16 An ancient olive tree growing on a relatively frost-free zone of the Yugoslav coast at Starri Bar.

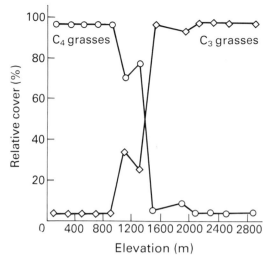

Fig. 2.17 The relative ground cover of C_3 and C_4 grass species, showing the abrupt replacement of C_4 species with increasing altitude in the Hawaii Volcanoes National Park. (Redrawn from Rundel, 1980.)

Eucalyptus species ecotypes from high altitude ranges develop greater frost tolerance than those from lower altitudes. A similar degree of altitudinal ecotypic variation in relation to frost tolerance is also found in certain tuber-bearing *Solanum* species (Palta and Li, 1979).

The hardening of plants to withstand freezing temperatures is a stepwise process. At temperatures between $+5°$ and $0°C$ plants attain a first level of frost hardiness which enables them to withstand a moderate frost. Progressive and persistent frost then leads to complete hardiness and in some boreal species can even protect some tissues against immersion in liquid nitrogen at $-196°C$. This highly developed degree of hardiness is lost when frost becomes less severe but can be re-established rapidly. During bouts of cold winter weather, frost tolerance in trees can increase noticeably in $1-2$ days and the full effect can develop within 10 days. In cereals hardening can be slower and may take several weeks to develop (Gusta and Fowler, 1979). Dehardening is a more rapid process than hardening and can take place within $2-4$ days.

Survival capacity at low temperatures in the field depends on more than just the ability of certain essential organs to withstand freezing injury. The timing of phenological events is essential to avoid injury both during the winter and on the resumption of growth in the spring. In areas with long, hard winters, leaf fall takes place early and the over-wintering buds develop their protective bud scales well in advance of the onset of winter. Similarly a late resumption of growth in the spring is a characteristic feature of high-latitude trees. This topic is considered in greater detail in Chapter 4. In woody plants both latitudinal and altitudinal distribution is limited by inadequate synchronization of development with climatic seasonality. The dormant buds of woody plants are the most resistant tissues to freezing injury. However, even within buds on the same tree there are differences in tolerance which vary in relation to the position of the bud on the tree. Lateral buds are normally more resistant than terminal buds (Mair, 1968). Woody stems are even more resistant than leaves and buds. In fully-hardened twigs and trunks the cambium is the most resistant tissue and the xylem the most sensitive (Larcher and Eggarter, 1960). The death of the cambium would lead inevitably to the death of the tree and hence for survival it is essential that trees are not ringed by a zone of frost-killed cambium. A certain number of terminal and lateral buds can be lost in a severe winter and the tree will still survive. This situation occurs frequently at the timber-line and can enhance the development of the crooked tree forms described as *Krummholz* (*see* Chapter 4 concerning the genetic predisposition to *Krummholz* forms at the tree-line). In woody plants the survival of the whole plant (as opposed to local dieback injury, inflicted on exposed branches and buds) is limited by the resistance of the roots to freezing injury. Root tolerance of freezing temperatures is considerably less than that of the aerial portions (Fig. 2.18) and the most vulnerable region appears to be the root collar.

As with most adverse factors, evolution has adopted avoidance as well as tolerance mechanisms in relation to low-temperature resistance. Many plants restrict the exposure of their tissues to frost by life form adaptations which minimize their exposure to feezing conditions. Buried tubers and rhizomes develop a freezing resistance to temperatures of no more than $-5°C$, but as they are usually buried this degree of tolerance is adequate for the plant to survive in regions where air temperatures are considerably colder.

Although in terms of degree of cold frost injury is distinct from chilling injury, the two cold effects have a number of similarities.

1   Injury by frost and chilling is quantitatively related to the extent of the exposure of the plant to low

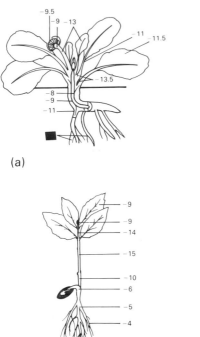

(a)

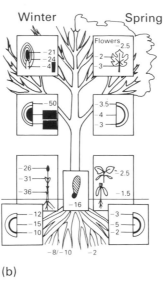

(b)

(c)

Fig. 2.18 Winter frost resistance of specific plant organs, (a) Organs and tissues of the common daisy (*Bellis perennis*), (b) the sycamore (*Acer pseudoplatanus*) and (c) a holm oak seedling (*Quercus ilex*). (Adapted from Larcher and Bauer, 1981.)

temperatures. Frost injury does not take place merely by the exposure of tissues to sub-zero temperatures but increases with the prolongation of freezing and with the depth of the freezing temperatures.

2   The speed of temperature change affects the extent of both freezing and chilling injury. After freezing the rate of temperature change on thawing is also related to the amount of injury caused.

3   Plants vary seasonally in their ability to withstand low temperatures (natural hardening cycles) and with proper temperatures conditioning can exhibit marked increases in their range of tolerance.

4   Water loss can be a major cause of injury in both chilling and freezing injury.

It is sometimes a surprise to those studying freezing injury for the first time to discover that the extent of injury to plants is not just a question of the leaf being exposed to sub-zero temperatures so that some of its water is in the frozen state. The extent of damage varies with both the depth and duration of the frozen condition. Ice formation in plant cells usually occurs outside the cell wall, as here the solute content is lower and there is a minimum depression of the freezing point. The ice crystals that form outside the cell, will continue to grow with continued freezing and this creates a gradient in vapour pressure which progressively removes water from the cell. The lower the freezing point the greater is the vapour pressure gradient. Thus with lower temperatures and prolonged freezing the cells are subjected to ever more severe dehydration. If the rate of freezing is rapid and if it water permeability of the cell membranes is low, then ice formation can take place both inside and outside the cell. The growth of ice crystals outside the cell is not in itself lethal; the extent of damage being dependent on the desiccation resistance of the cell. However when the ice crystals form within the cell death is inevitable. The ice does not physically rupture the cell like a burst pipe as was formerly believed, but seeds itself as crystals in the cell membranes which leads to the formation of membrane holes and thus causes irreparable damage.

Plants do not reach a eutectic point (a minimum freezing point when two or more substances in solution lower each others freezing point) and even when the ice crystals form in the cell the entire cellular contents are never frozen. Some portion of the cell water remains bound to macromolecules and in this state is protected from freezing. The composition of the solutes in this small portion of the cell water varies greatly from that of the unfrozen cell. It has been suggested that the concentration of salts and amino acids which is brought about in this way can cause direct toxic effects to the cell (Heber and Santarius, 1973). In addition changes in pH and salt concentration can bring about protein precipitation. A comparison of the freezing resistance of the various sub-components of cells shows that apart from structures such as membranes, proteins are the most susceptible to damage. Nucleic acids are undamaged by freezing (Mazur, 1969) and the polymer carbohydrates are likewise unaffected. Thus as membranes are in part made of protein, it is probable that the damage done to them by freezing may be caused by denaturation, salting out, dehydration or even pH precipitation.

In animal cells freezing injury is less if it takes place rapidly so that the cell becomes solidified with amorphous ice. Thawing has to be equally rapid for cell survival in animals as if the temperature remains in the sub-zero range for any time ice crystals will grow in size with water liberated from some constituents and membranes will become punctured. The same situation is not found in plants. It is a frequent observation that plants in the shade of a building may be frozen overnight to the same extent as unshaded plants, but when thawing takes place during the day it is the plants in the warmer sunlit areas that are killed and not the ones in shade. When leaves thaw rapidly the water permeability of the organelles within the cytoplasm is higher than that of the tonoplast. Thus the cell vacuole will expand only slowly and the greatest amount of water uptake will

be by the cell organelles, viz chloroplasts and mitochondria. Rupture of the membranes of these organelles by osmotic forces may then lead to the death of the cell. Thus in animals and micro-organisms which have no large central vacuole a rapid rate of thawing poses no osmotic problem and is in fact beneficial to the survival of the cells.

Examination of leaf physiology after freezing treatments shows that photosynthesis is the process most affected. From tracer studies (Heber and Santarius, 1964) photophosphorylation mechanisms are specifically affected. It is possible to isolate thylakoid membranes and examine the *in vitro* effects of freezing and cryoprotective substances. By working with isolated membranes it is possible to investigate the effects of cryoprotective substances in the absence of the effects of solutes on freezing point. A positive effect can be shown with sucrose while negative responses are obtained with sodium chloride (Fig. 2.19). In isolated membranes a cryoprotective effect of sucrose can be demonstrated at very low concentrations. This finding alone suggests that sugars apart from their action in depressing the freezing point have a specific membrane-protection role. It therefore follows that it is not the phase change of water itself that is the prime factor in damaging cells at sub-zero temperatures. It is the consequences of freezing on solute concentration in the cell which will determine whether or not injury takes place.

Exposure to a period of chilling before the onset of freezing has long been known to increase the amount of soluble sugars in a variety of plants. As the relationship between starch and soluble sugar is

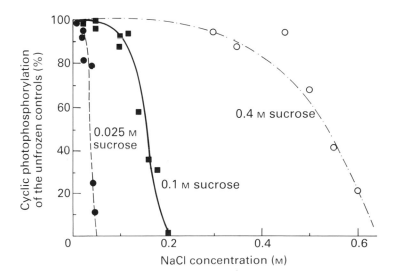

Fig. 2.19 The protective effect of sucrose on the cyclic photophosphorylation of isolated chloroplast membranes during freezing as a function of NaCl concentration. (Redrawn from Santarius and Heber, 1972.)

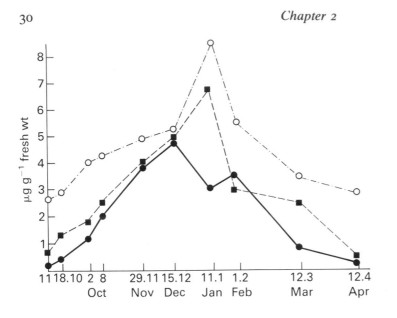

Fig. 2.20 Change in oligosaccharide concentration in leaves of *Cerastium arvense* during winter in plants that were cultivated for 3 years in the open; sucrose (O−·−·−O), raffinose ■−−−−■, lychnose ●−−−−●. (Reproduced with permission from Hopf *et al.*, 1984.)

exothermic there will naturally be an increase in the sugar content of the cell at lower temperatures. However, increase in sugar content alone is not sufficient to confer freezing tolerance on tissues. Potatoes convert large quantities of starch to sugar and do not develop a tolerance to frost. Similarly sugar cane which has throughout the year concentrations of soluble sugars higher than many species, like many other C_4 plants is sensitive to cold.

Successful protection against frost injury is not a function of the sugar content of the cell vacuole. The sugars that are found in relation to cold protection must be able to act to prevent frost damage to sensitive structures such as membranes. When isolated membranes are examined for freezing damage in

relation to sugars then it is found that mole for mole, trisaccharides are more effective than disaccharides and that these are more effective than mono-saccharides. Under artificial conditions sugar alcohols are also effective, namely mannitol, inositol, sorbitol, and glycerol. Glycerol is well known as a cryo-tective agent in insects but its occurrence in plants in the free state is relatively rare. Other sugar alcohols, mannitol and sorbitol, have been reported in high concentrations in *Gardenia* sp., *Malus* sp., *Sorbus aucuparia* and *Punica granatum* (Sakai, 1961). Figs. 2.20 and 2.21 show some typical changes in the levels of sugars which are commonly associated with the development of frost tolerance in higher plants. As well as sucrose, raffinose and lychnose are com-

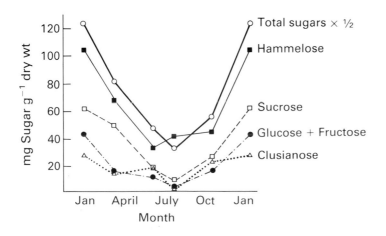

Fig. 2.21 The annual course of sugar content of young leaves of *Primula clusiana*. (Redrawn from Sellmair and Kandler, 1970.)

mon. In halophytic species, which overwinter in the green conditions, large quantities of raffinose are associated with frost resistance (Kappen and Ullrich, 1970). In *Primula clusiana* (Fig. 2.21) this same strategy in relation to sugar accummulation is also evident. Here there is demonstrated the common observation that frost hardiness is often associated with the accumulation of sugars that are out of the ordinary in that they are not part of the primary metabolism of plants. In *P. clusiana* the somewhat unusual oligosaccharaide, clusianose, makes a significant contribution to the total sugar content of the hardened plant. Other oligosaccharides frequently associated with frost hardiness include members of the raffinose family. The synthetic relationhips of the raffinose sugar family (Fig. 2.22) and their relationship with other sugar are shown in Fig. 2.23. Thus the accumulation of lychnose and raffinose as seen over winter in *Cerastium arvense* (Hopf *et al.*, 1984) (Fig. 2.20) can be seen as a pattern of progressive accumulation of ever more complex oligosaccharides.

Amino acids also play a role in the development of frost resistance, the best known example being the accumulation of proline. In cabbage leaves this is the only amino acid which shows significant ac-

UDP-α-D-galactose + *myo*-inositol $\rightleftharpoons$ galactinol + UDP

$$\text{galactinol + sucrose} \overset{\text{GST}}{\rightleftharpoons} \text{raffinose} + \textit{myo}\text{-inositol}$$

$$\text{galactinol + raffinose} \overset{\text{GRT}_1}{\rightleftharpoons} \text{stachyose} + \textit{myo}\text{-inositol}$$

$$\text{galactinol + stachyose} \overset{\text{GRT}_2}{\rightleftharpoons} \text{verbascose} + \textit{myo}\text{-inositol}$$

$$\text{galactinol + verbascose} \overset{\text{GRT}_3}{\rightleftharpoons} \text{ajugose} + \textit{myo}\text{-inositol}$$

Fig. 2.22 Biosynthetic relationships in the raffinose sugar family. GST = galactinol: sucrose $-$ 6 galactosyltransferase; GRT = galactinol : raffinose $-$ 6 galactosyltransferase, series 1, 3 for synthesis of higher homologues of the raffinose series. (Redrawn from Kandler and Hopf, 1984.)

cumulation rising from 2 to 4 per cent up to 60 per cent in fully-hardened leaves (Le Saint, 1969). Frost injury can even be prevented in cabbage by allowing leaves to absorb proline from dilute solution (5 mg l^{-1}).

2.3.3 LIMITING TEMPERATURE COMBINATIONS

In oceanic climates such as the west coast of Scotland and Ireland many exotic species can be grown in the

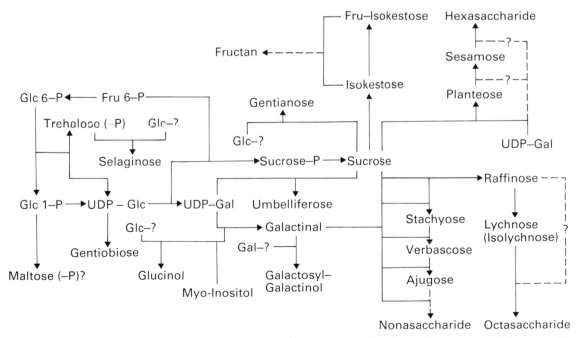

Fig. 2.23 Pathways of biosynthesis of primary oligosaccharides. Gal$-$? or Glc$-$? = unknown glycosyl donors; ($-$P) = phosphorylated intermediates; $---$?$---$ = suggested reaction. (Redrawn with permission from Kandler and Hopf, 1984.)

open due to the low incidence of frost. Similarly there is a native element in the British flora which is extremely oceanic and clings to the coastal fringes of the western littoral and rarely penetrates to the more frost-prone inland areas. A typical example of such a species is the wall navelwort (*Umbilicus rupestris*) shown with its distribution in Figs. 2.24 and 2.25.

Oceanicity as a climatic concept is a vague term comprising wet, cool conditions without marked temperature fluctuations. As it exists in the British Isles it is also distinct from the inland climate by being relatively frost free. Similar phytogeographical

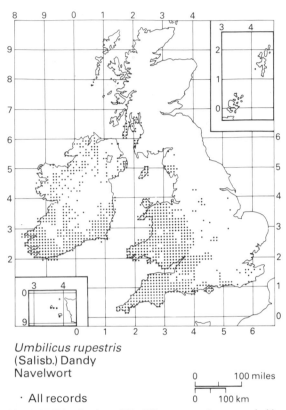

Umbilicus rupestris
(Salisb.) Dandy
Navelwort

· All records

Fig. 2.25 Distribution of *Umbilicus rupestris* as recorded in the *Atlas of the British Flora* (Perring and Walters, 1976).

Fig. 2.24 *Umbilicus rupestris* (wall navelwort) growing in a damp cliff on the west coast of Scotland.

boundaries also exist in southern Chile and the western seaboard of North America. Any attempt to relate plant distribution to just one climatic variable will be bound to produce a simplification with many exceptions and points of contention. Iversen (1944) produced a number of maps (Fig. 2.26) which clearly show the dual relationship that exists in the distribution of some species between summer and winter warmth. Mistletoe (*Viscum album*), ivy (*Hedera helix*), holly (*Ilex aquifolium*), and lime (*Tilia cordata*) (Fig. 2.27) all have their distribution limited by both summer and winter temperatures. In general the greater the degree of summer warmth the better is the plant adapted to survive low winter temperatures. The relationship however may not be direct in that it does not necessarily mean that plants which have more heat available in summer are better able to complete the protection of their overwintering tissues. Areas with high summer temperatures are usually

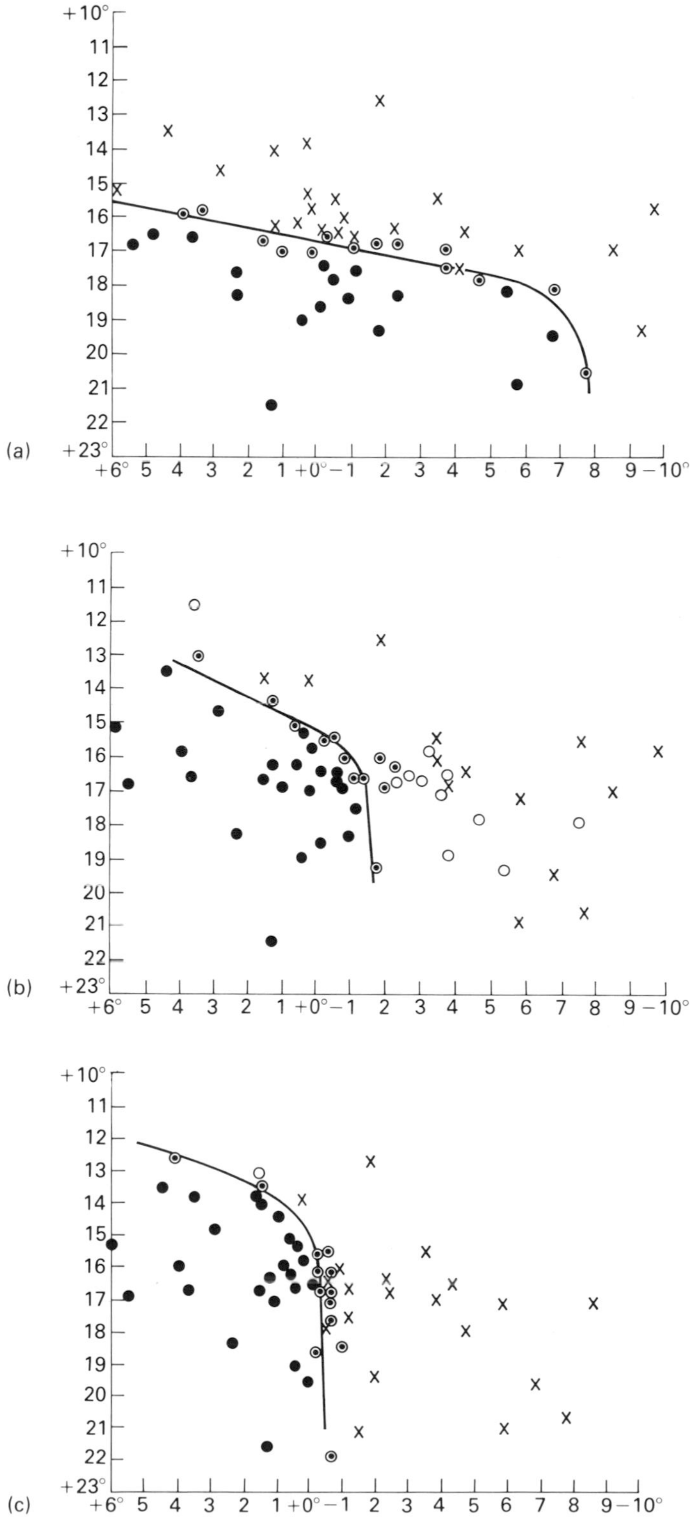

- ● *Viscum* within the station area
- x *Viscum* absent from station area
- ◉ Station on the area boundary of *Viscum*

(a)

- ● Develops normally within the station area
- ◉ Mostly sterile, only exceptionally bearing fruit
- ○ Never bears ripe fruit
- x *Hedera* missing within the station area

(b)

- ● *Ilex* within the station
- ○ Sterile only within the area
- ◉ Station on *Ilex* area boundary
- x *Ilex* missing in the station area

Fig. 2.26 Thermal limits to the distribution of three European species. (a) *Viscum album*, (b) *Hedera helix* and (c) *Ilex aquifolium*. (Taken from Iversen 1944 as adapted by Seddon, 1971.)

(c)

more continental. Thus in both North America and Europe ecotypes that evolve in habitats with high summer temperatures will automatically tend to be selected by winter exposure for endurance of low winter temperatures.

2.3.4 DEFICIENCES AND EXCESSES IN SUMMER WARMTH

The great variation in the thermal tolerances of higher plants is very clearly demonstrated by the experiments on related species of *Atriplex* and *Tidestromia* in which reciprocal transplants were made between a hot desert site (Death Valley) and cool maritime sites (Bodega Head) (Björkman *et al.*, 1974). The varying physiological behaviour of species in relation to high temperatures is illustrated in Fig. 2.28 which shows the growth and photosynthetic responses in controlled growth cabinets of species that are able to survive in a hot desert environment as compared with those that cannot survive in this habitat.

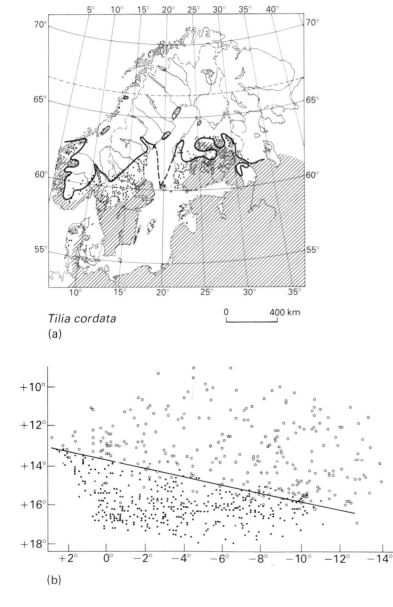

Tilia cordata

(a)

0   400 km

(b)

Fig. 2.27 (a) The Fennoscandinavian distribution of *Tilia cordata* (from Hulten, 1971; reproduced with permission from Esselle Map Service, Stockholm) and (b) the presence (●) and absence (○) of the species in relation to the mean monthly temperatures for the hottest and coldest months of the year. (From Walter and Straka, 1970, data of Hintikka).

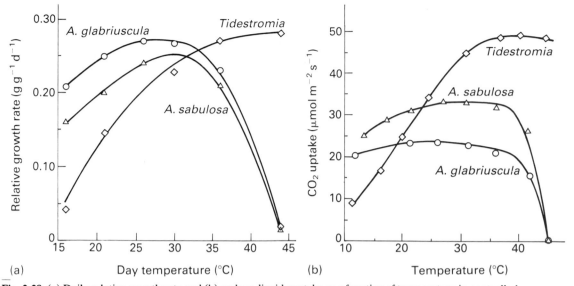

Fig. 2.28 (a) Daily relative growth rate and (b) carbon dioxide uptake as a function of temperature in controlled environment cabinets for *Atriplex glabriuscula* (C_3), *A. sabulosa* (C_4) and *Tidestromia oblongifolia* (C_4). The *Atriplex* species were unable to survive in the extreme heat of the Death Valley garden where *Tidestromia oblongifolia* showed rapid growth. (Adapted from data of Björkman and Osmond as quoted by Berry and Raison, 1981.)

However it is not only in the hot regions of the world that too much summer warmth can prove harmful to plants. Plants of cooler summers such as the Arctic and sub-Arctic can have their southern distribution limit controlled by too high a level of summer warmth. These southern temperature limits do not need to be high. Dahl (1951) was able to correlate the southern limit of distribution of 150 species of northern plants with maximum summer temperatures ranging from 22 to 29°C. These maximum summer temperatures were in general 10°C higher than the monthly mean temperatures. Coastal plants from Arctic and sub-Arctic shores also provide excellent material for studying the limiting effects of temperatures. Unlike mountain ranges coastlines in northern Europe present a continuous habitat type that can frequently be observed over a north–south direction for several thousand kilometres. Table 2.3 lists plants that reach their southern distribution limits on British coasts. These southern limits to Arctic species are not due to direct heat injury but appear to be related to a carbon imbalance (see Chapter 3). At the higher southern temperatures respiratory activity exceeds photosynthetic gain and the plants gradually waste away due to a lack of carbohydrate reserve. The critical period when this carbohydrate depletion acts on the plant appears to

be in the spring when the weakened overwintering organs attempt to resume growth (Stewart and Bannister, 1973; Bannister, 1981; Crawford and Palin, 1981).

2.4 Distribution limits to plant life forms

Ever since the beginnings of plant science the existence of plants of similar physiognomic appearance in different taxonomic groups has attracted attention of all those aspiring to put order into our understanding of natural phenomena. Theophrastus (372–288 BC) believed that all plants could be divided into three types, herbs, trees, and vines, and showed a degree of perception that matches closely the findings of twentieth century quantitative assessments of optimization of plant form in relation to environment. As plant exploration proceeded through the nineteenth and twentieth centuries the remarkable outward similarity of form in totally distinct species stimulated various attempts to provide a satisfactory morphological classification system. The classical example of taxonomically divergent floras that exhibit similar life forms is the evergreen sclerophyllous vegetation of mediterranean-type climates (Grisebach, 1872). Species of arid regions in varying parts of the world are also remarkable for their

Table 2.3 Coastal species with northern distributions in Britain; their habitat and world distribution.

Species	Habitat	Distribution and Notes
Cochlearia scotica DRUCE	Sands	Scotland, Northumberland and north and west Ireland. Taxonomy doubtful.
Ligusticum scoticum L.	Rocks, cliffs (sand)	Scotland, Northumberland, west Ireland. European coast from Denmark to north-west Russia. Circumpolar: Iceland, Greenland, North America. *L.s.* ssp. *hulteni* in Alaska and Siberia. Naturalized in New Zealand. Leaves formerly used as vegetable.
Mertensia maritima (L.) S.F. GRAY	Shingle (sand)	Lancashire and Aberdeenshire northwards. Formerly Northumberland. North and north-east Ireland. Denmark, Norway and north-west Russia. Circumpolar: Iceland, Greenland, North America to Alaska. Decreasing.
Juncus balticus WILLD.	Dune slacks	Lancashire and Fife northwards, but very local on west. Europe from Pyrenees to Faroes; Scandinavia and north-west Russia. Iceland, North and South America, Japan, New Caledonia.
Blysmus rufus (HUDS.) Link	Salt marsh	South Wales and Lincolnshire northwards. South-west and east Ireland. Circumpolar.
Carex marítima GUNN.	Damp hollows on fixed dunes	East and north Scotland. Hebrides, Orkney and Shetland. Also inland in Europe in the Alps.
Elymus arenarius L.	Active sand dunes	Local in south England, more frequent in north and east Britain. To 71°N in Europe from north France and Belgium. Circumpolar: Iceland, Greenland, North America and Siberia. Widely planted for dune stabilization.

Fig. 2.29 Species of *Euphorbia* growing with a 'cactus-like' life form on an arid mountain slope in the Canary Islands.

physiognomic similarity. The great similarity between some Old World *Euphorbia* species with the Cactaceae of the New World is shown in Fig. 2.29. The effect of similar conditions producing comparable physiognomic types is described as convergent evolution and is common in both plants and animals. Although the end result and its initial cause may be self-evident, it does raise the question of how, when, and where this selection acts on plant form.

The term life form stems from Warming (1895; English translation, 1909) who first used the term *Lebensform* in his book on the *Ecology of Plants*. A summary of various systems that have been in use is given in Table 2.4. Raunkiaer (1904) produced the most enduring system of life form classification that has been formulated this century. The simplicity of his classification which is based on the height of the perennating bud above the ground during the adverse or dormant season is shown in the footnote to Table 2.4. The simplicity of Raunkiaer's system led to its extensive use on practical grounds throughout most of this century and it still finds much popularity in survey work classifying poorly-studied regions for forestry plantation and other developments. Nevertheless ever since its publication there have been questions raised as to whether the position of an organ during the inactive season reflects the assumed evolutionary processes of optimizing plant form in relation to the selection forces operating in any particular habitat.

The Raunkiaer system presumes that the viability of the overwintering bud is the controlling factor for plant survival. Use is made of subclasses in relation to leaf size and longevity but these are employed only as secondary aids to the main divisions which are based on plant height. The correlation between form and climate is based on examining the nature of the stress during the adverse season. Temperature and precipitation have been the factors which have been principally considered. With the use of climatic

Table 2.4 A historical summary of the classification of higher plants in terms of life form. (Adapted from Schulze, 1982).

De Candolle (1818)	Grisebach (1872)	Raunkiaer* (1904)	Clements (1920)	Monsi (1960) Walter (1973)
Monocarpia	Woody plants	Phanerophytes	Annuals	Herbaceous plants
Annua	Succulents	Chamaephytes	Biennuals	Annual
Biennis	Climbers	Hemicryptophytes	Herbaceous perennials	Biennial
Perennis	Epiphytes	Cryptophytes	Sod grass	Deciduous perennial
	Herbs	Therophytes	Bunch grass	Evergreen perennial
	Grasses		Cushion herb	
Polycarpia			Mat herb	Woody plants
Rhizocarpia			Rosette herb	Deciduous woody
Caulocarpia			Carpet herb	Evergreen woody
Suffrutex			Succulents	
Frutex			Woody perennials	
Arbuscula			Half shrub	
Arbor			Bush	
			Succulent	
			Shrub	
			Tree	

* The Raunkiaer system is based on the height of the perennating bud above the ground during the adverse season.
Phanerophytes – the terminal bud is exposed freely, usually more than 0.5 m above the ground.
Chamaephytes – low growing plants with buds above ground surface, usually between 0.5 and 0 m.
Hemicryptophytes – aerial shoots degenerate to ground level during the adverse season.
Cryptophytes – buds and shoot apices survive the adverse season buried beneath the surface of the ground.
Therophytes – plants which survive the adverse season as seeds.

N.B. Raunkiaer also included numerous subclasses, for example phanerophytes, were divided into mega, meso, micro and nanophanerophytes; geophytes included geophytes (*sensu stricto*), hydrophytes, and helophytes depending on whether the bud and shoot apices were buried under earth, water, or in mud under water.

Fig. 2.30 A river bank scrub tree of the genus *Pithecellobium* growing on the banks of the Rio Negro 20 km north of its confluence with the river Amazon. Note the sclerophyllous leaf forms associated with the vegetation of this nutrient poor region and the pure sand of the river bank. Compare this tree with the trees seen in Fig. 2.31 where the temperature, rainfall and flooding conditions are identical but where the soil is nutrient rich.

diagrams in which temperature and rainfall are superimposed it is possible to assess the occurrence, length, and intensity of humid and arid seasons, the duration of cold weather, and the possibility of late and early frosts. An integrated representation of climatic stress plotted in this manner is available in the *World Atlas of Climatic Diagrams* (Jena, 1967). From this data it is even possible to construct homoclines (lines linking areas of similar climate).

Several authors have suggested that the life forms of cormophytic plants (plants that exist complete with roots, stems, and leaves) are not determined by the susceptibility of the terminal bud during the dormant or adverse season. Instead it is argued that plant form can be more readily understood on the basis of carbon gains and losses during the growing season (Schulze, 1982). Natural selection will act to distinguish individual variation in relation to carbon

gain as measured against carbon investment. The principal organ to react to this selection pressure will be the plant leaf and the expectation would be that greater fitness will automatically result from better leaves. This does not mean that it will benefit a plant to invest as much nitrogen as carbon as possible in producing the maximum amount of photosynthetic enzymes in leaves. In sun leaves it is rare for the photosynthetic apparatus to operate at more than 20 per cent effectiveness due to light, water, and other environmental limitations (Schulze, 1970; Moser, 1973). Thus in terms of carbon gained in relation to carbon invested, maximizing photosynthetic enzymatic content would not be a rewarding strategy. If carbon supplies, or the nutrients necessary for leaf growth are limiting, then prolonging leaf longevity would increase the potential return on investment. Therefore in arid regions of the world or areas

Fig. 2.31 Malacophyllous leaved trees growing in the flood-plane forests of the river Amazon west of Manaus. Compare the growth of these trees with the river bank vegetation of the neighbouring Rio Negro (Figs. 5.14 and 5.15).

deficient in nutrients it is leaf form and survival rather than the height of the perennating bud above the ground that would be expected to alter. Ecologically there is great variation in leaf longevity. In the genus *Araucaria* leaves may persist on the trees for up to 25 years. The home of the greatest *Araucaria* forests is in the arid belt that crosses South America between 37 and 40°S. Similarly, low availability of nutrients can induce sclerophyllous foliage without any other marked environmental change in temperature or water relations. In the upper regions of the Amazon at Manaus the confluence of the Rio Negro which is nutrient poor as its waters come from the hard granitic rocks of Venezuela shows just such a disjunction in leaf type (Figs. 2.30 and 2.31). The riverside vegetation in both the Amazon and its tributary the Rio Negro both exist under the same uniform tropical conditions and both suffer an ex-

tended flood season when the water of both rivers can rise 6–8 m above their low season level and stay at this height for up to 6 months. As is seen in the photographs the quartz sands of the river banks of the Rio Negro support sclerophyllous vegetation while on the Amazon, the malacophyllous (soft-leaved) type is predominant.

Similar changes in foliage rather than the height of plants can be observed in relation to precipitation. In the cool rain forests of Chile and western Patagonia a rainfall of 2500–3800 mm is needed to support the lush growth of the mixed deciduous and evergreen *Nothofagus* forests (Fig. 2.32). Sixty kilometres to the east the Andean rain-shadow has reduced the rainfall to less than 350 mm yet in spite of this trees can still exist. Fig. 2.33 shows a stand of *Libocedrus chilensis* which is remarkable for the size that this tree can attain even under arid conditions.

Fig. 2.32 *Nothofagus* forest in a high rainfall region on the frontier zone between Chile and Argentina. The evergreen species growing down to the edge of the lake is *N. dombeyii.*

Once again it is the leaf form that has responded to changing climatic conditions rather than the height of the terminal bud above ground level.

Apart from changes in leaf longevity, optimizing carbon gain in terms of investment will also affect the partitioning of carbohydrates into different organs within the plant. Here again, the partitioning will be influenced by the longevity of these organs. The ratio of photosynthetic to non-photosynthetic tissue differs greatly in different plant forms. Even within the same species different ecotypes can vary depending on environmental conditions. The suppression of non-photosynthetic tissue in favour of all-green plant forms is achieved best in lianas (the vines of Theophrastus) and in autotrophic parasites such as Mistletoe. The opposite extreme with no green photosynthesizing tissue is found in certain heterotrophic cormophytes such as *Monotropa*

hypopitys (Fig. 2.34a and b) and *Lathrea squamaria*.

Following the life form scheme of Monsi (1960) and Walter (1973) it is possible to compare the allocation of carbohydrate reserves in annuals, herbs, and woody plants.

2.4.1 ANNUAL SPECIES

In annuals the allocation of dry matter to seeds is higher than in other life forms. This should not be confused with numbers of seeds produced. Other life forms notably herbaceous biennials, such as foxglove, can produce large quantities of seeds as do certain perennials, for example *Salix*, but in terms of the proportion of the plant's annual production in dry matter the annual plants investment in seeds normally exceeds that of all other plant forms. Annual plants are characteristically rapid developers. The

Fig. 2.33 *Libocedrus chilensis* growing in the Andean rain-shadow in western Patagonia with an annual precipitation of only 350 mm and illustrating the ability of the tree form to survive in areas of low rainfall.

sunflower (*Helianthus annuus*) within 1 year can produce 600 g dry weight, whereas the beech tree seedling (*Fagus sylvatica*) attains only 2.5 g (Walter, 1981). Thus annuals can rapidly colonize bare ground and can complete their life cycles after intermittent rain in warm deserts. However in cold regions temperature inhibits the completion of their seed to seed cycle in one growing season. Thus in the flora of Arctic Greenland and Spitsbergen there is only one annual species *Koenigia islandica* (Polygonaceae) (*see* Fig. 3.14), a plant so small that it is easily overlooked if one is not previously acquainted with the species and its habitat.

Perennial herbaceous plants differ only from annuals in that they acquire their carbohydrate reserves over more than one growing season. Their morphological similarity in terms of life form comes from an absence of dead tissues. Both annuals and perennial herbs maintain living non-photosynthesizing tissues. Perennial herbs may flower only once when they are described as monocarpic. Examples of this type of be-

haviour are found in the bamboos and in the genus *Agave* where the carbohydrates reserve of 20–30 years accumulation may be used in supporting one season's flowering which is then followed by the death of the plant (Fig. 2.35).

2.4.2 PERENNIAL SPECIES

Perennial herbs are most successful in certain stressed sites. The accumulation of carbohydrate reserves can help to overcome limitations due to cold and short growing seasons. In the north *Saxifraga oppositifolia* typifies the success of herbs in the high Arctic being able to survive at 83° 39′ N. Perennial herbs are equally successful in shaded habitats. Their ability to survive for many years at levels only marginally above the compensation point allows them to accumulate enough reserves so that eventually they can use their assets to secure a place in the sunlight. In flooded habitats (*see* Chapter 5) the possession of reserve carbohydrates can enable cormophytic plants to withstand extended periods of oxygen deprivation. The limitation to continued accumulation of reserves in herbaceous perennials is due to the maintenance costs of supporting an ever-increasing mass of respiring non-photosynthesizing tissue. As the ratio of photosynthetic to non-photosynthetic tissues declines, net productivity decreases. Many perennial herbs however shed their older tissues. Thus in the genus *Iris* older parts of the rhizome rot away to the advantage of the plant. In *Primula scotica* no more than 2 years tissue persists although the plant can live for up to 20 years or more (Bullard *et al.*, 1987).

2.4.3 WOODY DECIDUOUS AND EVERGREEN SPECIES

Woody plants are distinguished by the production of secondary growth and its lignification. Thus the plant removes the need for maintenance costs for a large part of its structure. This dead tissue is not however functionless. As well as providing support and access to light it can provide storage of water and nutrients. Limitations to growth also exist in woody plants as well as in herbaceous perennials. Maintenance costs will eventually rise to reduce net photosynthetic gain. A disadvantage of the woody form, if it is in the shape of a single trunk, is that the senescence of this one stem can lead to the death of the tree. Beech trees are surviving well when they live to over 200 years. In the mature forests of North America typical ages for grand old trees are from 350 to 500

(a)   (b)

Fig. 2.34 The yellow bird's nest (*Monotropa hypopitys*) a totally achlorophyllous, saprophytic species of higher plant (a) growing on the roots of pine trees in eastern Scotland and (b) close up of flowering shoots of *Monotropa hypopitys* (Pyrolaceae).

years. However these ages are as nothing compared with the 11 000 years that have been estimated for *Larrea tridentata* (Vasek, 1980). Bush forms, like herbaceous forms, permit the fragmentation of the individual plant and thus shed accumulated unproductive tissues. The oldest known living tree, the bristle cone pine (*Pinus aristata*), owes its extreme longevity (4000 years) to its unique capacity for die back of the supporting stem and root which again reduces maintenance costs.

The interaction of woody form with varying leaf types is strikingly evident in the division of temperate trees into deciduous and evergreen. The evergreen habit has a bimodal distribution. The tropical rain forests are predominantly evergreen but to the north the broad-leaved forests of the temperate zone are largely deciduous. However to the north the circumpolar boreal coniferous forest is mostly evergreen.

This sequence of change can be seen again in the dwarf woody plants of the Arctic. In moving from the low to the high Arctic the evergreen habit becomes more frequent (*see* Chapter 3). Here also the gain from leaf longevity in a region of minimal resources has influenced life form. In temperate deciduous trees the most competitive plant form is achieved with leaves with short life-spans combined with large vessels to facilitate rapid translocation of water and nutrients to a height. As the growing season is reduced this advantage is lost to trees with tracheids and evergreen foliage. The evolution of the deciduous needle-bearing genus *Larix* appears to be of very recent origin and occupies more specialized habitats as in the mountain larch forests of central Europe or the extreme northern forests of *Larix dahurica* which are able to grow on Arctic permafrost soils.

Fig. 2.35 *Agave americana* the 'century plant' flowering in central Mexico. The species is perennial in that it lives for 20–30 years before flowering but is monocarpic in that it dies after flowering (height of wife 1.79 m).

2.5 Identity of views on plant form over 2000 years

Although more than 2300 years (Theophrastus fourth century BC) separate the first writings on plant form from those discussed above, (for example Monsi, 1960) there is a remarkable similarity in their view of the salient features of plant life forms. Theophrastus perhaps overlooked the ecological advantages of annual plants, but otherwise they are in overall agreement in that the main contrasts are found between the herbaceous and the woody plants, with the climbers (Theophrastus's vines), forming a third distinctive group. In all the above examples, whether they are based on organ longevity or ratio of photosynthesizing to non-photosynthesizing tissues or vascular architecture, it is possible to show a direct link between optimizing resource utilization and plant form. Even in the cushion plants such as the dwarf campion (*Silene acaulis*) and the dwarf azalea (*Loiseleuria procumbens*) it is not necessary to invoke the argument of the position of the terminal bud to understand the ecological advantages of this life form. Inside the cushion plant the temperature can be as much as 15°C above the ambient (Larcher, 1977). Humidity is also greater within the cushion and the carbon balance of the plant is likely to be improved by the supply of carbon dioxide from soil respiration due to the possession of the cushion form.

The Raunkiaer life form system has its advantage in that by relying principally on the position of the terminal bud, plants can be readily classified in the field. It is more difficult to decide from a visual inspection of a plant just how it allocates its carbohydrate resources. However in seeking to understand the forces that fashion the growth forms of plants it is no argument that a system is good merely because it is has practical merit. Attempts to elucidate the potential advantages of different carbon investment strategies appears to be a truer reflection of the natural selection process in relation to plant form and thus increases our understanding of the limits to plant distribution.

2.6 Conclusions

An enormous variety of barriers exist which act in numerous ways to limit plant distribution. The action of these barriers is due in part to the manner in which the species have evolved. The success of any biotype in a particular area is dependent on the successful interaction of many facets of its life cycle including the breeding system and the phenology of development with its bearing on the correct timing of flowering, seeding, and germinating. Physiological barriers which act directly on discernible attributes such as chilling resistance and frost hardiness are also a factor. Where limitations are due to variations in metabolic rates, then the limits to distribution are extremely subtle. As plants are autotrophic organisms they are responsible for the generation and consumption of their own foodstuffs and skeletal resources. Thus there is a very high degree of fine-tuning in re-lation to balancing production and consumption, which once achieved in a particular habitat, limits populations in their ability to spread to new areas with different environmental conditions. The weakest barrier is the geographical limit and when this is broken by man's agency it allows exotic species to be introduced into cultivation or spread as weeds, as in the introduction of the prickly pear into Australia and the spread of the water hyacinth throughout Africa. Fortunately the physiological and ecological limits to species distribution make these biological disasters not too commonplace.

Fig. 3.1 Arctic habitats near Mesters Vig north-east Greenland (72°N) with heathland in the foreground, morraines in the middle distance and nunataks in the background.

3

Plant survival in the Arctic

Tolerance of frost and low temperature at all times of the year and even during the growing season are obvious basic requirements for all species that can survive in the Arctic. Survival in the Arctic however does not just depend on being able to avoid frost damage. Adaptation to Arctic conditions requires plants to be capable of making a net carbon gain for the year in a very short growing season and also to be able to complete their developmental and reproductive cycles with minimal thermal input. The Arctic covers a large area and includes many different habitats (Fig. 3.1) and climatic types. As resources are limiting in both energy and nutrients it is to be

expected that Arctic plants should show a high degree of optimization in resource exploitation. Perhaps the most surprising feature of Arctic vegetation is the high degree of habitat specificity that exists in different species. It might have been expected that where conditions are universally minimal for plant growth that there would be correspondingly little variation between plants in their micro-habitat preferences. As any visitor to the Arctic can testify this is not the case and small differences in micro-habitats result in different species groupings in the varying plant communities. This chapter is an attempt to show the high level of specialization that has evolved

Fig. 3.2 Nunataks in Greenland (72°N). These mountains emerging above the snow and ice probably served in the Pleistocene period as they do today as refugia for higher plants thus preserving the distinct nature of Arctic biotypes of species which are also found in Alpine habitats to the south.

47

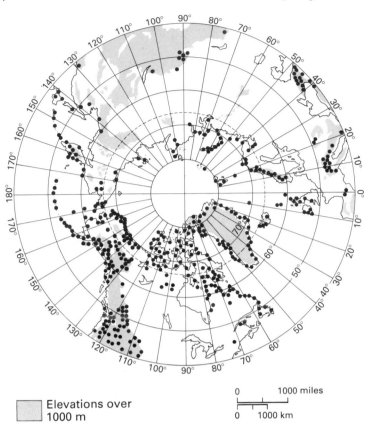

Fig. 3.3 The distribution of the typical Arctic–Alpine species *Oxyria digyna* (L.) Hill, based on collections in herbaria. The map extends to only 40°N but *Oxyria digyna* extends as far south as 28°N in the Himalayas. The Arctic forms of the species are however physiologically distinct from the Alpine races (Chabot *et al.*, 1972). (Map reproduced with permission from Billings, 1974.)

Elevations over 1000 m

0   1000 miles

0   1000 km

in Arctic vegetation in relation to both large and small variations in habitat type. Despite the small number of species in the Arctic and their diminutive size and limited range of life forms, there is a surprisingly large degree of specialization among the plants that compete for survival in this most limiting of all terrestrial habitats.

3.1 The Arctic flora

Thermodynamically, the earth is a cold habitat for all living creatures and the Arctic is but the chilly end of an unpromising thermal environment. Physiologically it is of interest in that it represents the limits of biological adaptation to low temperatures and short growing seasons. Ecologically, it is important for its extent (5.5 per cent of the earth's land surface) and as a last refuge where the processes of regulation in the natural environment proceed with minimal interference from man. Botanically the Arctic is usually considered as the area occupied by the

'tundra'. The original meaning of this word in Finnish or Lappish is 'a treeless hill' but has been absorbed via Russian to describe the dwarf shrub, herb, and moss vegetation that exist in polar regions too cold to support the growth of trees. The Arctic is also sometimes defined as the region of permafrost. Permafrost by reducing the depth of soil effectively prevents the growth of most trees. Noteable exceptions are the black spruce of North America (*Picea mariana*) and the dahurian larch of north-eastern Siberia (*Larix dahurica*). In these trees their shallow rooting systems can spread out above the permafrost zone sufficiently well to anchor the trees.

The restricted life forms and species numbers in the Arctic flora can create the impression that this inhospitable region serves as a last refuge for species in retreat behind the melting Pleistocene ice-sheets. Although it is true that many polar species once had a much wider range, it is incorrect to imagine that they have recently moved into the Arctic with the retreat of Pleistocene ice. The Arctic climate is not

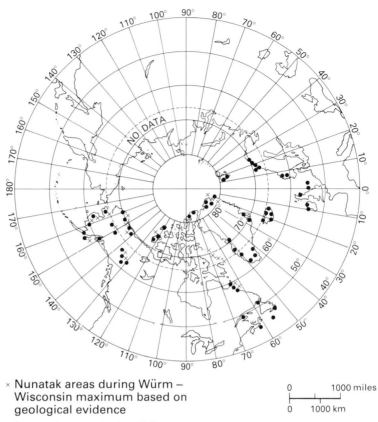

Fig. 3.4 Outline of presumed distribution of Arctic and high latitude refugia in the northern hemisphere as determined from biological and geological evidence. The most reliable indications are where geological and biological evidence agree as in Kodiak Island, interior Alaska, Yukon, Banks Island and Peary Land (the northern extremity of Greenland). The status of the Labrador sites is controversial. (Reproduced with permission from Ives, 1974.)

× Nunatak areas during Würm – Wisconsin maximum based on geological evidence

• Nunatak areas during Würm – Wisconsin maximum inferred from biological evidence but contested by some on geological grounds

0   1000 miles

0   1000 km

now thought to have been very different during the ice-age from what it is at present. From an analysis of the polar ocean sediments, it appears that the Arctic Ocean was not covered by an ice-cap 1000 m or more thick during the Pleistocene period (Clark, 1982). There were periods of iceberg activity during the ice-age and therefore there must have been some open water in the Arctic Ocean at this time. Radio carbon dating of timber flotsam found around the shores of Spitsbergen shows that there was free water circulation around this archipelago about 10 000 years ago (Hyvärinen, 1972). Therefore as the ice-age ended further south in Europe and North America there was clearly no need to wait thousands of years for mountains of polar ice to melt. The nunatak and coastal habitats that exist today in Greenland and Spitsbergen were very probably also available for plant colonization during the Pleistocene period (Figs. 3.2, 3.3 and 3.4).

An examination of the varying biotypes of species that are found at present in the Arctic with those further south shows that in North America and Europe the Arctic and Alpine races are genetically quite distinct. Cultivation of Arctic species outside their natural environment can present problems as not all species survive transplantation to warmer regions. With the construction of the Arctic greenhouse in the Copenhagen Botanic Garden it became possible to cultivate a range of varying biotypes from a number of different species under comparable conditions. The species included *Ranunculus glacialis*, *Juncus trifidus* (Figs. 3.5 and 3.6), *Saxifraga paniculata* and *Arabis alpina* (Böcher, 1972). The plants originating from southern regions such as the Alps invariably were richer in biotypes than those from the Arctic. *Ranunculus glacialis* although diploid over its entire range showed large differences between the Alpine plants and those originating from Green-

Fig. 3.5 *Ranunculus glacialis* the Arctic-Alpine species which holds the high altitude record in Scandinavia (2370 m Galdhøpiggen) and in the Alps (4275 m Finsteraarhorn).

land, Iceland, and Norway. The Alpine plants had between two and four flowers per stem and the leaves were narrow and stipitate. The northern plants by contrast rarely had more than one flower per stem (Fig. 3.5). As might be expected fewer biotypes appear to have survived the Pleistocene glaciation in polar regions. However, the difference in range of genetic types between polar and non-polar provenances is so great that there is little likelihood that any significant immigration took place into the Arctic at the end of the ice-age. Some of the enormous distribution disjunctions that are found in Arctic—Alpine species confirm the view that most elements of the Arctic flora survived the ice-age in the polar region. Fig. 3.7 shows the large gap between the polar and the more southern occurrences of *Poa alpina* and *Phleum alpinum*. Alopecurus alpinus occurs in Spitsbergen and Greenland with its next southern station in the southern uplands of Scotland. *Eriophorum scheuchzeri* (Fig. 3.8) occurs in Greenland and Spitsbergen and then again many thousands of kilometres to the south in Switzerland. In a survey of the geographical distribution of Arctic species Hadac (1960) concluded that many elements of the

Fig. 3.6 *Juncus trifidus* the three-leaved rush growing on a mountainside in the Jotunheimen (Norway). This was one of the species used by Böcher in his study which established the distinct nature of Arctic and Alpine biotypes for a number of circumpolar and Alpine species.

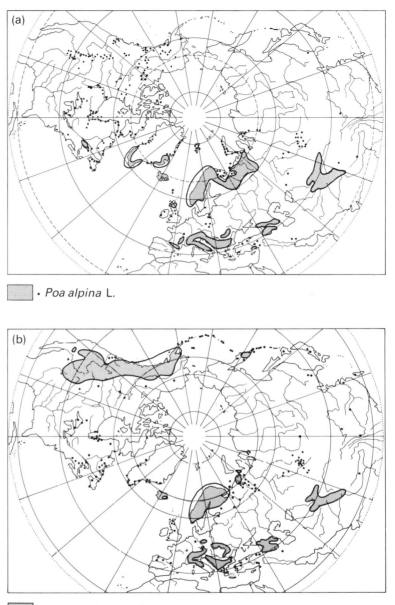

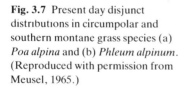

Fig. 3.7 Present day disjunct distributions in circumpolar and southern montane grass species (a) *Poa alpina* and (b) *Phleum alpinum*. (Reproduced with permission from Meusel, 1965.)

circumpolar flora were of considerable antiquity. Some of the species are thought to have been extant in the Arctic for up to two million years.

Thus our present Arctic vegetation is most probably descended directly from plants that survived the last ice-age in polar latitudes. These species will have been exposed over many generations to the most limiting conditions on earth in terms of temperature and shortness of growing season. The undisturbed nature of much of the Arctic even today provides an excellent opportunity for studying the responses of these long-established species to the environmental selection pressures of polar conditions. The plants of polar regions therefore belong

Fig. 3.8 *Eriophorum scheuchzeri* an Arctic and Alpine cotton grass with a disjunct distribution in Greenland, Iceland, Spitsbergen and Switzerland.

3.2.1 LOW AND HIGH ARCTIC

The tundra as well as accounting for a significant portion of the earth's land mass also extends through a number of different climatic regions. The north–south extent of Greenland is equivalent to the distance from south Norway to Sicily. It is therefore to be expected that within the Arctic zone many different vegetation types will occur. The principal phytogeographical distinction is made between the 'low Arctic', where shrub and graminaceous communities are characteristic and the 'high Arctic' in which the vegetation is sparse and contains many dwarf evergreen species.

This broad division covers a very wide range of vegetation types that occur throughout the entire extent of tundra vegetation. Low Arctic communities include relatively tall willow scrub with some birch as in the south of Greenland. Typically the graminoid form is also prominent and grasses and sedges comprise the bulk of the vegetation in sedge meadows, valley bogs, and grass slopes. Herb slopes are also well developed in low Arctic sites.

By contrast 'high Arctic' vegetation is poorer in graminoid forms than that from lower latitudes (Figs. 3.9 and 3.10). A typical 'high Arctic' shore-line and dune system lacks the usual grass species (*Leymus arenarius* and *Ammophila arenaria*), and instead rosette and mat forbs containing many saxifrage species are a principal feature of the vegetation. The disappearance of grasses as a dominant vegetation type from the 'high Arctic' is due to their pattern of shoot growth and development. The grass manner of growth involves the sequential production of leaves many of which do not last an entire growing season. Leaves that are produced sequentially do not make use of the entire growing season. Consequently this type of resource investment does not give an adequate return in the high Arctic and the amount of grass and sedge meadow is much reduced in both extent and species richness. Instead woody species with reduced exposure above ground such as the polar willow (*Salix polaris*) and evergreen heaths are more frequent (Fig. 3.10).

3.2.2 WIND EXPOSURE AND SNOW DEPTH

Thermal input, water, and nutrients are the principal limiting factors for plant growth in the Arctic and all three are influenced by the extent the plants are exposed to wind. Vegetation in the polar regions being diminutive in size does little to protect itself through the development of complex multilayered

to a much more ancient vegetation type than their immediate neighbours to the south, the conifers of the boreal forest. The interface between the tundra and the northern limits of the boreal forests is therefore not just a competition zone between two distinct vegetation types but the confrontation between the relatively new northern coniferous forests with long-established polar communities.

3.2 Arctic habitats and plant form

Arctic plants although small in size show much variation in both their above- and below-ground morphology. The sparse nature of plant cover does not generally permit the vegetation to alter significantly the environment in its own favour. Consequently the only amelioration that is provided in the habitat comes from topographical and microclimatic differences. Variations in plant form expose different tissues to varying degrees of climatic stress and are consequently very closely related to small changes in topography and microclimate.

Fig. 3.9 High Arctic vegetation with circular clones of the Arctic heather (*Cassiope tetragona*) in north-east Greenland. The size and regularity of the patches of this slow-growing species probably represent several centuries of undisturbed growth.

Fig. 3.10 Vegetation on a high Arctic shore. *Luzula spicata* with *Plantago maritima* and Alpine bearberry (*Arctous alpina*).

communities as in warmer climates. Microtopography has therefore a very marked influence on the pattern of vegetation development in Arctic habitats. Wind exposure results in thin snow cover, low soil water and reduced temperatures during the growing season. These adversities are heightened by the wind removing plant litter and thus depleting the system of nutrients. In addition such sites have well-drained soils which can lead to further acidification and the immobilization of phosphorus. Where snow cover is slight, high wind speeds (in certain parts of the Arctic) can affect plant growth. Abrasive damage to exposed leaves and buds favours prostrate growth forms. Exposed sites therefore have vegetation that is composed of drought-resistant plants with sclerophyllized leaves having a long life and a high resistance to the diffusion of water vapour and carbon dioxide. These evergreen plants are typified by the low pH tolerant heath species such as *Cassiope tetragona* (Fig. 3.9) and *Vaccinium vitis-idaea*. The Arctic heather (*Cassiope tetragona*) can also survive under considerable snow cover and both musk oxen and Arctic hares are in the habit of digging through snow in winter to gain access to the more nutritious stands of these species that occur on the lusher, snow-protected sites. With less wind exposure there is a deeper snow-lie, higher soil–water content and hence taller growth and greater nutrient storage in the above ground stems.

In sheltered habitats with minimal grazing the above ground plant cover will gradually increase. This inevitably reduces the amount of heat reaching the soil surface. Thus with increased plant cover there is a reduced thawing of the soil and the permafrost zone comes nearer to the soil surface. This in turn restricts root growth and reduces the cycling of nutrients. In the coastal experimental site at Barrow in Alaska (Miller *et al.*, 1980) it was noted in areas fenced off to exclude grazing that the vegetation became dominated by moss communities. Grazing and wind pruning are therefore essential for the continued survival of the larger plant types in the Arctic. At first sight it may seem surprising that an added burden in the form of herbivory and wind damage should increase survival chances.

3.2.3 SOIL DRAINAGE

Underground morphology of Arctic vegetation varies just as the exposed aerial portions of the plants do in relation to different habitat types. In all areas of the world vegetation changes in relation to soil drainage

with some species being restricted to well-drained sites and others appearing more successful in water-logged soils. In temperate regions the successional development of vegetation from xeric to mesic communities on one hand and from aquatic to terrestrial on the other does much to provide a range of mesic habitats where plants can have access to water resources throughout the year without enduring either frequent flooding or drought. The lack of vegetation development in the Arctic prevents any extensive development of mesic habitats and the absence of this buffering of the soil environment results in a marked polarization between dry and wet sites. In many parts of the high Arctic as in north-east Greenland the permanence of the polar high pressure zone produces very low amounts of precipitation which are frequently less than 127 mm per annum (5 inches). In such areas the hillsides, sand banks and screes become progressively drier after the disappearance of the winter snow. Meanwhile the bogs remain wet all summer as the cool climate does little to dry out their surfaces.

The differences between wet and dry sites for winter survival are evident in the photoperiodic responses of Arctic vegetation to the shortening days of late summer. In the dry exposed sites the plants become dormant and the foliage takes on autumnal colours well in advance of low-lying wetland vegetation. Exposed sites require the timely protection of above ground overwintering buds and the plants have to forsake the growth potential of the last days of summer in order to achieve a sufficient degree of winter bud-protection before the onset of winter. In the lowland areas the overwintering buds can be protected in the wet-muds of the better developed valley soils. Here plants such as the bog-cottons can continue growing until the very last days of summer without having to to allow time and resources for the development of special protecting bud scales.

3.2.4 TEMPERATURE AMELIORATION

Some morphological adaptations however do exist in Arctic plants which go some way to ameliorating the environment. One of the best known is the sun-tracking activities of certain flowers. Species such as *Dryas octopetela*, *Papaver radicatum* and *Ranunculus glacialis* keep their flowers facing the sun as it circles overhead during the perpetual Arctic day. This sun tracking combined with the parabolic shape of the flowers reflects the warmth from the sun's rays towards the centre of the flower. This can have a dual effect in that it makes the centre of the flower a warm spot and therefore attractive to potential pollinating insects. It also speeds up the process of growth and development for the sexual (and possibly in some cases asexual) production of seed.

The manner in which plants sense the direction of sunlight (heliotropism) has been the subject of debate for over 70 years. The suggestion made at the beginning of this century by Haberlandt (*see* Smith, 1984) that it is the widespread occurrence of pappilose epidermal cells acting as surface lenses that perceive the direction of light, still offers a promising line of investigation. Such cells would concentrate perpendicular light into a central zone of the inner wall. Oblique illumination of the epidermis would result in a reduction in illumination of the centre of the inner wall and an increase in the relative illumination of the flanking areas. It is still conjectural whether or not this gradient is sufficient to stimulate a growth movement of the leaf or petal so that the flower or leaf will reorientate itself with the sun's rays falling vertically on its surface. Paradoxically the largest flowered varieties of *Saxifraga oppositifolia* are found in the extreme north of the high Arctic (Fig. 3.11). It

is possible that the large petals may function as heat traps and give this form a thermal advantage in the extremely short growing seasons of northern Greenland and Spitsbergen.

Apart from leaf and flower orientation other plants use different methods to increase their tissue temperature. In the willows it has been observed (Krogg, 1955) that the temperature in the catkin surrounded by translucent down is several degrees higher than that of the surrounding atmosphere. It appears that the catkin hairs reflect solar radiation towards the inner surface and thus trap the infrared radiation (Fig. 3.12). The cotton grasses (*Eriophorum* spp.) seem equally well adapted to accomplish this same warming effect by the possession of their own self-developed glasshouses.

3.2.5 SLOPE AND ASPECT

The Arctic is remarkable in that it is one area in the world where the thermal conditions for plant growth frequently improve on going up the mountainside. This is due to two reasons. First, slope increases the angle of incidence of the sun's rays and contributes

Fig. 3.11 The large flower of the biotype of *Saxifraga oppositifolia* which is characteristic of the extreme northern populations of this species in Spitsbergen and northern Greenland.

Fig.3.12 A glasshouse effect produced by willow down over the developing catkins of *Salix arctica*. The temperature inside this canopy of down can be several degrees warmer than the surrounding atmosphere.

Fig. 3.13 High summer (July 14) in the high Arctic at Biskayerhuken Spitsbergen 79° 40′N. The bare patches in the foreground contain extensive stands of the Arctic willow *Salix polaris*.

greatly to the total energy input. Secondly, and even more importantly, is the escape from cold sea fogs that is enjoyed by mountain vegetation as opposed to vegetation at low levels near the coast. In inland areas there is even an improvement in temperature conditions on moving to the far north. Above 72°N the sun remains sufficiently high throughout the 24 hour cycle to permit the progressive attainment of considerable temperatures at ground level. It is this northern region that is sometimes referred to as the 'Arctic Riviera'.

Despite the climatic amelioration that can result from persistent insolation in high latitudes, species number and above ground exposure both decrease in northern regions. In the northern extremes of the high Arctic above ground storage organs are much reduced. In the woody plants the polar willow (*Salix polaris*) is one of the very few of its type to maintain a significant population as far north as the eightieth parallel (Fig. 3.13). The herbaceous perennial is the most successful life form in this region. Among annual plants there is only one species that survives in the high Arctic to demonstrate that this form of life cycle is just possible under such extreme conditions. The diminutive red-coloured annual species *Koenigia islandica* can be found throughout Greenland and Spitsbergen but becomes rarer in the north (Figs. 3.14 and 3.15.)

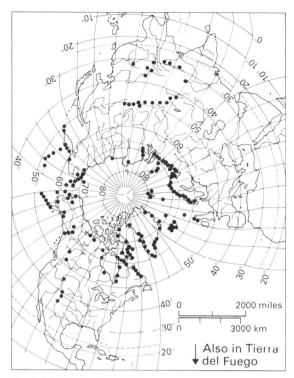

Fig. 3.15 Distribution of *Koenigia islandica* L. (Reproduced with permission from Löve and Löve, 1974.)

Fig. 3.14 A full-grown specimen (scale divisions = 10 mm) of the annual Arctic species *Koenigia islandica* (Polygonaceae).

3.3 Carbon balance

3.3.1 CARBON ALLOCATION AND NORTHERN LIMITS TO DISTRIBUTION

Arctic plants are remarkable for the high rates of photosynthesis that can be attained at low temperatures. In a survey of the photosynthetic rates for some Arctic species in Alaska, Tieszen, and Welland (1975) recorded values ranging from 7 to 31 mg CO_2 dm^{-2} h^{-1} which are comparable with rates achieved by many temperate species. Not only can plants of low temperature habitats match their photosynthetic rates with species from warmer climates but they can achieve these rates at lower temperatures. In northern species the limiting factor to the rate of dioxide fixation is closely related to the metabolic carboxylation power of the leaf (r_m = mesophyll resistance) and not to the limitations imposed by physical restrictions such as stomatal resistance (r_s) (Treharne, 1972). When ribulose bisphosphate carboxylase (the enzyme responsible for carbon dioxide fixation) is compared in northern and southern grass provenances then the northern form is always found to have the higher activity (Fig. 3.16). Acclimatiz-ation can also alter the carboxylation potential of Arctic species. When kept at low temperatures *Oxyria digyna* increases its ribulose bisphosphate carboxy-lase activity (Fig. 3.17) (Chabot *et al.*, 1972). The ability to increase metabolic rates at low temperatures by means of increased amounts of active enzymes is referred to as metabolic compensation. As a means of adaptation it is found in fish (Hochachka and Somero, 1973) as well as in plants. In poikilohydric organisms the maintenance costs of increased enzyme complements at low temperatures will be kept low as long as they remain within the low-temperature environment. As a general phenomenon low-temperature adaptation by metabolic compensation is one which although advantageous in the low-temperature habitat will be likely to be a disadvantage due to increased maintenance costs in other tempera-ture regimes and will therefore contribute to the

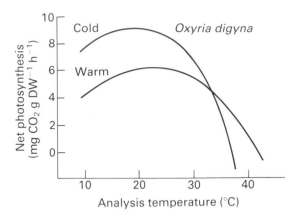

Fig. 3.17 Facility for temperature compensation in the Arctic form of *Oxyria digyna*. These plants from Ellesmere Island (north of Greenland) showed a higher net photosynthetic rate when grown under a cold temperature regime. (Redrawn from Billings, 1971.)

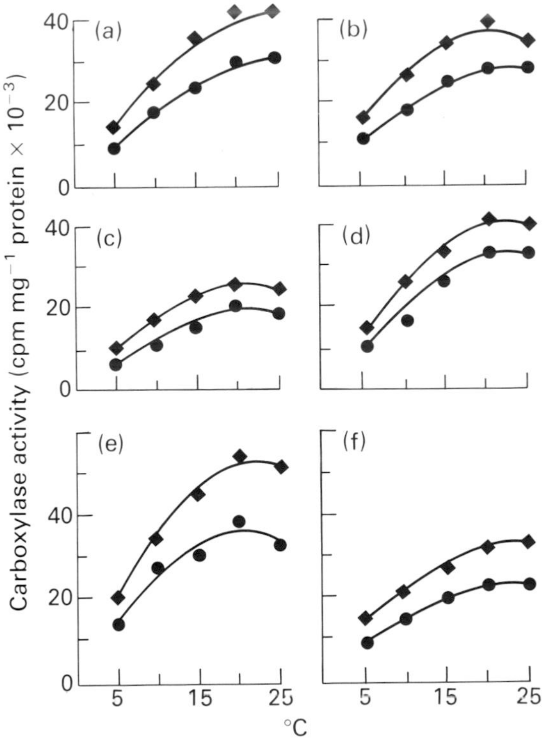

restriction of the adapted species to the Arctic habitat. As photosynthetic rates in the Arctic can match those of temperate plants it is not surprising therefore that comparable growth rates can also be observed (Fig. 3.18). The time over which these maximal rates are maintained is however much shorter in the Arctic.

If northern and southern provenances of grasses are placed in the dark together in the glasshouse in order to deplete their carbohydrate reserves it is possible to compare the rate at which they make good their losses on being re-exposed to light. The results of such an experiment are shown in Fig. 3.19 and again the superiority of the northern provenances is evident. As far as is known Arctic plants are exclusively C_3 plants. High ribulose bisphosphate carboxylase activity can act as a compensation mech-anism which increases the rate of carbon dioxide fixation at low temperatures. Under these conditions the inherent dangers of increased oxygenase activity and loss of carbon due to photorespiration will not be serious as temperatures are generally low. How-ever this type of adaptation will cease to be effective in increasing the carbon gain of plants in warmer habitats due to increased photorespiratory losses. It is therefore understandable that this form of adap-tation will be restricted to plants in the colder parts of their range and thus metabolically distinguishes Arctic from temperate biotypes of the same species.

The short growing seasons and low thermal input into Arctic habitats require plants that live in these

Fig. 3.16 Ribulose bisphosphate carboxylase activity in extracts of grass leaves (▲) northern and (●) southern provenances of (a) *Dactylis glomerata* (b) *Lolium perenne*, (c) *Holcus lanatus*, (d) *Festuca arundinacea*, (e) *Deschampsia cespitosa*, and (f) *Poa pratensis* (Crawford, 1982a).

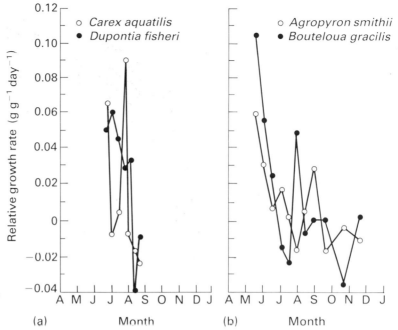

Fig. 3.18 Mean relative growth rates of graminoids observed *in situ* in (a) tundra and (b) temperate grassland. (Reproduced with permission from Chapin III *et al.*, 1980.)

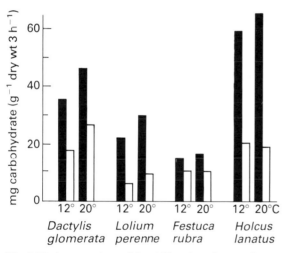

Fig. 3.19 A comparison of the ability of northern and southern provenances of grasses to replenish the carbohydrate content of their leaves after 48 hours of darkness. The northern provenances show consistently higher rates of carbohydrate replenishment than the southern provenances, at both 12 and 20°C.

areas to control carefully the allocation of their carbon resources. In addition to the short period normally available for photosynthesis, the Arctic has also a very variable climate. In bad years the

accumulated temperatures in day degrees can be so low as to result in no net photosynthetic gain being made by the plants. Table 3.1 shows the effect of a cold growing season on the Arctic grass *Dupontia fisheri* growing in Alaska. A fall of just over 1 degree was sufficient to reduce the net photosynthesis by 27 per cent. Therefore in maintaining an adequate partitioning of their carbon resources between growth and reserves there will have been selection for an ability to conserve sufficient carbohydrate reserves to last not just from 1 year to another but to tide over an entire missing growing season if need be.

The possession of large carbohydrate reserves also means that northern plants differ from those in the south in that a much greater proportion of the carbon dioxide that they fix is allocated to the maintenance of underground organs. The large root systems of Arctic plants and their considerable carbohydrate reserves while enabling these plants to survive long periods with little photosynthetic gain require nevertheless a considerable energy input just for their maintenance. The bog rosemary (*Andromeda polifolia*) growing in the Swedish tundra translocates 75 per cent of its fixed carbon below ground. Similarly in an overall estimation for grass species in Alaska 59 per cent of the carbon dioxide fixed was found to be translocated below ground where half was used just for maintenance respiration

Table 3.1 Response of the Arctic grass *Dupontia fischeri* to warm (1972) and cold (1973) growing seasons. (Adapted from Chapin III *et al.*, 1980.)

| | Temperature (°C) | | | Biomass (g m^{-2}) | Net annual photosynthesis (g CO$_2$ m^{-2} yr^{-1}) | Phosphorus in *Dupontia* leaves | | Phosphorus uptake (μmoles hr^{-1} [g fresh wt root]$^{-1}$) |
	Air	Canopy	Soil			(% dry wt)	(mg tiller^{-1})	
1972	5.1	8.1	5.0	125 ± 19	620	0.163 ± 0.017	0.105	0.041
1973	4.0	5.5	3.1	93 ± 13	450	0.526 ± 0.068	0.105	0.039
Difference	1.1	2.6	1.9	32	170	0.363	0.0	0.002
% difference	−22	−32	−38	−25	−27	+223	0	−5

and the other half contributed to new biomass (Chapin III *et al.*, 1980). By comparison, prairie grasses translocate 57 per cent of the carbon fixed below ground but use 85 per cent of this portion to contribute to new growth and only 15 per cent for maintenance. The policy of the northern plants in maintaining adequate reserves to combat the thermal uncertainties of their environment places them at a disadvantage in terms of growth potential with less well-insured species. Thus it is not in their photosynthetic capacity that Arctic plants differ from those further south but in the manner in which they allocate their reserves.

Frequent reference is made in the literature on Arctic vegetation to the role of lipids as storage foodstuffs either in the stems or evergreen foliage of Arctic and Alpine plants. The analyses which led to these suggestions were based on observing total lipid content and calorific values (Hadley and Bliss, 1964; Larcher *et al.*, 1973). Such generalized estimations of lipid content do not distinguish between membrane phospholipids or glycolipids, and structural or storage neutral lipids such as waxes and the acyl-glycerols. These acyl-glycerols are the only lipid components which are likely to contribute directly to the energy reserves of plants. An examination with careful fractionation of all lipid classes of leaf lipid contents throughout a 12 month period in lowland and montane forms of *Empetrum nigrum* (Hetherington *et al.*, 1984a) has shown that although lipid content can be high in this species in both the lowland and montane varieties of *E. nigrum* it is not in the form that contributes to the energy turnover of the plant. The triacylglycerols in the two subspecies of *Empetrum nigrum*, *nigrum* and *hermaphroditum* contained respectively only 4.5 and 1.4 per cent of the total lipid present. Attractive as the idea is, therefore, there is no evidence in this species that high lipid content will provide any contribution to the over-wintering carbohydrate supplies. The high lipid values for such plants are most probably due to their extensive cuticular wax production (Fig. 3.20). It would appear that only in seeds do lipids act as storage compounds in higher plants.

3.3.2 CARBON ALLOCATION AND SOUTHERN LIMITS TO DISTRIBUTION

More enigmatic than the northern limits to distribution are the southern limits that are found to the dispersal of many Arctic plants. The nature of this enigma is illustrated strikingly by the spider plant (*Saxifraga flagellaris*) which does not occur south of 75°N in either Greenland or Spitsbergen. In both areas this saxifrage grows in open shingle habitats and there are no grounds for believing that it is competition with other more aggressive species that limits its southern spread. When visiting this plant at 75°N (Fig. 3.21) it is difficult to imagine what there might be in more southern climates that is inimical to the survival of this species. Limitation in southern distribution will be caused by competition in many species but there is a significant number of examples where this explanation is not adequate. Coastal habitats with their open communities which frequently have to be renewed after storms give ample opportunity for the establishment of fresh colonists. The continuous nature of coastlines also provides a habitat which extends through changing climatic zones without imposing any topographical barrier to plant dispersal. Starting on the shores of Greenland and Spitsbergen there are several species which enjoy an uninterrupted distribution as far south as the British Isles and even northern France. Examples of such plants include the herbaceous perennials *Ligusticum*

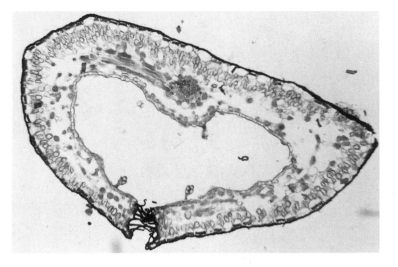

Fig. 3.20 Transverse section of leaf of *Empetrum nigrum* ssp. *nigrum* stained with sudan IV showing the prominent wax cuticular layer which accounts for the higher lipid content reported in the leaves of some Alpine and montane species. This lipid will not contribute to the carbohydrate reserves of this species.

scoticum and *Mertensia maritima* (Figs. 3.22, 3.23 and 3.24).

The unique advantage of the herbaceous perennial as a life form in the Arctic stems from its ability to save up carbohydrate from one year to another and to use the accumulated reserves to initiate rapid growth in the spring. Plants such as the Scottish lovage (*Ligusticum scoticum*) are remarkable for the speed with which they extend their foliage when growth recommences after winter. Such rapid growth, exploiting the reserves of previous carbon gains requires high respiration rates. A typical response of *L. scoticum* to temperature in its respiration rate is shown in Fig. 3.25. The rapid increase in dark respiration rate with temperature is also found in *Mertensia maritima* (the oyster plant) and contrasts with the lower respiration rates of two comparable coastal species of more southern distribution.

The depletion of carbohydrate reserves by high respiratory activity at warm temperatures has often been suggested as a limiting factor in the southward extension of northern species (Björkman and Holmgren, 1961; Mooney and Billings, 1965; Stewart and Bannister, 1973). An examination of the carbohydrate content of various *Vaccinium* species growing in Scotland showed that the species which suffered the severest depletion of carbohydrate reserves at warm temperatures *V. uliginosum* was the most restricted in its southern extension (Fig. 3.26). Thus for perennial herbs the consequence of having high respiration rates to exploit carbohydrate reserves in short cool growing seasons places the plants at a disadvantage in warmer climates.

Table 3.2 shows the length of time required to produce a 50 per cent depletion of carbohydrate reserves in four maritime species that differ in their

Fig. 3.21 The spider plant *Saxifraga flagellaris* growing on an Arctic shore at 75° 30′N in Spitsbergen. This species occupies open habitats in Arctic shores with apparently no barriers to its southern migration yet is not found south of 73°N.

(a)

(b)

Fig. 3.22 Arctic species that reach the southern limit to their distribution on the shores of the British Isles and Denmark (a) Scots lovage (*Ligusticum scoticum*) on the shore of a Norwegian fjord (b) The oyster plant (*Mertensia maritima*) on a pebble beach in Orkney.

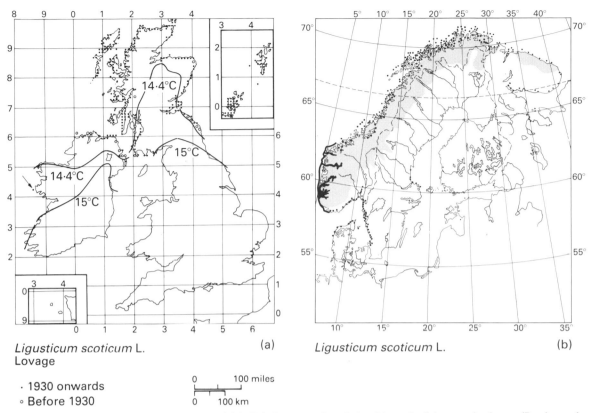

Fig. 3.23 Distribution of *Ligusticum scoticum* (a) in Britain – note the relationship to the July mean isotherms (Perring and Walters, 1976) and (b) in Scandinavia (Hulten, 1971; reproduced with permission from Esselle Map Service, Stockholm.)

north–south distribution limits. The northern species will lose 50 per cent of their reserves in 14–17 days at 25°C while the southern species can endure this temperature twice as long before dark respiration produces a similar depletion in their reserves. For the southern species warm winter temperatures will therefore be less likely to lead to carbohydrate starvation after a mild winter than in northern species. The damage caused by these warm winter temperatures is unlike heat injury in that no visible sign or tissue lesion is observed. Plants transplanted to the south of their natural range may grow for a season or two until they have exhausted their carbohydrate reserves. Spring-starvation will then prevent the resumption of growth and the plant will just disappear. This will not be the case for all Arctic plants as some can be successfully grown in the open in temperate gardens. Böcher (1972) noted that it was the plants of dry habitats that persisted best outside his Arctic greenhouse. It is often assumed that species of dry habitats possess rhizomes and tap roots more frequently than plants from Arctic wetlands where adventitious roots are more usual. Tap roots and rhizomes will conserve greater carbohydrate supplies and this may be a contributory factor to their persistence in the open in temperate environments.

3.4 Nutrition

Arctic soils fall into two broad classes: (1) skeletal soils of screes, dunes, and fellfields and (2) the peat accumulating types of the bogs and sedge meadows. In both types the climatic conditions of the Arctic severely limit the rate of nutrient input. Permafrost by reducing the available soil depth restricts the area of potential weathering and accumulation of nutrients. The low-temperature regimes of polar regions also mean that chemical weathering of the soil is negligible. Thus the main sources of nutrients

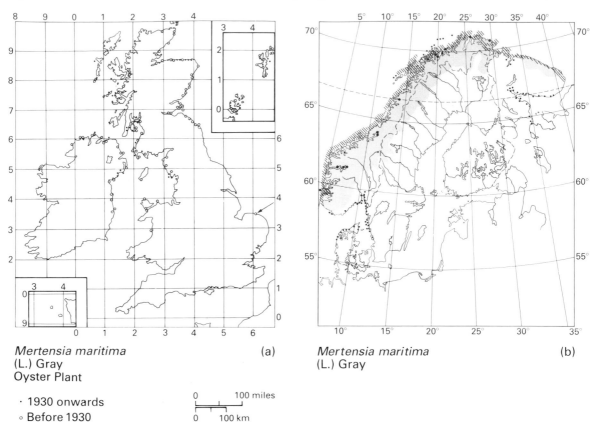

Mertensia maritima
(L.) Gray
Oyster Plant
(a)

Mertensia maritima
(L.) Gray
(b)

· 1930 onwards
∘ Before 1930

Fig. 3.24 Distribution of *Mertensia maritima* (a) in Britain (Perring and Walters, 1976) and (b) in Scandinavia (from Hulten, 1971; reproduced with permission from Esselle Map Service, Stockholm.)

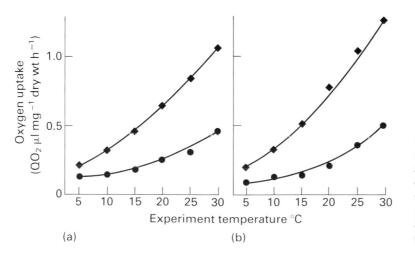

(a)   (b)

Fig. 3.25 Respiration rate at a range of temperatures of (a) *Ligusticum scoticum* (■) a northern species and *Crithmum maritimum* (●) a southern species; (b) *Mertensia maritima* (■) a northern species and *Limonium binervosum* (●) a southern species. (Crawford and Palin, 1981.)

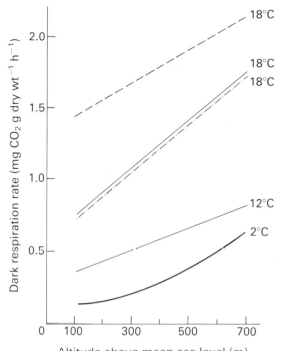

Fig. 3.26 Relationship between respiration rate (y) and altitude (x) elucidated by multiple regression analysis in *Vaccinium uliginosum* (– – – – –), *V. myrtillus* (———), and *V. vitis-idea* (–·–·–·). (Reproduced) with permission from Stewart and Bannister 1974.)

the short summer period. As a result the input of nutrients from the atmosphere is an order of magnitude less than that found in temperate ecosystems.

In addition to these low input rates, the Arctic has the lowest nutrients cycling rates of any ecosystem with typically only 1 per cent of the ecosystems nutrients occurring in the living biomass. In this 'locking up' of nutrients in the non-living material the Arctic reaches yet another extreme end-point in comparison with other world ecosystems (Chapin III *et al.*, 1980). Where there is an adequate supply of moisture as on wet, mossy sites there is active growth of blue green algae which account for the bulk of nitrogen fixation in most Arctic habitats. Although some nitrogen comes into the system through snowfall and rain this is more than lost by the run-off that takes place with snow melt in the spring. In the low temperature conditions of the Arctic, nitrogen fixation is severely limited by temperature and the input even in favourable coastal sites in Alaska is only 5 per cent of the nitrogen that cycles through the vegetation annually (Chapin III *et al.*, 1980). This low contribution to nitrogen turnover can be compared with a figure of 30 per cent for shortgrass prairie (Woodmansee *et al.*, 1978).

Much attention has been attracted to the discovery that in northern Canada *Dryas drummondii* possesses nitrogen fixing nodules (Lawrence *et al.*, 1967). During the period that the *Dryas* occupied a morraine in a successional series it was estimated that the nitrogen content of the soil rose by 305 kg per hectare (272 lb per acre). The nitrogen fixing capacity of the nodules was confirmed using N[15] and has excited speculation as to whether during the immediate post-glacial *Dryas* period this ability to fix nitrogen was more widespread throughout the genus. If it were, it would do much to explain the extraordinary pioneering capacity of this plant. However

for the vegetation have either to be through precipitation or from biological sources. The presence of the Arctic high pressure zone means that precipitation is low. The persistence of sea-ice for the greater part of the year also reduces the nutrients that come from ocean spray to the few storms that may occur during

Table 3.2 Calculated loss of carbohydrate (as glucose) expressed as per cent loss of original (total) dry wt. The time for loss of 50 per cent of dry wt was calculated directly from these values.

	1 day		1 week		Time to lose 50% of dry wt in days	
	5°C	25°C	5°C	25°C	5°C	25°C
Ligusticum scoticum	0.7	3.0	5.0	20.9	70	17
Mertensia maritima	1.0	3.5	7.0	24.3	50	14
Crithmum maritimum	0.4	1.3	2.7	9.4	131	37
Limonium binervosum	0.5	1.5	3.5	10.5	100	33

as yet there is no evidence that the *Dryas* populations at present found within the Arctic (Fig. 3.27) are in any way different from those that existed 12 000 years ago in the general absence of nitrogen fixing capacity. In addition to free-living blue-green algae there are also those living in association with lichens which contribute to the nitrogen gains of the Arctic ecosystem. The lichen genus *Peltigera* in association with the blue-green alga *Nostoc* is one of the most active combinations.

Soil phosphorus is also extremely low in Arctic soils. The only widespread source is from snowfall and precipitation. In the IBP study at Barrow in Alaska it was estimated that the level of available soluble phosphorus was so low that it would have had to be replenished in the soil system 220 times in a growing season or three times per day to account for that taken up by the plants. On a similar basis available nitrogen would have had to be replenished 60 times during the growing season. So slow is the accumulation of nutrients in the Arctic that it has been calculated that the standing crop of nitrogen and phosphorus in some sites at Barrow in Alaska would take respectively 10 000 and 70 000 years to regenerate at the present rate of input. In certain habitats such as bird cliffs abundant supplies are available and it can be a perpetual astonishment to Arctic travellers who have seen nothing but small and stunted plants to arrive at a bird cliff and come face to face with luxuriant vegetation. Fig. 3.28 shows the growth achieved by early July in *Alopecurus alpinus* growing below a bird cliff at 79° 50′N in Spitsbergen a mere 630 nautical miles from the North Pole.

The importance of animals in recyclying nutrients has therefore very profound influences on Arctic ecology and no discussion of plant nutrition in the Arctic would be complete without considering the question of fluctuations in lemming populations (Fig. 3.29). The rise and fall of lemming populations in the cycles that are so characteristic of Arctic habitats appear to be linked to both the extent of predation

Fig. 3.27 *Dryas octopetala* at its seeding stage in late summer on a hillside in north-east Greenland enjoying the warmth above the the cold sea fog that fills the fjord below.

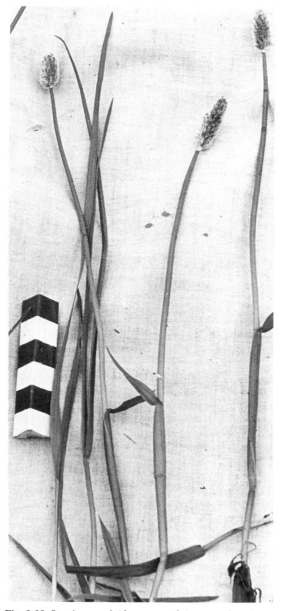

Fig. 3.28 Specimens of *Alopecurus alpinus* growing at 79°N beneath bird cliffs in Spitsbergen. These plants were photographed at the beginning of July and show the rapid growth that is possible at only 11°S of the North Pole provided nutrients are not limiting.

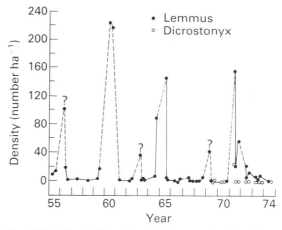

Fig. 3.29 Estimated lemming densities averaged for all habitats in coastal tundra at Barrow for a 20-year period. The question marks indicate numbers based on observations other than trapping. (Reproduced with permission from Batzli *et al.*, 1980.)

and the forage quality of the vegetation (Schultz, 1969). Significant rises in lemming populations take place when their is sufficient winter forage to permit breeding under the protection of snow cover. Under these conditions the lemming population can increase rapidly free from the attacks of predators. Thus three factors, suitable forage, adequate snow pack, and low predator density all contribute to the build up of high lemming populations. Dramatic reductions in populations take place when sustained winter breeding has produced populations that can no longer be supported. Such population peaks due to a 100-fold population increase occur every 3–6 years. The pattern is not regular as many modifying factors including weather appear to operate. However the very high populations that develop from time to time do much to recycle the nutrients that are locked up in the standing vegetation biomass. The effect of plant nutrition on the growth and maintenance of lemming populations has stimulated a number of studies comparing field observations with computer simulation models (Batzli *et al.*, 1980). The preference of lemmings for the Arctic grass *Dupontia fischeri* may be related to the tendency of this grass to grow in nutrient rich areas. Relatively large amounts of calcium and phosphorus are retained in the forage and these nutrients may be limiting for lemming reproduction as well as for plant growth. From computer simulation studies of nutrient uptake from forage it could be estimated that in years when nutrient content in the forage was low then adult females (120 days old) could barely support litters of seven during midsummer, four during winter and two during late August. It would appear that in lemmings calcium and phosphorus deficiencies in the diet are more limiting than energy supply. This may perhaps explain the paradoxical situation in lemmings compared with temperate mircotines in that they show a lower digestive efficiency than their counterparts from warmer climates. Lemmings process a large volume of forage not so much for its calorific value as for its use as a source of mineral nutrients. The Arctic environment in this way acts on plants and lemmings alike in that both growth and reproduction are limited by the availability of mineral nutrients. Lemmings eat considerable quantities of moss but if their diet is restricted to moss alone they reduce their intake and rapidly starve. The mosses appear to act as a mineral supplement as they are 10−20 per cent higher in phosphorus and 200−300 per cent higher in calcium than graminoids. In winter lemming reproduction requires higher energy input due to the greater thermal demands of reproductive activity at low temperatures.

Due to locking up of nutrient in forage the extent of the consumption of plant material by lemmings will have a profound influence on plant growth as well as the subsequent nutritional status of the forage after locked-up nutrients have become available again from lemming urine and faeces. Thus after a period of low grazing the vegetation will be relatively nutrient poor although extensive amounts of forage material will have accumulated. As lemming populations rise, this forage material may be reduced but the quality of the subsequent plant growth in terms of nutrient status will be higher thus supporting higher lemming reproduction. The population can then rise very rapidly if the correct conditions of snow cover and predator evasion are possible. Large populations eventually collapse due to the simultaneous action of large predator populations and the exhaustion of adequate energy supplies in the forage material.

In the larger grazing animals such as reindeer, caribou and musk oxen (Figs. 3.30, 3.31 and 3.32) population size is also related to food supply. Differences in life history however produce entirely different population effects in relation to the potential changes in population size. Assuming 100 per cent survival (Batzli *et al.*, 1980) it can be shown that lemmings have the capacity to increase their population by a factor of 1300 in 1 year while the factor for caribou is only 1.5. In addition due to the small body size, energy flow through the lemming is considerably greater than that of caribou. Lemming respiration rate (weight for weight) as estimated in populations at Barrow is about 50 times that of caribou. Caribou migrate and rely on increasing

Fig. 3.30 An Arctic hare in north-east Greenland. (Photograph by courtesy of E.F.B. Spragge.)

body weight in the summer to survive the following winter. The reindeer population that is native to Spitsbergen can double their body weight during the summer grazing season and therefore concentrate on using high quality forage for a short period rather than low quality forage throughout the year as is the case with the lemming.

Because of the above differences in grazing pattern it is the lemming which has the greatest influence on vegetation among Arctic herbivores. Exclosure of lemmings from trial plots in studies in Alaska have shown that in well-drained sites carpets of mosses and lichens developed and graminoids became less plentiful. In low wetland sites graminoids continued to develop although the standing crop increased and productivity declined. Apparently heavy lemming grazing disrupts moss growth which only recovers slowly after grazing. Graminoids recover more readily due to the availability of underground reserves in rhizomes. Soil analyses also show decreased phosphorus in enclosed areas protected from grazing.

3.5 Reproduction

A comparison of the metabolic cost of producing a new grass shoot by sexual reproduction via a seed or

Fig. 3.31 Reindeer in Spitsbergen. The short-legged form of animal that is native to Spitsbergen can double its body weight during the summer grazing season.

by the growth of a new tiller based on rough calculations of the carbon costs of inflorescence production, percentage seed set, seedling survival, etc, suggests that sexual reproduction is some 10 000 times more costly than vegetative reproduction (Chapin III *et al.*, 1980). In regions where carbohydrate supply is limiting it is therefore not surprising that asexual reproduction is very common. It has been estimated that 80 per cent of all the angiosperm species in certain Scandinavian floras possess some means of asexual reproduction (Gustafsson, 1947). The form of asexual reproduction can be very varied. Stolons and rhizomes are probably the commonest form of asexual reproduction. As has been already pointed out carbohydrate storage in stolons and rhizomes is favoured as a means of perennation

in many Arctic species. Apomixis (agamospermy, the production of embryos without fertilization) is also common in Arctic genera such as *Taraxacum, Antennaria, Hieracium* and *Potentilla*. However despite the high metabolic costs and the many other uncertainties of success due to lack of pollinators and the variability of the seasons, sexual reproduction is not abandoned even in the high Arctic. A number of Arctic species have adaptations which allow them to reproduce both sexually and vegetatively. Noteable among these species is the highly successful Alpine bistort (*Polygonum alpinum*) which produces both flowers and vegetative bulbils on its flowering axis (Fig. 3.33). The same strategy of having both vegetative and sexual reproduction taking place on the same flowering stem is found in

Fig. 3.32 Musk oxen in north-east Greenland. These relatively large herbivores survive largely by grazing the Arctic willow *Salix arctica* which they can find even in the months of darkness of the polar night.

Fig. 3.33 The viviparous Alpine bistort (*Polygonum alpinum*) growing on a mountainside in the Jotunheimen (Norway). This species uses the flowering stalk to produce both sexual and asexual propagules. Vegetative bulbils grow on the lower stalk while seeds are produced in the flower at the top of the stalk.

Saxifraga cernua. Other species overcome the problem of short growing seasons by completing the process of sexual reproduction over several seasons. In 1 year flower initials may be laid down followed by the development of pollen in a subsequent year, flowering then taking place the year after, and finally seed maturation and dispersal in the next growing season. In a study of the developmental time sequences (phenology) of Arctic plants Sorensen (1941) showed that in *Pyrola grandiflora* (Fig. 3.34) and *Pedicularis palustris* as well as several other species it could take up to 7 or 8 years from the initial formation of a shoot until flowering and then fruiting were finally accomplished.

3.5.1 POLYPLOIDY AND APOMIXIS IN COLD CLIMATE VEGETATION

Angiosperm reproduction varies with the degree of ploidy and as the development of polyploidy varies geographically (Fig. 3.35) this in turn superimposes geographical differences on the mode of reproduction. In many plant groups the degree of polyploidy increases in colder climates (Haskell, 1952). The degree of polyploidy in the monocotyledons is higher than that found in dicotyledons and Haskel has argued that as the ancestors of modern monocotyledons were aquatic plants then this selection for cold wet habitats would have favoured polyploidy. Ecologically a relationship between soil temperature and polyploidy has been found in a study in a single valley in Alaska (Johnson and Packer, 1965). The valley at 68°N was in a region that was free of ice during the Pleistocene period. The valley bottom was always wet and cold and subject to intense frost heave, with the permafrost boundary only 0.5 m below the surface. The valley sides were both warmer and drier than the ground below with less frost action and the permafrost zone at least 3 m below the soil surface. They found that the extent of polyploidy was greatest in both monocotyledons and dicotyledons in the cold lowland sites. A number of explanations have been advanced to account for this phenomenon. Stebbins (1971) believes that climatic cooling itself is not directly responsible but that it is the disturbance of the vegetation with species combinations being brought into proximity in unusual combinations that is the root cause. These species juxtapositions have then produced various outcrossing, backcrossings and chromosome doublings that have led to polyploid complexes which have produced rich gene pools enabling certain plants to adapt to invade new habitats when the ice-sheets retreated. A putative example of

Fig. 3.34 The Arctic wintergreen (*Pyrola grandiflora*) growing in north-east Greenland at 72°N. Species such as this can take several years to complete the reproductive cycle from the initial formation of a flowering shoot to the dispersal of seed.

a species originating in this way is the rare *Saxifraga nathorstii* which occurs only in north-east Greenland in cold wet sites with an extremely short growing seasons (*see* Fig. 3.11). In this area it has been suggested that the putative parents *S. oppositifolia* and *S. aizoides* are induced to flower at the same time thus giving rise to the hybrid. Visits in recent years (Balfour, personal communication) to northeast Greenland record different flowering times for the two parents. However given the great seasonal variation that takes place in relation to temperature

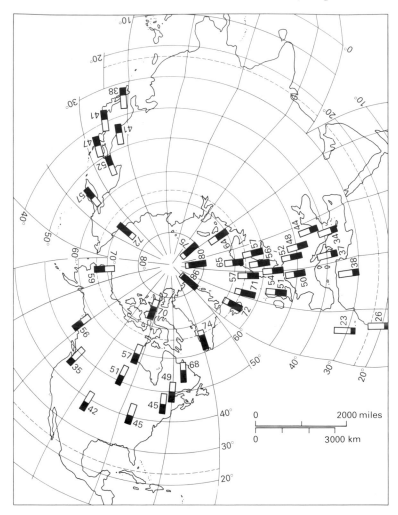

Fig. 3.35 Frequency of polyploids in the floras of the northern hemisphere. Full column denotes 100 per cent. (Reproduced with permission from Löve and Löve, 1974.)

and hence plant development it may be that the breeding barrier between these species has been penetrated from time to time in the long history of Arctic species.

Given the great antiquity of the Arctic flora and the absence of any direct evidence that new polyploid biotypes have immigrated into the Arctic since the end of the ice-age there is no reason to neglect the alternative view to that of Stebbins, namely that polyploid species are better adapted physiologically to cold climates. In a comparison of diploid and tetraploid rye (*Secale cereale*) it has been shown that the tetraploids are less able to withstand high temperatures (Hall, 1972). At high temperatures it appears that the higher respiration rates of root meristems are limited by the increased thickness of the root which restricts the rate at which oxygen can diffuse to the respiring tissues. The conclusion from this

experiment is not that polyploids are better adapted to cold conditions but that they are less adapted to warmth.

Possibly more significant than any direct link between polyploidy and physiological tolerance is the greater degree of heterozygosity that will arise due to gene duplication in polyploid species. The degree of heterozygosity will also remain fixed in those species that reproduce vegetatively or by apomixis. The Arctic flora is rich in apomictic species. Given the increased presence of polyploid species in the Arctic this is not unexpected as polyploidy and apomixis are commonly associated with each other in plants. There are however a number of other factors which favour the production of apomictic species in the Arctic. The first is the ability to produce embryos without the uncertainty of waiting for a suitable insect pollinator. Secondly in any disturbed habitat where the opportunities for establishing a new cohort only arise from time to time then the preservation of a suitable genotype suited to a particular microhabitat will be favoured in any system which abandons the hazards of sexual recombination.

It is probable that the existence of apomictic species in the Arctic is a phenomenon of great antiquity. In the genus *Taraxacum* there are at present approximately 2000 apomictic species and 50 species with normal sexual reproduction (Richards, 1973). The present abundance of apomictic species in Europe is thought to be due to the sexually reproducing species coming in contact with the polar apomictic species with the retreat of the Pleistocene ice-sheets. The apomictic species are not entirely isolated genetically as they can produce pollen which will transfer genetic material to sexually reproducing individuals. In this way the genes favouring apomictic reproduction have spread out of the Arctic into the northern temperate regions since the last ice-age (Richards, 1973). In the United States dandelions are presumed to have been introduced from Europe by early settlers and are devoid of sexual reproduction. There are fewer microspecies all of which are triploid with no diploids providing residual sexuality as in Europe (Solbrig and Simpson, 1974).

3.5.2 DIOECISM IN ARCTIC VEGETATION

Specialization usually increases efficiency and with regard to sexual reproduction it is of interest to note the conspicuous position of dioecious plants in the circumpolar flora. In the world flora as a whole dioecious species only account for about 4 per cent of all flowering plants (Bawa, 1980). On a species basis Arctic floras also show a similar percentage occurrence of dioecious species. However, in relation to their relative contribution to the vegetation dioecious species are very conspicuous in the Arctic. Apart from the willows which can be a dominant feature of the vegetation cover in the extreme north there are many other dioecious species, for example *Oxyria digyna, Rubus chamaemorus, Sedum arcticum,* and *Silene acaulis.*

Dioecious species probably evolved from the monoecious condition. Self-incompatible species could have become self-compatible and then dioecious. The enforced outcrossing may have generated sufficient genetic variability to allow these species to colonize a wide variety of Arctic habitats.

In a study in Spitsbergen of the distribution of the sexes in the most northerly willow *Salix polaris*, which involved examining over 1300 plants taken over a wide range of locations a predominance of female over male plants was found at most sites. Over the entire survey the female plants averaged 59.1 per cent of the willows in flower (Crawford and Balfour, 1983). A similar survey in Iceland on *S. herbacea* showed 59.3 per cent of the plants to be female. This tendency for female plants to outnumber male plants in Arctic willows has subsequently been confirmed by other expeditions and Table 3.3 summarizes their results. After noting the female predominance in *Salix arctica* in north-east Greenland

Table 3.3 Summary of percentages of female willows recorded by various researchers in populations of northern and Arctic willows of varying species.

Species	Location	Latitude	Per cent Females	Reference
S. polaris	Spitsbergen	79°N	59.1	Crawford and Balfour (1983)
S. herbacea	Iceland	67°N	59.3	Crawford and Balfour (1983)
S. arctica	Denver Island, Canada	75°N	60.5	Dawson (personal communication)
S. arctica	North-east Greenland	73°N	57.8	Crawford and Balfour (unpublished)
S. arctica	North-west Greenland	77°N	61.2	Crawford and Balfour (unpublished)
S. glauca	West Greenland	67°N	68.6	Crawford and Balfour (unpublished)
S. myrsinifolia × *S. phylicifolia*	Umea North Sweden	63°N	57.8	Danell *et al.* (1985)

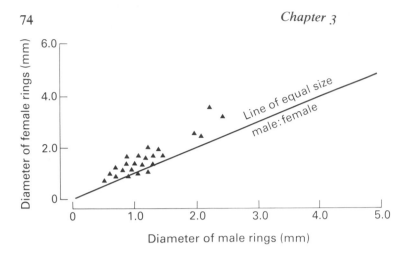

Fig. 3.36 Plot of mean annual growth ring width of female versus male stems of *Salix arctica* in north-east Greenland. A paired *t* test shows female mean annual ring width to be significantly greater than the male annual ring width (*P*<0.05).

an analysis of leaf form and growth rates in male and female plants showed that the females had larger leaves and grew faster as estimated from their annual rings (Fig. 3.36).

The coincidence of these values in two similar species over a widely scattered range of sites in different regions prompts speculation as to the selective forces which could conceivably achieve this degree of equilibrium between male and female plants. Individual selection will favour equality of the sexes at least in seeds (Fisher, 1930). Differential mortality between male and female plants could modify this and would have the greatest opportunity to do so in long-lived species such as dwarf willows. In the Arctic, regeneration by seedlings can be expected to take place by cohorts when conditions are favourable either for seed production or seedling establishment. Such conditions can also arise from sudden perturbation in the environment when landslip or frost-heave produces a new site for colonization. Under such conditions populations that have a greater quantity of seed to disperse are more likely to establish themselves than those of lower fecundity.

Genetically, the effective size of any population is not the total number of individuals of all ages but the breeding population (Wright, 1931). As populations fluctuate in size or as the sexes depart from equality the effective size approximates to the lower number. More precisely the effective population size (N_n) of any population made up of N_m males and N_f females is given by:

$$N_n = \frac{4N_m \times N_f}{N_m + N_f}$$

Deviations from the equilibrium ratio of 50 : 50 in sex distribution have little effect on the effective

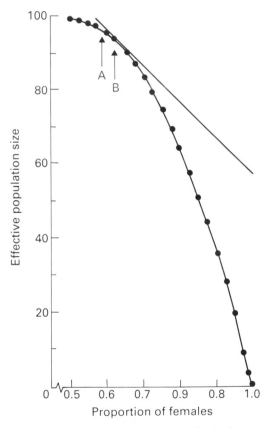

Fig. 3.37 Relationship between the variation in the sex distribution from equality and the effective breeding size of any population (Wright, 1931). The arrow A represents the observed population of female plants of *S. polaris* in Spitsbergen and *S. herbacea* in Iceland. The arrow B is the hypothetical equilibrium point between increase in seed bearing population and decrease in effective breeding population size (Crawford and Balfour, 1983).

population size (N_n) until the percentage of female plants exceeds 62.5 per cent (Fig. 3.37). Any increase in the proportion of females above 62.5 per cent by reducing the effective populations size (N_n) will reduce progressively the level of heterozygosity in the population and lead to severe inbreeding depression effects. Some balance appears therefore to have been struck in all these Arctic willows which maximizes the fitness of the individuals when the sex ratio approximates to 60 : 40, female to male. The exact mechanism by which this was achieved is not clear. However it is possible that differential mortality in the different sexes may play a role.

In a study of *Salix polaris* in Spitsbergen (Crawford and Balfour, 1983) it was also noted that leaf resistivity differed between male and female plants. Paired observations of male and female leaves frequently showed that female leaves had higher resistance values than the male leaves. This suggested that on these occasions the female plants were more efficient than the males in restricting water loss. As water deficits are frequent in the skeletal soils of the Arctic this could be a factor which might allow female plants to predominate. Although there is much bare ground in the Arctic, sites suitable for willow growth are usually occupied. In these favourable sites competition between male and female plants will be lessened if there is increased mortality in males due to less effective stomatal control. Thus populations that developed this degree of physiological differentiation between male and female plants would be able to establish the most fecund stands of willow and would be in the best position to perpetuate this trait when the opportunity presented itself for seedling establishment. In *Salix arctica* it has been found that female plants both grow faster and produce larger leaves than male plants. In this species therefore vegetative reproduction alone might account for the excess of female plants over males. However it is still not clear how the constancy of the 60 : 40 female to male ratio is attained with such regularity in all the willow species so far examined in the Arctic. It is more usual to note that in dioecious species female predominant sex ratios are rarer than male predominant sex ratios. In tropical or temperate conditions there are many examples of male predominance. Any habitat which hinders the distribution of pollen will put male plants at a genetic advantage. Thus in areas of lush vegetation where much pollen is inevitably lost selection for parents favouring male offspring can be expected. Conversely in the treeless exposed habitats of Spitsbergen, Greenland, and Iceland conditions are reversed and ovules rather than pollen are at a premium and thus female biased sex ratios prevail.

3.6 Conclusions

This chapter has outlined a number of features where Arctic plants show positive adaptations which can be interpreted as increasing their survival capacity. In all the major biological activities discussed, namely, respiration, photosynthesis, and reproduction it is possible to discern attributes that are a potential advantage in the Arctic but would place the plants at a disadvantage in warmer habitats with longer growing seasons. The vegetation of the Arctic cannot therefore be dismissed as an assemblage of passive survivors with slow growth rates and sluggish metabolic activity which can only survive in polar refugia. The plant life of the Arctic is a highly specialized group of plants of great antiquity with a variety of metabolic, developmental, and reproductive mechanisms which have evolved in direct response to maximizing their efficiency in low temperature environments.

Fig. 4.1 Rowan (*Sorbus aucuparia*) growing at 610 m (2000 feet) above sea level in Glen Clova, Scotland. The rowan ascends readily to 975 m (3200 feet) in the British Isles, higher than any other tree.

4

Polar and altitudinal limits to tree survival

4.1 Timber-line geography

The definite line that is frequently produced by the upper limit to tree growth on mountainsides is a visible boundary to plant survival and consequently has attracted much ecological speculation. So definite is this line in certain situations that it has prompted many investigators to search for a common factor that limits the altitudinal establishment of trees. Before attempting to relate tree survival to any particular consequence of higher altitude or latitude it is important to look at the characteristics of the trees at different timber-lines to see whether or not they can be compared with one another. The climatic conditions which exist at the timber-line clearly differ between temperate and tropical mountains as well as between the Arctic and Alpine boundaries to tree growth. In all these areas geology and climate may also interact to produce different soil types. In the Arctic the existence of a permafrost zone will produce limitations that differ from those that restrict tree growth on temperate and tropical mountains. In some regions altitudinal timber-lines may not be due to climate. The upper regions of the mountains of Java are treeless due to the hostile nature of the young volcanic rocks (Walter, 1962). In different parts of the world there are not only different species at the timber-line but different types of tree. In some regions the timber-line is exclusively evergreen conifers while in others it is made up entirely of deciduous broad-leaved trees. The common ecological factor is merely that this boundary represents a limit to the successful survival of the tree form. This chapter attempts to examine the varying climatic factors that produce this boundary and their action on the growth, development, and regeneration of trees.

4.2 Tropical timber-lines

In the tropics the upper limit to tree growth is at the altitude that coincides with the rapid decline in rainfall at the top of the troposphere. The troposphere is the lowest of the atmospheric zones and in tropical regions the cloud zone that frequently lies in its upper regions produces a humid climate with minimal temperature fluctuations. In mountainous areas, particularly in warm temperate to tropical regions with limited rainfall at lower altitudes, the cloud zone often supports a distinctive and well-developed forest. Above the cloud zone precipitation falls off rapidly. This generally does not produce such a definite timber-line as is seen in temperate mountains. Instead there is a rapid reduction in tree size with altitude which culminates in the production of dwarf shrubs and finally mountain desert. In Central America, *Pinus* and *Quercus* species are predominant at the timber-line. In Mexico as precipitation declines with increasing altitude the timber-line is marked by a thinning out of the stands of *Pinus hartwegii* rather than by a reduction in size. In Asia numerous species of *Lithocarpus*, *Vaccinium*, and *Rhododendron* form the transition zone between forest and mountain desert or permanent snow. In the Himalayas the timber-line at 4000 m is the native habitat of the birch *Betula utilis* and the large so-called tree-rhododendrons which although predominantly bush-like in having more than one stem can produce massive trunks up to more than 18 m (60 feet) in height. *Rhododendron arboreum* can grow to up to 24 m (80 feet).

In Africa the sides of Kilimanjaro have a cloud forest with an upper limit at 2800–3100 m. The best development of the forest is on the south-west side of the mountain facing the direction of the monsoon arrival. The north-east side of the mountain lies in a rain-shadow area where its development is much reduced. The upper tree boundary here consists of species of *Podocarpus* and *Hagenia* and heather bushes (*Erica arborea*) which can be over 6 m in height. The size of this shrub vegetation then gradually decreases as precipitation falls with increasing altitude. In the Canary Islands, the Canary pine (*Pinus canariensis*) maintains a relatively high timber-line at 1900–2000 m due to its ability to intercept and trap small drops of moisture out of the north-east trade winds on the enormous surface area of the pine needles contained in the forest canopy (Fig. 4.2). Rain gauges placed in the natural precipitation zone on the sides of El Teide, the gently smouldering volcano that produced the island of Tenerife, record only 510 mm (20 inches) of precipitation. When

Fig. 4.2 Forest of *Pinus canariensis* emerging above the cloud zone at 1800 m (5900 feet) in Tenerife, Canary Islands. Note the long needles which give the forest the ability to condense dew sufficiently to quadruple the annual precipitation.

these gauges are placed within the forest the precipitation can be as much as 2030 mm (80 inches). The needles of the canary pine are borne in threes in densely crowded clusters up to 270 mm in length, the longest of any species of pine. This great needle length increases the surface area of foliage available for the condensation of dew.

It is a curious feature of tropical mountains that as the true tree species disappear, due to inadequate precipitation, they are replaced by tree forms of genera that are normally herbaceous plants. Thus on the upper slopes of African mountains are found the famous tree *Lobelia* and *Senecio* species, while in South America the large erect tree-like flowering stalks of the genus *Puya* and *Echium* are a striking feature (Figs. 4.3 and 4.4). This development of the

tree form by genera that are normally herbaceous is found in many families in different parts of the world. These so-called pachycaul plants have been the subject of some debate as to whether they are primitive forms that have survived due to their ecological isolation (Corner, 1964) or whether their existence is due to parallel evolution in isolated mountain areas in response to particular climatic conditions. The primitive relict view was first put forward by Darwin in his *Origin of Species* as most of the first discoveries of pachycaul plants were made on remote oceanic islands. However, further botanical exploration has found this particular life form in many other areas including the Andes where the north–south orientation of the mountain chains allows a ready migration route and thus does not

Fig. 4.3 *Puya* sp. a pachycaul plant with Professor Cesar Vargas of the University of Cuzco above the Andean timber-line at 4300 m (14 000 feet).

support the isolation-relict theory. In addition there is no evidence from wood anatomy to support the primitive-relict theory in these plants. It is more likely that the habit of these Alpine Afro-Andean plants has arisen as an adaptation to wide diurnal temperature ranges. The dense rosette of leaves closes at night and this combined with thick leaves and many hairs serves to protect the growing point from frost (Hedberg, 1964). In the higher regions of tropical mountains plants have to tolerate a frost risk all the year round and are unable to exploit seasonal differences for the development of frost resistance combined with a cessation of growth. If growth is to take place, then the plant has to maintain actively growing tissues, ready to expand and develop by day and yet have them protected from frost by night. This physical rather than cellular protection

mechanism developed by the pachycaul plants is well suited to this particular temperature situation.

In a re-examination of the natural climax vegetation of the high altitude inter-Andine steppe lands (altiplano) of Peru, Ellenberg (1958) concluded that the natural timber-line lay at an altitude of 4500 m and that the present line of 4000 m is artificially low. Much of the land over 4000 m in Peru and Bolivia has been treeless for a long time. The vast grasslands of the altiplano made a strong impression on the earliest Spanish *conquistadores* (Prescott, 1847) so that the antiquity of this feature, together with the high altitude led ecologists to consider the area a natural steppe formation. However it seems probable that the grazing of these upland areas has been intensive ever since the rise of the Inca empire in the eleventh century AD. We know from historical

Fig. 4.4 *Echium wildpretti* a pachycaul species growing at 2750 m (9000 feet) in the montane arid zone of El Teide, Tenerife (Canary Islands).

accounts made at the time of the Spanish conquest (1528–31) that the herds of llamas and alpacas were so large that grazing could not be found for them (Garcilaso de la Vega, 1608). This example of depression of the upper timber-line due to grazing is a general feature of all upland areas and is not confined to the drought-prone areas of tropical mountains. The Peruvian altiplano is however one where the antiquity of grazing pressure is historically well documented (Fig. 4.5).

4.3 Temperate timber-lines

In temperate forests it is possible to group timber-lines into two broad classes irrespective of the actual species that are present. Either there is an abrupt end to tree cover or else the upper limit for the growth of well-formed trees is followed by a zone of stunted or twisted trees referred to as *Krummholz* (German, *krumm* = twisted, *Holz* = wood).

4.3.1 ABRUPT TIMBER-LINES

When no *Krummholz* develops and where there is a definite timber-line without a zone of outlying individual trees it appears that the sensitive stage in tree establishment is when the seedling starts to emerge from the short ground vegetation. Young tree seedlings outside the cover of parent trees become progressively more vulnerable to exposure and fluctuations of the climatic conditions on the upper regions of the mountain. Where a timber-line has advanced up a mountainside during an earlier more favourable climatic period the adult trees can occupy a post-climax position. In this situation exposure, short growing seasons, drought, and frost-heave can all act to eliminate young seedlings as rapidly as they attempt to establish themselves. In shade-tolerant deciduous trees the young seedling however can generate under the cover of the parent trees. The forest thus perpetuates itself despite climatic deterioration but the post-climax position results in few seedlings being found above the established timber-line which therefore has a very abrupt appearance.

Such a situation is found in the southern Andes where for over 2000 km in a north–south direction the timber-line is made up of almost pure stands of a deciduous southern beech *Nothofagus pumilio* (Fig. 4.6). These beeches survive as erect trees to the highest point of their establishment without the production of any twisted *Krummholz* forms. In New Zealand another species of southern beech *N. solandri* also forms a distinct timber-line without the production of scrub forms. These trees at the upper limits of their altitudinal distribution appear to have established themselves at one particular time, possibly during periods of favourable weather conditions. Once established these mature trees have greater environmental tolerance than their seedlings. During

Fig. 4.5 The treeless altiplano of Peru (altitude 4200 m) where deforestation has resulted from centuries of overgrazing by llamas and alpacas.

Fig. 4.6 The well defined timber-line of the southern beech (*Nothofagus pumilo*) in south-western Patagonia. This timber-line extends north–south along the Andes for 2000 km.

(a)

(b)

Fig. 4.7 Natural timber-line of the evergreen silver beech (*Nothofagus menziesii*) at 1200 m at Haast Pass, New Zealand. (a) General view of extent of natural tree boundary. (b) Close up of marked community boundary between silver beech and sub-Alpine scrub. (Photograph by courtesy of Professor A.T. Mark.)

periods of climatic deterioration the mature trees may survive but no further advance of newly established trees will take place outside the canopy of the parent trees. In shade-tolerant trees, such as the southern beech, regeneration can take place under the canopy of the parent trees. Thus, the established stand can perpetuate itself by regeneration of protected seedlings in habitats too harsh for the colonization of new territory by seedlings. Trees, such as the southern beech, as well as other species which form a timber-line with no *Krummholz* are usually deciduous and shade tolerant (p. 88) (Wardle, 1974).

4.3.2 ECOTONE TIMBER-LINES

The clear timber-lines of these *Nothofagus* forests of New Zealand (Fig. 4.7) and South America (Fig. 4.6) are in striking contrast with the extensive development of tree scrub or *Krummholz* that is found in the boreal forests of Europe and North America. In these regions the mature tree is more severely affected by the adverse conditions of high altitude or latitude than the seedling. Seedlings can usually establish themselves in these regions above the altitude or latitude of the parent plant. However, with time they become increasingly affected by adverse climatic conditions as they raise their branches and terminal buds into zones of increased exposure and away from the level of protection afforded by winter snow cover. This results in the frequent die-back of the terminal bud producing trees of stunted growth with strongly twisted stems. Not only is the growth of the trunk misformed, but the shoots are often chlorotic and poorly cuticularized. In the western United States needles of species such as *Picea engelmannii* show all these symptoms at the timber-line together with a predisposition to desiccation by winter winds (Wardle, 1968). Exposure similarly affects the uppermost populations of *Betula pubescens* in Scandinavia and Scotland. The scrub form of the *Krummholz* vegetation also minimizes the effects of exposure on the plant and in some species there is a genotypic fixation of the twisted *Krummholz* form (*see* p.84). Where *Krummholz* develops it is usually possible to distinguish three distinct boundaries. The first is the upper limit to the existence of closed forest which is referred to as the *forest limit*. This is then followed by a region of isolated trees whose altitudinal or latitudinal limit is termed the timber- or tree-line (German = *Baumgrenze*). Above this line is the *Krummholz* zone where only stunted tree forms are found. Between the upper limit of the closed forest and the top of the *Krummholz* zone there is a transition or ecotone zone which is described by German-speaking ecologists as the *Kampfzone* (*Kampf* = struggle). In English this is more usually termed the timber-line ecotone (Fig. 4.8).

In Scandinavian timber-line birch forests, *Betula pubescens* is represented by a genetically distinct form, *B. pubescens* var. *tortuosa*. This type of birch

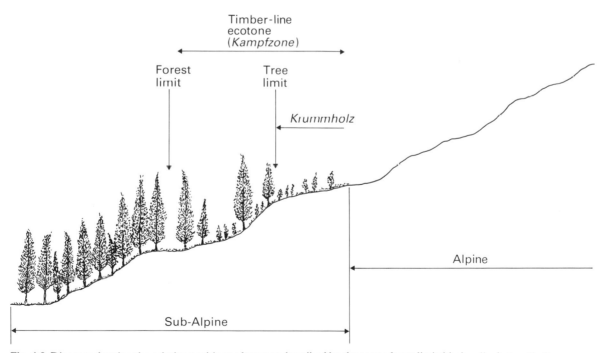

Fig. 4.8 Diagram showing the relative positions of trees as described by the terms forest limit (timber-line), tree limit, timber-line ecotone and *Krummholz* (*see* text).

Fig. 4.9 *Betula pubescens* var. *tortuosa* growing at the timber-line in the Hardangervidda region of Norway.

keeps its twisted *Krummholz* form even when transplanted to lower and more favourable habitats (Fig. 4.9). Presumably the twisted and lower-growing form of this variety of birch has had a greater survival capability at the timber-line and has therefore been preferentially selected. A similar situation is found in pines. In eastern Siberia and Japan, *Pinus pumilo* forms scrub communities which apart from their stunted growth form are taxonomically identical with the erect *P. cembra* which is a characteristic tree in the upper regions of montane forests in central Europe and Asia (Fig. 4.10). The reverse situation geographically also occurs as *P. mugo* which is found as a timber-line bush in the Alps (Fig. 4.11) is considered by some authorities to be conspecific with the eastern *P. uncinata* which grows as a well-formed tree.

The permanence of these stunted forms suggests that the production of scrub *Krummholz* is not entirely due to the direct pruning effects of exposure. It is possible that *Krummholz* formation is a direct growth response regulated by internal growth factors which are produced as a result of poor growing conditions. *Krummholz* specimens of *Picea engelmannii* have a higher content of the growth regulator

scopulin than normal trees (Love *et al.*, 1970) which may be related to the stunted appearance of the trees. In the southern Andes both tree and stunted forms of *Nothofagus antarctica* grow side by side with each other (personal observation). If the sites are wet then there is usually a predominance of scrub trees. Poor soil aeration (*see* Chapter 5) can also alter internal concentrations of growth regulators which can be related to changes in tree form.

There are also many instances where there appears to be a genetic predisposition to low growth forms. In areas of cool temperatures, where heat is a limiting factor plants that keep their foliage near the ground will achieve higher leaf temperatures than those with more exposed foliage (Grace, 1977) and thus reduced size will have a selective advantage. There is also physiological evidence to support the view that there is genetic selection for specialized tree populations at the timber-line. In a study of root respiration in *Abies lasiocarpa* and *Picea engelmannii* at high and low elevations the plants from near the timber-line were shown to have slightly higher respiration rates than those collected at low altitude sites (Sowell and Spomer, 1986). In rowan (*Sorbus aucuparia*) winter-resting buds col-

Fig. 4.10 A timber-line stand of *Pinus cembra* at an altitude of 2300 m near Zermatt, Switzerland.

lected from higher altitudes have markedly higher respiration rates from those collected at sea-level (Fig. 4.28). However in this latter case the differences may be phenotypic.

In areas free from too much human interference as in the mountain areas of Scandinavia there is a characteristic birch forest zone above the level of the pines and spruces. Apart from the permanent *Krummholz* already discussed, the form of the birch trees that make up this upper forest is extremely variable. In the genus *Betula* there is much hybridization and introgression. In northern regions frequent introgression between the dwarf birch *B. nana* and the common birch *B. pubescens* produces a range of intermediate types. This may be a contributory factor to the production of the subspecies *B. pubescens*

ssp. *odorifera* which occurs frequently in the northern and upland regions of Scotland. This form of birch always produces shrubby specimens with several stems arising form the base of the bush. Such a growth form minimizes exposure and increases the ratio of productive (photosynthetic) tissue to non-productive tissue in the plants. By contrast, the closely related species *B. pendula* the silver birch, does not adopt the shrub form and is limited to lower altitudes in both Britain and Scandinavia. In Scotland the growth of the silver birch requires a minimum of 1100 day degrees, whereas *B. pubescens* can survive in areas where the accumulated temperature is as low as 600–670 day degrees (Forbes and Kenworthy, 1973). The tendency to increase the investment in productive tissues is found in other

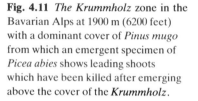

Fig. 4.11 *The Krummholz* zone in the Bavarian Alps at 1900 m (6200 feet) with a dominant cover of *Pinus mugo* from which an emergent specimen of *Picea abies* shows leading shoots which have been killed after emerging above the cover of the *Krummholz*.

Fig. 4.12 A last remnant of natural *Krummholz* formation of *Pinus sylvestris* in Scotland at 640 m (2100 feet) in the Cairngorm Mountains above Loch an Eilean.

woody species of montane habitats. In the crowberry (*Empetrum nigrum*) there are two forms; ssp *nigrum* which has a lowland distribution and is dioecious and a montane or northern form ssp *hermaphroditum*. In the lowland form, typically 67 per cent of the annual production has been noted as being taken up by the leaves, whereas in the vicarious mountain form 85 per cent of the productivity is used in leaf growth (Flower-Ellis, 1973).

In the British Isles very little forest remains at the

level of the true potential altitudinal limit for the growth of trees. In the Cairngorms there is a small stand of pine growing at 640 m (2100 feet) (Pears, 1967) which has all the typical characteristics of *Krummholz* and which appears to be genetically distinct from other pines growing in the area and which do not develop this extreme contorted growth form (Fig. 4.12). The present timber-line lies at 530 m (1750 feet) but this level is largely artificial being depressed by constant grazing and periodic burning (Fig. 4.13).

4.3.3 DECIDUOUS AND EVERGREEN TIMBER-LINES

The question of the relative advantage of the deciduous to the evergreen habit at the timber-line has to be answered in relation to the specific climatic conditions that exist in any particular area. In large areas of Europe and North America the timber-line is occupied by evergreen forest. Yet on the southern side of the European Alps in the Apennines and Pyrenees deciduous species such as the beech (*Fagus sylvatica*) together with various species of *Acer*, *Sorbus*, and *Populus* are also found at the timber-line. Fig. 4.14 is taken at just below the timber-line on southern side of the Spanish Pyrenees where at 41°N there is a mixed forest of *Fagus sylvatica*, *Fraxinus excelsior*, *Buxus sempervirens* as well as *Pinus sylvestris* and *Abies alba* growing at 1450 m. In northern Argentina a deciduous timber-line that extends southwards along the backbone of the Andes

begins in the north with *Alnus jorullensis* and then continues south from 40 to 55°S with the deciduous upland beech *Nothofagus pumilo* (*see* Fig. 4.6). In Tasmania another species of southern beech *N. quinnii* forms the timber-line but in New Zealand it is the evergreen species *N. solandri* var. *cliffortioides* that is the uppermost tree species. It is only in the extreme south of the Andean range near Tierra del Fuego that the evergreen species *Nothofagus betuloides* replaces *N. pumilo* at the timber-line.

The relative advantages of leaf longevity over annual renewal raise the issue of optimization of resource utilization (*see* p.41). In lowland areas where temperature is not limiting the advantage of the evergreen habit is generally supposed to be its value in nutrient conservation. However, at the timber-line the relationship between summer and winter warmth appears more important regarding the relative success of the deciduous versus the evergreen habit. If there is sufficient winter warmth for active photosynthesis then the evergreen habit will potentially have an advantage. However, this advantage is realized only when summer warmth is a limiting factor for tree growth and winter photosynthesis can make a significant contribution to annual productivity. Thus in oceanic areas with mild winters and cool summers evergreen species will have a productivity advantage over deciduous trees. Consequently in the British Isles in the large areas of upland and marginal land planted for forestry this century the evergreen sitka spruce (*Picea sitchensis*) has been the favoured species. In the oceanic climate of the British Isles this tree is particularly productive

Fig. 4.13 A non-regenerating remnant of upland forest of *Pinus sylvestris* in Easter Ross, Scotland, showing mature trees which have survived the burning of the heather (*Calluna vulgaris*) undergrowth.

Fig. 4.14 *Fagus sylvatica* near the timber-line on the southern side of the Spanish Pyrenees near Pamplona at 1450 m (4750 feet) at 41°N.

as it is capable of making a net photosynthetic gain in every month of the year. However in areas where the winters are cold and the summer relatively warm as on the southern side of the Alps this advantage is lost and the deciduous *Fagus sylvatica* is able to survive at the timber-line. Similarly the harsh winters and warm summers of the Patagonian Andes favour the deciduous forms of the southern beech. Only in the coldest regions of northern Europe and Asia are the evergreen genera *Pinus* and *Picea* replaced by the deciduous larches (*Larix europea, L. dahurica,* and *L. gmelinii*).

The existence of a deciduous or evergreen timber-line has a consequence on whether the timber-line possesses a *Krummholz* zone or not. The deciduous trees are generally more shade tolerant than conifers and thus permit the establishment of their seedlings under the shade and climatic protection of the parent trees. In this manner the trees are able to maintain themselves as discussed above with a timber-line that is perched above the level that would be reached if such protection was not available. Deciduous tree timber-lines are therefore more likely to represent the maximum altitude where at some time in the past it was possible for saplings to establish themselves in the open. Thus the timber-line in these areas represents a post-climax vegetation type that is able to maintain itself due to the fact that it alters the microclimate in favour of its own regeneration. By contrast, in evergreen taxa, phenotypically or genetically induced *Krummholz* forms occur with timber-lines that represent the current altitude or latitude at which the establishment of trees is climatically possible.

4.4 Climate and timber-lines

4.4.1 TEMPERATURE

On a world basis, when timber-lines are plotted against altitude it is possible to show a clear relationship between latitude and timber-line. Daubenmire (1954) has calculated that in northern temperate zones there is a reduction of 110 m in the timber-line for every 1 degree increase in latitude. In addition the position of the timber-line coincided with the altitude where the temperature of the warmest month was 10°C, thus confirming the general applicability of Köppen's rule that tree growth requires a minimal summer warmth represented by a mean maximum temperature for 1 month of 10°C (Figs. 4.15 and 4.16). As already discussed (p.81) in the Andes between 38° and 55°S there is a unique timber-line made up of one species *Nothofagus pumilo*. This one species is the principal tree for a north–south distance of over 2150 km. Fig. 4.17 shows the close linear regression of latitude against altitude that is found with this species. For every degree of latitude polewards the timber-line is reduced by 80 m.

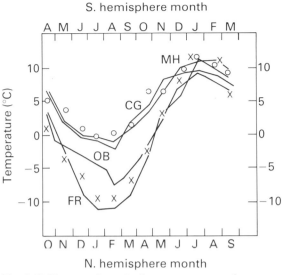

Fig. 4.15 The annual march of temperature near the natural timber-line at several geographical locations. OB, Obergurgl, Austria (2070 m); FR, Front Range, Colorado (3350 m); o, Craigieburn Mountains, New Zealand 1340 m); x, Haugastøl, Hardangervidda, Norway (988 m); MH, Moor House, England, (558 m). (Reproduced with permission from Grace, 1987; *see also* Grace, 1977).

The southern hemisphere contains relatively less land and more ocean than the northern hemisphere and the polar regions are covered by a much more extensive ice-cap than that of the Arctic Ocean and

Fig. 4.16 Treeless landscape in Orkney at 59°N. This archipelago has been practically treeless since the arrival of neolithic man over 5000 years ago. Tree regeneration is slow in oceanic climates and burning and grazing have a more deleterious effect than in areas with warmer summers. Wind itself, although severe and causing severe pruning is not thought to be a natural limiting factor to tree regeneration.

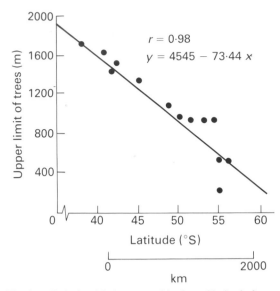

Fig. 4.17 Relationship between altitude and latitude for the South American deciduous beech *Nothofagus pumilo* between 40° and 56°S.

its surrounding lands. This is mirrored in the lower altitude of permanent snow cover in the southern hemisphere. Consequently in the southern hemisphere the timber-line is depressed below the level at which it would be expected if comparisons with the northern hemisphere were made on a basis of summer warmth only. Taking 47°N and 47°S as a reference point the timber-line in the Rockies lies at 2300 m and in the Austrian Tyrol at 2150 m. In the Andes however at this latitude tree growth is not found above 1100 m. In Norway the birch forest timber-line at 62°N on the Dovrefjell is found at about 900 m. Comparing this altitude with that found in the Austrian Alps represents a reduction of 80 m for every degree of latitude, a figure that matches the South American timber-line depression rate for the southern beech. Europe has a more oceanic climate than that of North America and therefore those European forests exposed to Atlantic weather have lower timber-lines. Thus in western Europe the timber-line in the Scottish Cairngorm Mountains at 57°N lies at 640 m and only rises to 1450 m at 41°N in the western section of the Spanish Pyrenees. This represents an increase of only 50 m per degree of latitude, only half the value recorded for the North American Rockies (110 m for each degree of latitude).

It can therefore be expected that relatively small changes in mean annual temperature will alter the level of the timber-line. It has been estimated that a fall of one degree would be sufficient to move the cultivation of the vine south by 450 km (Lamb, 1967) and depress the timber-line in central Europe by 100–200 m (Firbas and Losert, 1949). There is not sufficient palaeoecological evidence to substantiate the accuracy of these predictions. However, recent studies in the Sierra Nevada (Scuderi, 1987) have combined dendrochronological evidence from living trees and standing snags with radiometric data from relict logs of *Pinus balfouriana* to reconstruct the changes in the elevation of the timber-line over the past 6300 years. The data shows that the upper forest level of this species is sensitively adjusted to climatic variations. Elevations of 70 m above the present timber-line were found at this early period, followed by the establishment of a relatively stable timber-line between 3200 and 2800 yr BP. In the warmer period that existed at the beginning of this millenium 950–850 yr BP there was a rise of 10 m in the elevation of the timber-line. The subsequent onset of the 'little ice-age', however, did not appear to have had any significant effect on this timber-line. In southern Sweden a comparison was made of the position of the birch tree-limit at 213 sites between 1970 and 1975 and compared with data collected from similar sites between 1915 and 1916 (Kullman, 1979). In 75 per cent of the sites a rise in the timber-line was recorded with no change being noted at the remaining sites. From an age study of the stems the rise in the position of the birch-line was due to colonization that took place in the 1930s when a series of warm summers caused early snow thaw and rapid drying of the soil. This colonization had come to a halt by 1950. Birches once established during the favourable period continue to grow even although the climatic conditions are less favourable than when they became established.

It is difficult to assess the relationship between available heat and tree growth in terms of monthly mean temperatures. A quantitative measure of total heat available during the growing season is best obtained by summing the daily temperature above a threshold value and obtaining a figure for day degrees. However, even this calculation presents problems in making direct use of meteorological data in relation to plant physiology. The climatic data that comes from meteorological stations is taken from instruments that lie within weather screens and do not reflect the conditions that affect exposed foliage. This discrepancy between meteorological data and the conditions affecting exposed plants is particularly marked in mountain areas where tissue (leaf and meristem) and air temperatures can deviate substantially (Fig. 4.18) (Grace, 1987; Tranquillini and Turner, 1961). Thus, when attempts are made to correlate timber-line limits with the accumulated temperature above a base-line of 10°C, a figure of 600–700 day degrees is found for polar timber-lines while Alpine timber-lines coincide with the 200–300 day degree limit (Davitaja and Melnik, 1962). However, when the calculation is based on leaf temperatures both polar and Alpine timber-lines lie where the annual accumulated temperature reaches approximately 800 day degrees. The marked increase in leaf temperature above the ambient in the Alps is caused by the greater intensity of radiation in Alpine areas as compared with polar regions.

4.4.2 WIND, OCEANICITY AND TREE SURVIVAL

Early studies by Rydberg (*see* Grace, 1977) noted that the timber-line was higher in extensive mountain areas where it was argued that wind speeds would be less. Similarly numerous observations have suggested that in wind-swept areas (Figs. 4.16, 4.19 and 4.20) tree establishment only takes place with difficulty. In the Shetland and Orkney Islands to the north of Scotland it is often assumed that their treeless nature is due solely to wind. It is a common conclusion among those attempting to grow trees in extreme oceanic conditions with high wind exposure that it is wind rather than temperature that limits tree establishment. The Orkney Islands to the north of Scotland have probably been treeless for over 5000 years and are frequently considered to be naturally treeless. It has been argued that at about the time that Orkney became treeless there was a deterioration in the weather. The evidence for this stems mainly from increased sand deposits blown into lochs indicating greater wind speed (Keatinge and Dickson, 1979). However, it is equally possible that these increased ground wind speeds were due solely to the removal of tree and scrub vegetation. There is considerable evidence to connect the disappearance of trees from this northern region of the British Isles with the arrival of neolithic man. In the Northern Isles of Orkney and Shetland the extensive peat cover that now exists only began to develop with the arrival of neolithic man.

In the island of Hoy in a small valley there can still be found a relict of what was probably a much more extensive natural woodland (Fig. 4.20). At present it

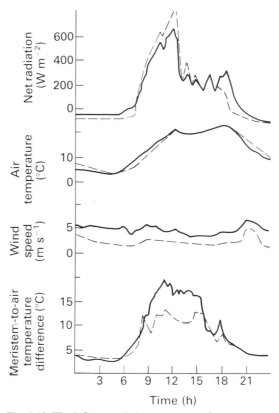

Fig. 4.18 The influence of plant stature on the temperatures of apical meristems near the timber-line in the Cairngorm Mountains; (– – – –) 2–3 m tall stand of *Pinus sylvestris* at the timber-line at 650 m; (———) less than 10 cm tall dwarf shrub community of *Arctostaphylos uva-ursi* at 700 m. Note that the highest apical temperatures are recorded within the dwarf shrub community. (Reproduced with permission from Grace, 1987.)

is the most northerly natural woodland in the British Isles. Due to the depopulation of Hoy, grazing pressure has been removed from this area and this relict woodland is now expanding and regenerating. A comparison of the growth rates of the birch trees in this wood with similar forests in Norway shows that these Orkney trees have a growth rate that compares favourably with that found in similar forests further north in Norway (Fig. 4.21). Here again it would be erroneous to conclude that the nature of the Orkney landscape is due solely to climatic causes. Trees will grow and regenerate despite the exposed and windy conditions. The process of regeneration is

Fig. 4.19 Papa Stour a windy exposed island of the west coast of Shetland that has been continuously settled by man for over 4000 years. The island has been denuded of all woody scrub vegetation and all peat has been completely removed. This remote, yet over exploited island has also the unenviable reputation of having fewer species of flowering plant than any equivalent area in the British Isles.

Fig. 4.20 Britain's most northerly natural woodland at Berriedale in Orkney (59°N) containing mostly birch, rowan and aspen (right foreground).

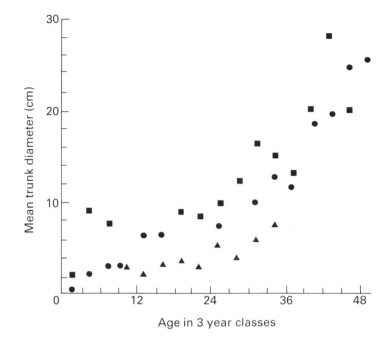

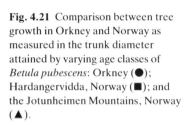

Fig. 4.21 Comparison between tree growth in Orkney and Norway as measured in the trunk diameter attained by varying age classes of *Betula pubescens*: Orkney (●); Hardangervidda, Norway (■); and the Jotunheimen Mountains, Norway (▲).

however extremely slow where existing cover has been removed. Oceanic climates are very deficient in summer warmth particularly in northern regions. The cooling effect of wind on surface temperatures further reduces the heat input to plants in already cool oceanic climates. It is these temperature effects which limit the rate of tree growth rather than the physical force alone of wind. Given time and freedom from human disturbance trees can build up their own protection against wind. However if there is some human disturbance, through burning, grazing, and felling, these disturbances will more rapidly produce a treeless landscape in an oceanic climate than in areas with greater regeneration potential. In both Orkney and Shetland it is a significant observation that the growth of peat started to take place actively only with the arrival of man. The removal of natural scrub and tree cover would reduce evapotranspiration and thus facilitate the growth of peat and hinder the regeneration of trees.

4.5 Tree physiology at the timber-line

Physiological studies into the reasons for the definite limit to tree growth in relation to temperature are divided into two basic approaches. The first which has been much studied by Tranquillini (1979) exam-

ines the carbon balance of the trees at the timberline to determine if it is the ability to maintain a positive carbon balance at high altitudes that determines the position of the timber-line. The second approach was that initially pioneered by Michaelis (1934) who considered that it was the length of the growing season in relation to the phenology of the

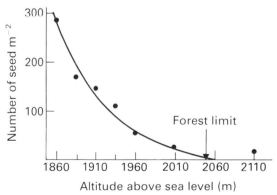

Fig. 4.22 Seed fall in spruce (*Picea abies*) on to the snow surface in late autumn and winter at a range of altitudes in a sub-Alpine forest near Davos, Switzerland. (Redrawn from Tranquillini, 1979 quoting data of Fischer *et al.*, 1959.)

Fig. 4.23 *Sorbus aucuparia* showing a good development of tree form in the *Krummholz* zone of *Pinus mugo* in the Bavarian Alps at 610 m (2000 feet).

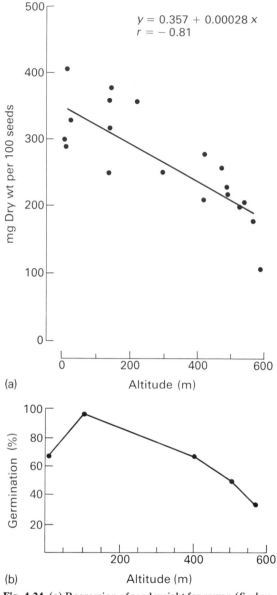

$$y = 0.357 + 0.00028\,x$$
$$r = -0.81$$

Fig. 4.24 (a) Regression of seed weight for rowan (*Sorbus aucuparia*) trees sampled over an altitudinal gradient of 600 m at Ballachulish on the west coast of Scotland. (b) Plot of percentage germination of rowan seeds (*Sorbus aucuparia*) as sampled over a 600 m altitudinal gradient at Ballachulish on the west coast of Scotland.

trees that was crucial to survival. Only trees that could complete their development in time to protect themselves against the stresses of winter weather would survive at the timber-line. The remainder of this chapter attempts to take a standpoint on this controversy and examines the effects of altitude, latitude, wind, and exposure on tree physiology and their role in determining the curious abrupt nature of many timber lines.

4.5.1 REPRODUCTION

Seed production in many trees takes place typically at irregular intervals which are called 'mast years'. Such years in which the tree has to support a large production of flowers, fruits, and seeds will place an exceptional demand on energy reserves and the extent to which these are depleted is likely to vary with altitude. In the Norway spruce (*Picea abies*) mast years occur at favourable sites every 4–5 years, whereas at higher altitudes this frequency is reduced to every 6−8 years and at the timber-line once in every 9−11 years (Tschermak, 1950). The quantity of seed produced per hectare of spruce forest also varies with altitude. At the timber-line (1990 m) in good years this can reach 0.4−2.3 million seeds per hectare. Down hill at 1600 m seed production rises to between 3 and 25 million seeds per hectare (Kuoch, 1965; cited by Tranquillini, 1979). This exponential drop in the level of seed production is shown in Fig. 4.22. The weight of the seeds also declines with altitude as does the germination capacity which can fall to less than 5 per cent at the timber-line. In New Zealand in the evergreen southern beech *Nothofagus solandri* var. *cliffortiodes*, flowers are produced every year, but good seed years occur only once in every decade (Wardle, 1971). Here again the amount of seed and its vitality decreased with altitude from 20 million seeds per hectare with a

germination capacity of 57 per cent at 1000 m to less that 0.9 million seeds per hectare with a germination capacity of only 7 per cent at 1350 m. The weight of the seeds also decreased with altitude.

The rowan (*Sorbus aucuparia*) is the tree that occurs at the highest altitudes on the mountains of northern Europe. When other species are reduced to *Krummholz* forms the rowan is able to maintain an erect, single, and undamaged trunk (Fig. 4.23). Despite this remarkable growth ability at high altitudes the rowan also shows a similar decline in reproductive capacity with increasing altitude. Fig. 4.24 shows the reduction in both seed weight and viability that occurs with increasing altitude on a Scottish mountainside (Barclay and Crawford, 1984). In the Scottish Cairngorm Mountains Scots pine at the timber-line has also been found to produce only infertile seed (Millar and Cummins, 1982).

Curiously, seed dispersal by wind above the timber-

Fig. 4.25 A colony of aspens (*Populus tremula*) growing in Britain's most northerly natural wood at Berriedale in Hoy, Orkney. These trees have never been observed to produce seed and appear to present an example of long-term vegetative survival.

line is very poor. Therefore, the failure to produce quantities of viable seed in the upper regions of the forest will be an additional factor in the dynamic equilibrium between the upward advance of the timber-line and its retreat downwards into regions more suitable for sexual reproduction. Where regeneration in high level forests is hindered by grazing and burning the lack of seed dispersal up the mountainside will aggravate the damage done to the maintenance of a natural timber-line. A number of tree species that persist at the timber-line in spite of disturbance by grazing and burning do so largely because they are capable of vegetative reproduction.

Fig. 4.25 shows a colony of aspens (*Populus tremula*) in Britain's most northerly natural woodland in the Orkney Islands. These trees have never been observed to set seed yet maintain themselves over the years by sending up new trunks from underground runners.

In Utah large clones of aspen (*Populus tremuloides*) are found extending over considerable areas. One such clone reported by Barnes (1966) occupied 107 acres and included 47 000 individuals. Under present conditions of low summer rainfall, seed germination of aspen in this region is negligible. As the present drought conditions are thought to have begun 8000

Fig. 4.26 An isolated timber-atoll of spruce (*Picea abies*) growing above the timber-line and beneath the north face of Eiger, Switzerland.

years ago it is possible that some of these aspen clones are as much as 8000 years old. A number of other species of upland areas also spread vegetatively. In spruce (*Picea abies*) the lateral branches of older trees develop adventitious roots which can eventually produce colonies of trees sometimes referred to as timber atolls (Fig. 4.26). In the deforested areas of the British upland such timber atolls are found with hawthorn (*Crataegus monogyna*). Like the timber atolls of spruce at the timber-line these hawthorn colonies help to ensure their survival by creating their own environmental enclave in an otherwise treeless landscape.

4.5.2 TREE GROWTH IN RELATION TO ALTITUDE

Detailed examination of tree growth at the timber-line shows that temperature acts upon plants in a variety of ways depending on the particular species concerned. In some species there is a direct correlation between temperature conditions during the growing season and the growth of the tree in that year. Thus in the birch (*Betula pubescens*) growth in height near the timber-line at 580 m (1900 feet) in northern England is directly related to the accumulated air temperature above 6°C (Millar, 1965). However in *Pinus sylvestris* the correlation between the growth of the terminal shoot and temperature holds when the current year's growth is related to the temperatures that prevailed during July and August of the previous year. This retrospective correlation is caused by the growth of conifers being determined by the number of needle-nodes laid down in the pre-formed terminal bud. Therefore, the temperature during the time of bud formation which starts as day length decreases, is crucial in determining the growth potential of the trees in the following year.

Growth in height in trees differs from production of seed in that there is no exponential decline as the timber-line is approached (Fig. 4.27). The most usual situation is for the forest to maintain the average height for mature trees and only at the timber-line does this rapidly decrease with the production of dwarf or *Krummholz* forms. Thus with larch (*Larix decidua*) and spruce (*Picea abies*) no change in height was observed in stands in the Bernese Alps up to 1800 and 1900 m respectively. Above these altitudes however tree height decreased by 3–5 m with every 100 m increase in altitude (Ott, 1978).

The genotypic differences that are found between lowland and upland provenances of the same tree

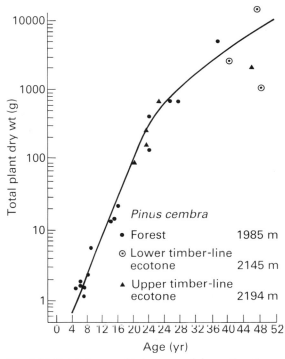

Fig. 4.27 Total above and below ground dry matter of *Pinus cembra* with respect to age near the timber-line in Austria. Weight development is not noticeably different near the timber-line ecotone. (Reproduced with permission from Tranquillini, 1979.)

species appear to affect the phenology rather than the intensity with which the trees use the growing season. In both lowland and upland provenances growth ends about the same time. It is the timing of bud break in spring which appears to be the main variant with upland forms showing a later development. Transplants into areas with shorter growing seasons from more favourable habitats can lead to severe frost damage in spring. As the timber-line is approached trees have less time to complete the growth of new shoots and in spruce the new needles and overwintering buds have less time to mature and harden for winter than in the valley trees. The same phenomenon is seen in the evergreen forests of southern beech in New Zealand where the seedlings at the timber-line approach winter with their terminal buds unripened and are largely destroyed (Wardle, 1972).

In a study of the growth rates of rowan (*Sorbus aucuparia*) seedlings germinated from seeds collected at different altitudes it was found that the seedlings

that germinated from the trees from higher altitudes
had the greatest relative growth rates (Table 4.1).
This difference was all the more surprising as, has
already been mentioned, the seeds from the higher
altitudes were lighter than those collected at lower
levels (Barclay and Crawford, 1984). The terminal
resting buds of the rowans from higher altitudes also
showed a very marked increase in their respiration
rate (Fig. 4.28). This difference was not seen in the
respiration rates of the seedlings grown from seed
collected at different altitudes so it is not clear
whether this is a genotypic or phenotypic response
to the effects of altitude. However it seems a regular
phenomenon in early spring as it has been observed
over the same altitudinal range in different years.
Whether or not the difference has any genetic basis,
such active respiration rates in early spring would be
likely to facilitate the prompt resumption of growth
that is such a feature of rowan trees when the spring
temperature reaches the threshold value necessary
for their development.

The differences in bud ripeness for overwintering
can be quantitatively assessed from measurements
of cuticle thickness. In the buds of the rowan cuticle
thickness is maintained up to the very uppermost
limits of tree establishment (Barclay and Crawford,
1984). The numerous studies that have been carried
out in various parts of the world from the European
Alps and Scottish mountains to the mountains of
New Zealand and the North American Rockies all
point to one limiting factor in tree growth and
development at the timber-line, namely the inability
of the young shoots, particularly those of seedlings,
to complete the necessary ripening process before
the onset of winter. This was the conclusion reached
by Michaelis (1934) in his pioneering work on the
osmotic pressure of tree sap on approaching the
timber-line. He observed a large increase in the
osmotic potential of tree sap at the upper limits to

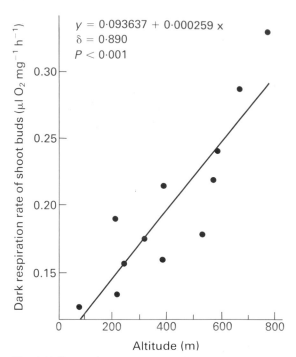

Fig. 4.28 Regression of dark respiration rate of terminal
buds of *Sorbus aucuparia* sampled in January over an
altitudinal gradient of 700 m in north Angus on the east
coast of Scotland.

tree growth and associated this with the winter desiccation of the buds due to their unripened condition.

4.5.3 CARBON BALANCE

As conditions become more severe with increasing
altitude or latitude it is logical to enquire as has been
done by Tranquillini (1979) if there is a limit above
which the tree fails to make a net gain in carbon over
the year. There are numerous difficulties in obtaining
an integrated estimation of total photosynthetic and
respiratory activity of trees throughout the year in
the severe conditions that exist at the timber-line.
Apart from the logistic problems of maintaining
observations with sophisticated instruments under
difficult conditions there are the added difficulties
which arise from tree size and variability. Trees are
obviously too large to be enclosed completely in
cuvettes for infrared gas analysis of their photosynthetic and respiratory activities, they need to be
sampled in some manner which gives an integrated
estimation of the carbon exchange rates in young
and old needles, young stems and old branches, as

Table 4.1 Relationship between the altitude of origin of
Sorbus aucuparia seeds and the mean relative growth rate
(R) of their seedlings grown under standard conditions.

Mean altitude (m)	n	R (g g^{-1} day^{-1})	95% C.L.
8	4	0.129	0.016
102	15	0.131	0.002
400	10	0.139	0.005
503	5	0.142	0.006
567	3	0.149	0.017

well as measure the respiration rates of roots. Further, the distribution of carbon resources varies with age in trees and estimations based on one particular age of tree or tree seedling may not be representative of the forest as a whole.

Instead, a simpler approach based on the analysis of timber production can be used to give an integrated measure of total productivity (Fig. 4.27). Estimations of dry matter production along altitudinal or latitudinal gradients do give an indication of the ability of a plant to make a net carbon gain. During the early stages of tree establishment an adequate biomass increment is essential for survival although, as pointed out by Tranquillini, at later stages only a minimal increment can suffice to keep the tree alive.

In a study which compared the growth of potted seedlings placed at the timber-line at 1950 m and in the valley at 650 m dry weight production decreased by 42 per cent in *Pinus mugo*, by 54 per cent in *Picea abies* and by 73 per cent in *Nothofagus solandri* (Benecke, 1972). By comparison 75-year-old trees of *Picea abies* in the Massif Central in France showed a 90 per cent reduction in timber-volume increment at the higher altitude of 1650 m as compared with

that between 1000 and 1300 m. Similar figures are found for varying forests in different parts of the world. However, it is only when an analysis is made of the various factors that are associated with this decline that it is possible to ascertain if it is due to the failure of the trees to maintain a positive carbon balance due to respiratory activity using up all the available photosynthetic gains.

In the beeches (*Fagus crenata*) of the cool-temperate broad-leaved forest zone of Mount Naeba (2145 m) in central Japan, Maruyama (1971) was able to relate the reduction in dry matter production with photosynthetic activity as affected by temperature, growing period, weather and leaf-area index. In these trees only 9 per cent of the production could be attributed to reduction in photosynthesis. A reduction in leaf-area index from seven to five accounted for only 17 per cent of the decline in productivity. By far the greatest factors in reducing production at the higher altitudes were shortening of the growing season from 185 to 122 days and reduction in radiation due to increasing cloud at high altitudes. Similar results were found in the Cragieburn range in New Zealand (Wardle, 1971) where potted

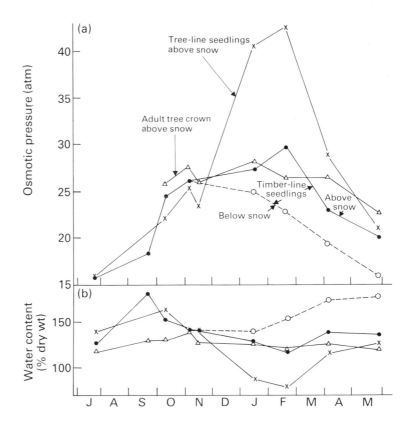

Fig. 4.29 Seasonal changes in (a) osmotic pressure and (b) water content (as percentage dry weight) of needles from *Pinus cembra* trees at the timber-line near Obergurgl, Austria. Determinations were made before sunrise. The three youngest age groups of needles were pooled to form a sample. Strongest desiccation was found in young plants from sites in the timber-line ecotone (*Kampfzone*) with little winter snow. (Reproduced with permission from Transquillini, 1979.)

seedlings of the mountain beech (*Nothofagus solandri*) placed 300 m above the natural timber-line (c. 1300 m) produced more dry matter than natural seedlings within closed stands near the timber-line which were subject to intense root competition. It would appear from these studies that there are no grounds for believing that the elimination of trees above the timber-line is due to inadequate dry matter production. Thus the possibility that tree growth is not possible above the timber-line due to an inability for photosynthetic capacity to match respiratory activity is not valid for Alpine timber-lines. No experimental data is available for latitudinal limits to tree growth such as are found on reaching the Arctic.

4.5.4 DESICCATION RESISTANCE

Many conifers of northern and montane habitats restrict photosynthesis by stomatal closure in dry air. *Picea sitchensis* is a typical example of a spruce which is adapted to cool moist oceanic conditions and is more sensitive to low air humidity than most other conifers. As this effect is due to atmospheric moisture conditions it imposes less of a restriction on tree growth as altitude increases. As a result the trees are able to compensate for the reduction in photosynthesis that is caused by reduced temperature and carbon dioxide concentration and increased wind speed. Soil moisture deficits at high altitudes are rarely severe enough during the growing season

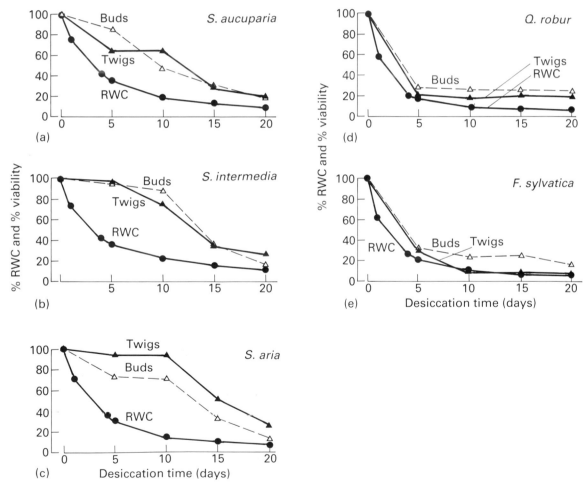

Fig. 4.30 Relationship between desiccation and cell vitality as measured with tetrazolium vital staining in (a) *Sorbus aucuparia*, (b) *S. intermedia*, (c) *S. aria*, (d) *Quercus robur* and (e) *Fagus sylvatica*. Note the retention of vitality in twigs and buds of the *Sorbus* species as compared with *Q. robur* and *F. sylvatica*. (Data from Barclay and Crawford, 1982a).

to reduce tree growth significantly. The studies of Pisek and co-workers in Austria (*see* Tranquillini, 1979) have shown that the osmotic pressure of *Pinus cembra* cell sap rarely rises above 30 atmospheres (−3 MPa). This is well below the maximum permissible value for this species of 40 atmospheres (−4 MPa) (Fig. 4.29).

In winter, however, with the prevention of water uptake by low soil temperature and freezing of the soil around the rooting zone, trees are subjected to an ever-increasing drought risk. Terminal buds differ between species as to their capacity to withstand progressive desiccation. In the genus *Sorbus* there is a remarkable ability for the buds to endure desiccation and still remain viable (Fig. 4.30). Fig. 4.30 also shows the effect of reduction in water content (RWC) on cell vitality in three *Sorbus* species as they were exposed to progressive desiccation. Compared with beech and oak the *Sorbus* species were capable of retaining much greater cell vitality at low cell water contents.

In 1934 Michaelis was the first to report the sudden rise in osmotic pressure of cell sap at the timber-line. Since this investigation many measurements have been made in different parts of the world and have shown comparable findings (Figs. 4.31 and 4.32). In stands of *Picea engelmannii* growing in the Medicine Bow Mountains in Wyoming, Lindsay (1971) showed that the water potential of the needles commonly remained above −20 bar (2 MPa) in the forest stand but in branches projecting above the snow this dropped to a minimum of −35 bar (3.5 MPa) in January. The level of desiccation that can be tolerated by coniferous needles varies with the season. In *Pinus cembra* 1−2 per cent of the needles are damaged when the water content drops to 73 per cent of dry weight in December. In April an equivalent damage is caused by the water content dropping to only 90 per cent of dry weight.

The factors controlling cuticular transpiration of timber-line trees during winter were examined by Sowell (1985) in an experiment to assess their relative

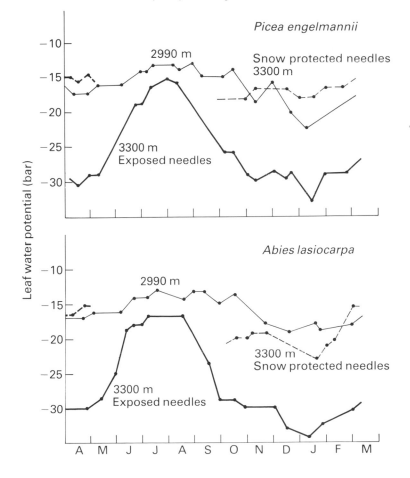

Fig. 4.31 Leaf water potential determined by the Schardakow method for *Picea engelmannii* and *Abies lasiocarpa* during a course of 1 year in the Medicine Bow Mountains in Wyoming at 3300 m (*Krummholz* zone) and 2990 m (near forest limit). Samples taken from below the snow during winter at the highest site showed no evidence of unbalance in their water relations. (Adapted from Lindsay, 1971.)

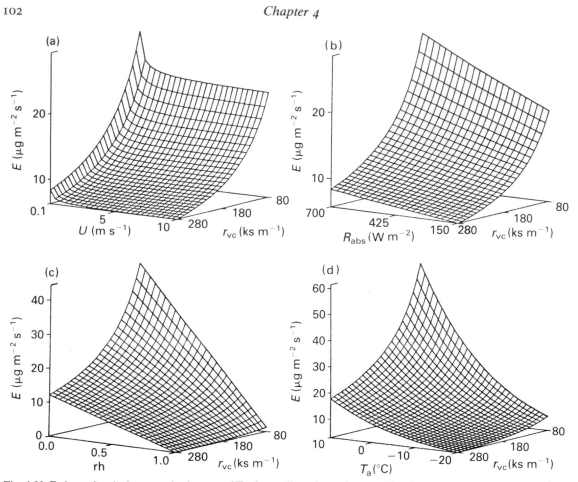

Fig. 4.32 Estimated cuticular transpiration rates (E) of a needle under various combinations of cuticular resistance (r_{vc}) and wind speed (U), absorbed radiation (R_{abs}) relative humidity (rh) and air temperature (T_a). When a parameter was not varied it was held constant: $U = 2$ m s^{-1}, $R_{abs} = 400$ W m^2 -1, rh $= 0.5$ and $T_a = -5$°C. (Reproduced with permission from Sowell, 1985).

influence on the desiccation of foliage exposed above the winter snow. Variations in wind speed, absorbed radiation, air temperature, relative humidity, and cuticular resistances were all examined for their effect on transpiration rate (Fig. 4.32). The results confirmed the earlier hypothesis of Michaelis (1934) that transpiration was strongly influenced by cuticular resistance and that inadequate cuticle maturation at the timber-line leads to excessive water losses. In Scots pine at the timber-line a high rate of water loss has been linked with an inability to close the stomata completely (Grace, personal communication).

Shoots likely to suffer most from desiccation damage can be predicted (Wardle, 1968) from their pale colour in the previous autumn and summer. Such shoots enter the winter period with a higher

water content than hardy shoots and lose this water more rapidly than more resistant tissues. In Finland after cool seasons the needles of *Pinus sylvestris* have been noted to be poorly differentiated with reduced resin canals and a thinner outer epidermal layer (Holtmeier, 1971). In spruce (*Picea abies*) it has been calculated that 3 months are necessary in lowland sites and longer in mountain situations for the development of needles from bud-burst to a final cuticle thickness that is sufficient to ensure survival (Lange and Schulze, 1966). Thus at the timber-line there will be an interaction between the state of maturation of the overwintering buds and needles together with the degree of exposure that they have to endure. Trees in the shelter of parent plants and protected by snow will not suffer the same degree of stress as plants that are isolated in the open on wind-exposed ridges with little snow. Further, once the establishment stage is passed, larger trees seem to have a greater resistance to desiccation possibly due to the water reserves contained in their vessels (Tranquillini, 1979).

4.6 Conclusions

The above account has illustrated that many facets of tree physiology from reproduction to the protection of overwintering buds will have an effect in altering the ability of trees to survive at the timber-line. Despite the definite timber-line that can often be observed in different parts of the world, no single factor on its own can be directly related to the actual position of the timber-line on any particular mountainside. The importance of the timing of phenological events does however seem to be a predominant feature in sites with cool temperatures and short growing seasons. On the one hand bud-break must not take place too soon or else frost damage in late spring will ensue, while on the other hand, delay will make it impossible for tissues to ripen to a suitable state of winter hardiness. In addition the exposure factor that is combined with any state of winter readiness is also crucial. In this way the very existence of a timber-line is in itself one of the main features that ensures its survival. Protection afforded at the timber-line will be sufficient to ensure the regeneration of trees that would not otherwise take place. Although this is particularly relevant for deciduous, non-shade-tolerant trees it is also an important factor in reducing desiccation damage to conifers. Where extensive timber-lines exist as in the remoter regions of North and South America and New Zealand they represent an ecological boundary of great complexity that has probably only arisen over a long period of plant and environment interaction.

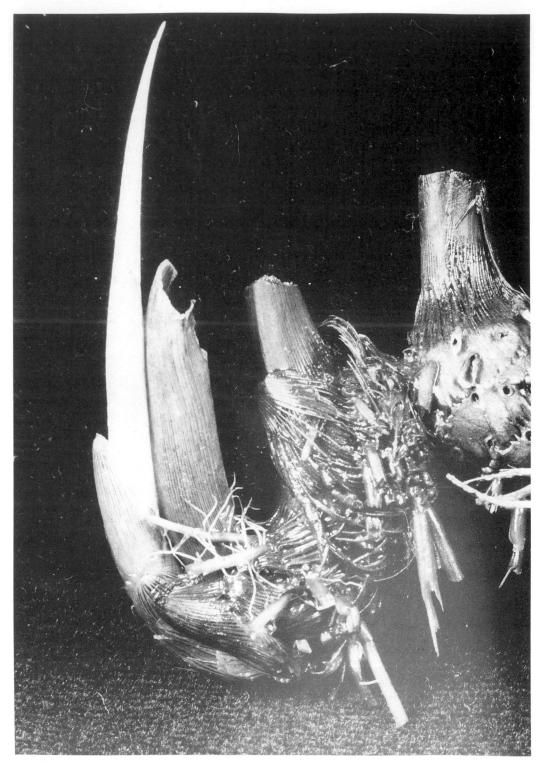

5

The anaerobic retreat

5.1 Ecological advantages of surviving without oxygen

The increased tempo of aerobic life which has resulted from the greater energy supply it affords, has made most aerobic organisms highly dependent on an adequate supply of oxygen. This is most clearly seen in the effect of anaerobic conditions on cell division. Mitosis, a major cytological feature which divides eukaryotes from prokaryotes, with the exception of a few species such as the yeasts, is inhibited in most species in anaerobic conditions. Most eukaryotic species therefore require some oxygen for a substantial part of their life cycle. Nevertheless this does not mean that all aerobic species require a constant supply of oxygen. Both plants and animals display a range of tolerance in their ability to survive for a period without oxygen. Being able to exist temporarily without something that is constantly essential for your competitors is one of the great keys to survival. The extent to which some aerobic plant species can survive without oxygen and the advantages that this confers in habitat availability (Fig. 5.1) has prompted this chapter on the ecological advantages of the anaerobic retreat.

The evolution of the cytochrome oxidases and their great ability to react with oxygen at low concentrations increased the efficiency of resource utilization and banished anaerobiosis as a way of life on earth to those regions where the oxygen concentration is less than about 1 per cent. The oxygen concentration which divides the aerobic from the anaerobic world is often referred to as the Pasteur point. The measurement of a critical oxygen pressure (COP) for the maintenance of aerobic respiration became easier with the development of the oxygen electrode which allows the simultaneous estimation of oxygen depletion and oxygen concentration from aqueous solutions. However as the technique developed it became clear that the exact oxygen concentration needed to sustain an undiminished uptake (zero

order kinetics) varies both with species and the conditions under which they are growing. Even in monocellular organisms the concentration at which oxygen becomes limiting for aerobic metabolism varies depending on whether the medium is static or moving and on the presence or absence of boundary layers. Their state of growth can also profoundly influence the critical oxygen pressure (COP) that is necessary for maintaining an undiminished oxygen uptake. Thus if yeast is fed with glucose a COP is reached at a partial pressure of 2.5 mm Hg. In the absence of glucose this point is reached only at 0.2 mm Hg (for review *see* Harrison, 1976).

Given the above difficulties in defining exactly a critical oxygen concentration for micro-organisms it is not surprising that the situation is even more complex in higher plants where tissue bulk and density are additional barriers to the supply of oxygen to the respiring cell. A typical trace for oxygen uptake in relation to concentration for the root system of a higher plant is shown in Fig. 5.2. Estimation of the critical oxygen pressure has however practical problems. Measurements made in Clarke-type oxygen electrodes are liable to overestimate the COP as in addition to the barrier produced by the external bathing liquid to the diffusion of oxygen there is the artificial inundation of the root air spaces with solution which will further impede the diffusion of oxygen to the cells. Armstrong and Gaynard (1976) have compared data obtained in this manner with measurements which estimate the changes in oxygen concentration in roots of intact plants by surrounding the root with a cylindrical platinum electrode. When COP is measured without flooding the cell spaces, values as low as 0.024 atmospheres of oxygen are obtained which is an order of magnitude less than when such precautions are not taken.

As evolution has produced larger organisms the critical oxygen pressure necessary to prevent aerobic metabolism from being inhibited reaches values much higher than the 1 per cent Pasteur point. This is particularly marked in warm-blooded animals but

Fig. 5.1 (*opposite*) Shoot growth of the bullrush (*Schoenoplectus lacustris*) that took place during a 14−day anaerobic incubation. The ability to regenerate new shoots in the total absence of oxygen allows this species to spread into anaerobic muds and thus secures a habitat that is denied to plants that require oxygen for shoot extension.

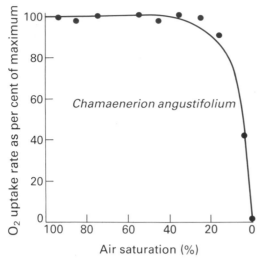

Fig. 5.2 Root oxygen uptake as a function of the air saturation of the bathing medium. Measurements were made using a Clark oxygen-electrode containing six root tips 1 cm each in length and immersed in a phosphate buffer pH 6.0 (0.1M) with 2 per cent sucrose and kept at 20°C. The figure shows that oxygen concentration does not limit the rate of uptake in roots of fireweed *Chamaenerion angustifolium* until the concentration is less than 10 per cent of air saturation.

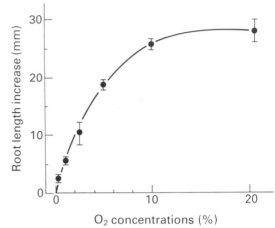

Fig. 5.3 Effect of oxygen concentration on growth in length of rice roots of seedlings grown in air for 3 days and subjected to different oxygen concentrations for 24 hours. Note that even in the water-loving rice plant root growth is reduced as soon as the oxygen concentration falls to below half of that normally present in ambient air. (Data from Bertani and Brambilla, 1982a.)

in all plant and animal phyla the evolution of large and complex organs increases the dependence of the organism on adequate supplies of oxygen. Figs. 5.3 and 5.4 illustrate the reduction in growth and protein synthesis that occur in wheat and rice as the oxygen concentration is progressively reduced below the ambient level of 21 per cent. It can be seen that even a reduction to 10 per cent oxygen in the gas mixture to the roots, significantly reduces growth and protein synthesis. Thus although the experiments with polarographic techniques can suggest a very low internal

COP for intact roots, the oxygen concentrations in nature that are needed to maintain these values require in some cases that the ambient oxygen concentration does not fall by more than half of that normally present in the atmosphere.

In spite of this basic oxygen dependence in all eukaryotic organisms, certain species can suspend their aerobic metabolism for a period and survive under anoxia for varying lengths of time. These species can partially withdraw from the aerobic world to an *anaerobic retreat*, free from disturbance by totally oxygen-dependent organisms. The ecological advantages of this adaptation must be very real as this mode of behaviour is found in nearly all classes of plants and animals. All diving animals, whether

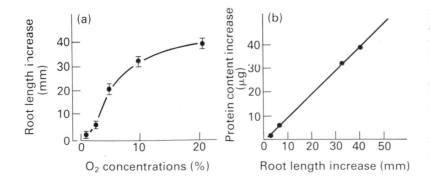

Fig. 5.4 (a) Effect of oxygen concentration on growth in length of wheat roots of seedlings grown in air for 2 days and then subjected for 24 hours to different oxygen concentrations. (b) Relationship between increase in length and increase in protein content in wheat roots grown for 2 days in air and then subjected for 24 hours to different oxygen concentrations. (Data from Bertani and Brambilla, 1982b.)

they are reptiles, birds, or mammals are terrestrial oxygen-requiring species which not only gain protection from diving but also benefit from access to food supplies that are denied to more oxygen-dependent species. Although fundamentally aerobic animals, the turtle and the marine iguano can graze on the algae of the sea-floor by being able to endure prolonged periods of anaerobiosis. The tolerance of the red-eared turtle (*Pseudemys scripta*) to anaerobic conditions is sufficient to allow it to remain submerged for 2 weeks at 16–18°C. The ability of this turtle to dispense with aerobic respiration is so well developed that it cannot be killed by poisoning with cyanide (an inhibitor of aerobic respiration). Similarly, molluscs that can survive periods of natural anaerobiosis out of the water with their shells closed when the tide ebbs, also gain an ecological advantage in that they gain access to a littoral habitat that is less exposed to predation by aquatic species.

Likewise in land plants, a return to water allows the exploitation of a new ecological habitat. For a higher plant the ability to root in a wetland site can confer a number of advantages. Competition from flood-intolerant species is avoided, grazing by herbivores is reduced, and the burial of overwintering roots and rhizomes in anaerobic mud is an excellent protection against winter frost desiccation (*see* p.54). Plants that can survive inundation are also likely to be in sites that are less prone to drought and will thus enjoy a prolonged growing season in areas of uncertain rainfall. Seeds that can germinate under water or in waterlogged soils, where the oxygen supply is much reduced are able to establish themselves in habitats where competition for space is less. Very few species are able to germinate in the total absence of oxygen. Rice (*Oryza sativa*) and some graminaceous weeds of paddy fields in the genus *Echinochloa* are the only well-documented examples of seeds that can germinate in a total absence of oxygen (Rumpho and Kennedy, 1981).

These examples of germination without oxygen are exceptional. There is however a considerable number of species where the seeds can germinate at very low oxygen concentrations. In some species in which the seed reserves are in the form of carbohydrate, as in the Gramineae and Leguminosae, germination can take place slowly at oxygen partial pressures as low as 0.01 KPa with germination speed increasing up to the oxygen concentration of ambient air, 21 KPa (Al-Ani *et al.*, 1982; 1985). By comparison seeds in which reduced carbon is stored as lipids showed little germination until the partial

pressure of oxygen was close to 2 KPa. Seeds in this category included cabbage, lettuce, flax, turnip, soybean, radish, and sunflower. Lipids might appear at first to be the most suitable energy store for seeds as they would contain the greatest amount of energy in the smallest volume of matter. However the early stages of germination are often accompanied, particularly in leguminous seeds such as pea, by low oxygen (hypoxic) conditions within the seed. Before the rupture of the testa oxygen diffusion to the respiring tissues is often insufficient to meet the increasing metabolic needs of the germinating seed and consequently there is frequently a period of natural anaerobiosis. Frequently this can be prolonged due to the onset of cold or wet conditions in early spring or any other factor which reduces soil aeration. Consequently lipid-containing seeds which have a demand for high oxygen concentrations are likely to be frequently at a disadvantage.

5.2 Plant survival in oxygen-deficient soils

Different species have adopted very different solutions to the problem of survival in oxygen-deficient soils. The stresses include not just a shortage of oxygen for the metabolism of roots and other underground organs (for example rhizomes) but the avoidance of toxic effects from products of the reducing conditions in the soil such as ferrous and manganous ions, sulphide, and nitrite. As with all stresses plants can be divided into 'avoiders and toleraters' (Levitt, 1980). For 'avoiders' the region of the soil that the plant exploits can be restricted to surface layers where the consequences of flooding are minimal both in terms of soil toxins and oxygen deficits, or alternatively the timing of root growth can be restricted to those seasons when water-tables are at their lowest. 'Toleraters' on the other hand can survive prolonged periods of inundation or burial in anaerobic mud without access to oxygen and can either resume growth anaerobically or wait undamaged until the oxygen supply is restored (Table 5.1). Although this distinction can be made experimentally in many species there is a variety of mechanisms some using the tolerating principle and others the avoidance mechanisms. Thus in the bullrush (*Schoenoplectus lacustris*) shoot extension growth can take place in the total absence of oxygen (Fig. 5.1) but if last year's stalks remain and can provide a 'snorkel' for the provision of oxygen to the submerged rhizome then shoot emergence from anaerobic mud is more rapid.

Species	*Anoxia endurance in days	Shoot elongation
Carex rostrata	4	none
Juncus effusus	4–7	none
J. conglomeratus	4–7	none
Glyceria maxima	7–21	occasional
Ranunculus lingua	7–9	none
R. repens	7–9	none
Mentha aquatica	4	none
Eleocharis palustris	7–12	none
Filipendula ulmaria	7–14	none
Carex papyrus	7–14	none
C. alternifolius	7–14	none
Spartina anglica	>28	none
Iris pseudacorus	>28	none
Phragmites australis	>28	none
Typha latifolia	>28	frequent
T. angustifolia	>28	frequent
Scirpus americanus	>28	frequent
S. maritimus	>90	frequent
S. tabernaemontani	>90	frequent
Schoenoplectus lacustris	>90	frequent

Table 5.1 Length of anaerobic incubation that can be endured in detached rhizomes without loss of regenerative power. For experimental methods *see* Barclay and Crawford (1982b).

* These figures represent the minimum time that the species were able to survive the anoxic treatment; longer periods of anoxia survival may be possible in those species that survived 90 days or more. It should be noted that tolerance can vary with season and the state of development of tissues. Normally the tolerance is greater in spring and less in summer when carbohydrate reserves are small. Similarly well-formed terminal buds in *Glyceria maxima* can elongate under anoxia but this is not seen in smaller lateral buds.

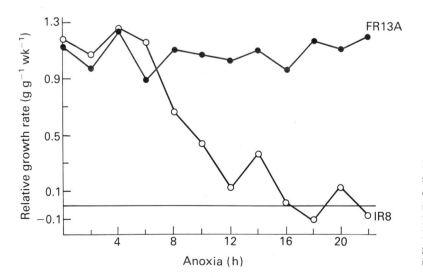

Fig. 5.5 Relative growth rate of seedlings of two varieties of rice during recovery periods after anoxic incubation lasting from 0 to 24 hours. Note that the improved fast growing IR8 variety has subsequent aerobic growth severely reduced by anaerobic incubations of 8 hours or more.

A noticeable feature of a comparison of relative tolerances of anoxia is that closely related species have very different capacities to survive anaerobic stress. One of the most striking differences is in the genus *Iris*. The common yellow flag *I. pseudacorus* will survive 2 months anaerobic incubation and produce new shoots immediately on being returned to air. Its close relative *I. germanica* which has a similar

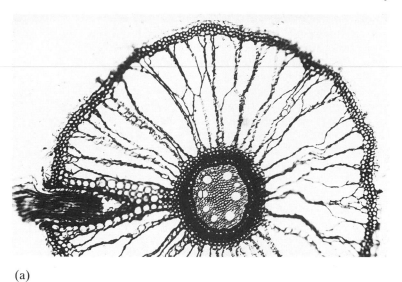

(a)

(b)

Fig. 5.6 Transverse section of roots taken at least 2 cm behind the tip showing structure of the cortical aerenchyma in (a) *Juncus effusus* and (b) *Narthecium ossifragum*. (Reproduced from Smirnoff and Crawford, 1983.)

but slightly more dense rhizome is killed by only 1−2 weeks anaerobic incubation. Table 5.1 records the length of time that rhizomes can be kept under anoxia and still be able to resume growth when restored to air. *Cyperus alternifolius* one of the papyrus sedges will survive a week of anoxia while *C. diffusus* a species from the same region but confined to drier river bank margins is killed. Species that are morphologically similar can also be quite different in their tolerance of anoxia. The common rushes *Juncus effusus* and *J. conglomeratus* always fail to survive anaerobic incubation, both being killed by as little as 5 days anoxia. Morphologically they appear similar to the *Scirpus* and *Eleocharis* species that are highly tolerant of anoxia. Examination of these *Juncus* species in winter shows that they always retain a green basal region in their shoots and this combined with a relatively shallow rooting system may allow them to survive in bogs and marshes as oxygen stress avoiders rather than tolerators.

Even in the typical flood-loving rice crop there are varietal differences in their ability to withstand anoxia. Fig. 5.5 shows the effect of 0 to 24 hours flooding on the new fast growing IR8 variety as measured by its subsequent growth in air as compared with a more tolerant variety. It can be seen that 8 hours of anoxia is sufficient to reduce the subsequent growth potential of this variety. These results emphasize one of the dangers of crop improvement aimed only at increasing yield, namely a reduction in tolerance of adverse conditions.

5.2.1 AVOIDANCE OF ANOXIA

Habitats that are prone to flooding usually support a distinctive flora. Rushes, sedges, and reeds are vernacular terms which convey an image of a wetland plant community even to the non-botanist. The predominance of monocotyledons in wetland plant communities does not mean however that this group of species has any special ability to survive without oxygen. Grasses are particularly characterized by their ability to grow an entirely new root system each year from a perennating shoot base. This strategy obviates the need to maintain a deep-rooted perennial rooting system during periods of winter flooding. Thus as the water-table subsides in spring new root growth can follow its descent and the progressive aeration of the upper horizons of the soil. Many monocotyledonous plants have a well-developed aerenchyma (Fig. 5.6) which allows them to penetrate the water-table due to the facilitation of oxygen diffusion that is provided by large and continuous air spaces (Armstrong, 1979). Within the same species or even within the root system of an individual plant, roots that grow into the water-table usually increase their development of aerenchyma more than those that are formed on 'dry' land. This is particularly marked in the Gramineae but is also to be observed in dicotyledon species. In a study of aerenchyma development in the field in *Senecio aquaticus* it was shown (Fig. 5.7) that there was a

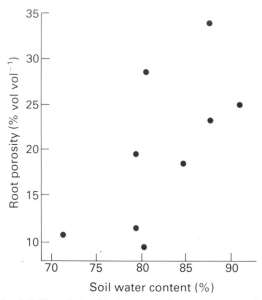

Fig. 5.7 The relationship between soil water content and the air space of root systems of *Senecio aquaticus* plants growing in an Orkney valley mire. (Reproduced from Smirnoff and Crawford, 1983.)

positive correlation between soil water content and the amount of aerenchyma possessed by the plant roots.

The structure of the aerenchyma tissues varies between different species. In *Caltha palustris* and *Filipendula ulmaria* aerenchyma is formed by the enlargement of intercellular spaces (schizogeny). In the case of *Caltha palustris* this is well developed and produces a root with a spongy appearance. Very commonly in wetland species aerenchyma is formed by cell disintegration (lysigeny) to give large lacunae. On the basis of the arrangement of the cells forming diaphragms between lacunae it is possible to distinguish two types of lysigenous aerenchyma production (Smirnoff and Crawford, 1983). In the first the lacunae are separated by longtitudinal diaphragms, consisting of broken cell walls and varying numbers of intact cells stretching from the inner to the outer cortex. Species in this group include *Mentha aquatica*, *Ranunculus flammula*, *Potentilla palustris*, *Juncus effusus*, *Glyceria maxima*, *Narthecium ossifragum* and *Nardus stricta* (Fig. 5.6). In *Mentha aquatica* the diaphragms contain up to several rows of intact cells while at the other extreme *Juncus effusus* has almost no intact cells in its diaphragm. The second type of lysigenous aerenchyma develop-

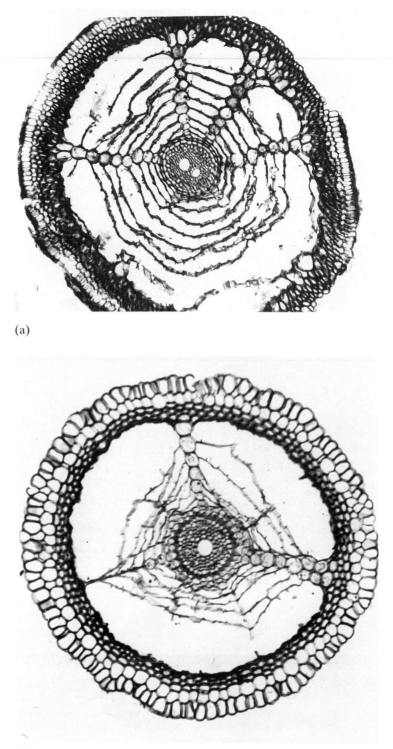

(a)

(b)

Fig. 5.8 Transverse sections of roots, taken from at least 2 cm behind the tip showing the structure of cortical aerenchyma in (a) *Eriophorum vaginatum* and (b) *Carex curta*. (Reproduced from Smirnoff and Crawford, 1983.)

ment is seen most characteristically in the Cyperaceae. In this type the diaphragms consist of entirely intact cells arranged radially from the inner to the outer cortex. Stretched between these radial diaphragms are tangentially arranged diaphragms consisting entirely of broken cell walls. This arrangement results in a spider's web-like structure and is shown in Fig. 5.8 in *Eriophorum vaginatum* and *Carex curta*.

Ethylene has been clearly shown to stimulate the production of aerenchyma in maize (Jackson *et al.*, 1985a) but not in rice (Jackson *et al.*, 1985b). However, in those cases where ethylene is the stimulant for aerenchyma formation, its synthesis from its precursor (ACC) is an oxygen-dependent reaction (Young and Hoffman, 1984) and will therefore necessitate an adequate level of tissue aeration. Shoot extension is also generally assumed to be

ethylene controlled (Osborne, 1984). Nevertheless, the most anoxia-tolerant species described above are capable of shoot extension in complete absence of oxygen and must therefore receive a stimulus for bud extension that is not dependent on *in situ* ethylene production.

Ethylene can also play a vital role in growth regulation of the floating or deep water rice (*Oryza sativa* L., cv. Habiganj Aman II). This deep water cultivar is only tolerant of inundation if its leaves reach the water surface. The rice seeds are broadcast after the first rains and the plants are established on dry land. As the flood waters rise the rice leaves grow apace and show remarkable rates of internode elongation. This growth response appears to be due to the flooded conditions stimulating the production of the ethylene precursor 1-aminocyclopropane-1-

Fig. 5.9 Swamp cypress trees (*Taxodium distichum*) with pneumatophores growing in flooded bottom-land forest in Southern Louisiana. Pneumatophore development is most marked in sites with fluctuating water-tables. The ability of the trees to survive long periods of waterlogging with or without pneumatophores gives them access to sites with fewer competitors than are found in dry land forests.

carboxylic acid (ACC), which is then converted to ethylene. Inhibition of ethylene biosynthesis by amino-oxyacetic acid and aminoethoxyvinylglycine inhibits internode elongation (Metraux and Kende, 1983).

In all plants the supply of oxygen from shoot to root is purely a physical process prone to leakage from the root into the soil with gaseous diffusion following the path of least resistance. Consequently, small plants with shallow adventitious rooting systems are likely to be in the best position to avoid oxygen deficits when flooded. The short distance from shoot to root facilitates oxygen diffusion downwards and the dispersed and sub-surface distribution of the adventitious roots places them in the soil region that is least likely to suffer a severe oxygen deficit. Nevertheless, many larger plants, including trees (for example the swamp cypress *Taxodium distichum*, Fig. 5.9) can survive in sites that are regularly flooded. Some of these species appear to be low-oxygen stress avoiders. The swamp cypress like the mangroves (Fig. 5.10) has some characteristics of a stress avoider in that it produces specialized roots termed 'knees' which bend upwards and emerge above the soil surface. These above ground roots are well provided with ventilating tissue and promote gas exchange down into the submerged roots. Curiously these 'knees' are only produced in areas where the water-level fluctuates. Swamp cypress growing in areas with a permanently high water-level do not produce knees. This is somewhat puzzling as it is just in these areas that 'knees' would be of the greatest advantage if they functioned as efficient eration organs.

In the spruces (*Picea*), shallow root systems spread out above the wetter anaerobic regions of the soil and form the bulk of the anchoring and water and nutrient absorbing organs of the tree with only a few sinker roots penetrating to a greater depth (Fig. 5.11). Some trees develop their own characteristic oxygen conducting tissue by developing large cavities in their xylem. Lodgepole pine (*Pinus contorta*) when grown in waterlogged conditions produces large stelar cavities (Fig. 5.12). The efficiency of these cavities in transporting oxygen is shown in Fig. 5.13 where it can be seen that roots of lodgepole pine unlike those of spruce can transport oxygen downwards even when the outer cortex of the root has been removed (Coutts and Philipson, 1978). The development of lenticels at the base of the stem or trunk appear to be important for the entry of air for if flooding takes place above the level of the soil the damage to the root system is more severe.

Fig. 5.10 Mangrove (*Avicennia* sp.) pneumatophores. Like the swamp cypress pneumatophores the organs are found in sites with fluctuating water-levels which in the case of the mangrove take place daily with tide movements.

5.2.2 TOLERANCE OF ANOXIA

Tolerance of anoxia is only a relative term when applied to higher plants. As mentioned above (p. 105), all eukaryotes require oxygen especially for mitosis and are therefore dependent on oxygen for their ultimate growth and survival. However the speed with which they die when deprived of oxygen differs

Fig. 5.11 Root system of a felled stand of *Picea abies* exposed by a peat fire which burned for several months removing the peat and leaving the stumps with their root systems intact. Note the lateral spread of the roots and the relatively low production of sinker roots. Spruce trees are commonly found on soils that are poorly aerated and their ability to root in the shallow upper layers makes them stress 'avoiders'. This shallow rooting also renders them liable to premature wind-throw in conifer plantations in upland sites with wet soils.

greatly. An indirect indication of varying sensitivity to oxygen starvation is seen in the length of time that certain tree species will survive when flooded. With the construction of reservoirs, data is sometimes available for how long certain trees will survive constant flooding (Gill, 1970). Table 5.2 shows some of this data with tree survival varying from 3 years to 4 months. Fig. 5.14 shows a riverside forest on the banks of the Rio Negro (Brazil) where an unusually extended period of flooding has caused the death of a large number of trees that normally survive a

predictable seasonal flooding of a few months. So predictable is the flooding that freshwater sponges live in the branches of the trees as for a substantial part of the year the trees are regularly under water (Fig. 5.15). Death under natural flooded conditions however may not be due directly to the removal or reduction in oxygen supply; anaerobic soil toxins and pathogenic micro-organisms may also play a role in determining the potential viability of flooded root systems.

A more direct indication of anoxia tolerance is

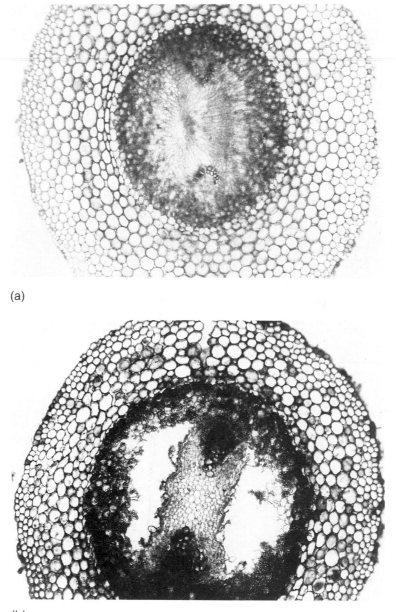

(a)

Fig. 5.12 Transverse section of *Pinus contorta* roots cut at a point 40 mm from the apex. (a) Root growing in a freely-drained soil above the water-table. No large cavities are present. (b) Root growing in a waterlogged soil, showing large cavities in the stele. (Photographs reproduced with permission from Coutts and Philipson 1978.)

(b)

obtained when plant tissues are incubated under strictly controlled anaerobic conditions where their viability can be tested in relation to duration of anoxia. Table 5.1 shows the enormous range of tolerance of varying plant species to survive incubation in a strictly controlled anaerobic environment. Although this is a highly artificial test system it shows the relative tolerance of plant organs to strict anoxia for prolonged periods without any complications due to oxygen diffusion or the susceptibility of different species to soil toxins. The results of these experiments on anaerobic survival in wetland species demonstrate that it is possible to subdivide them into the following distinct physiological categories:

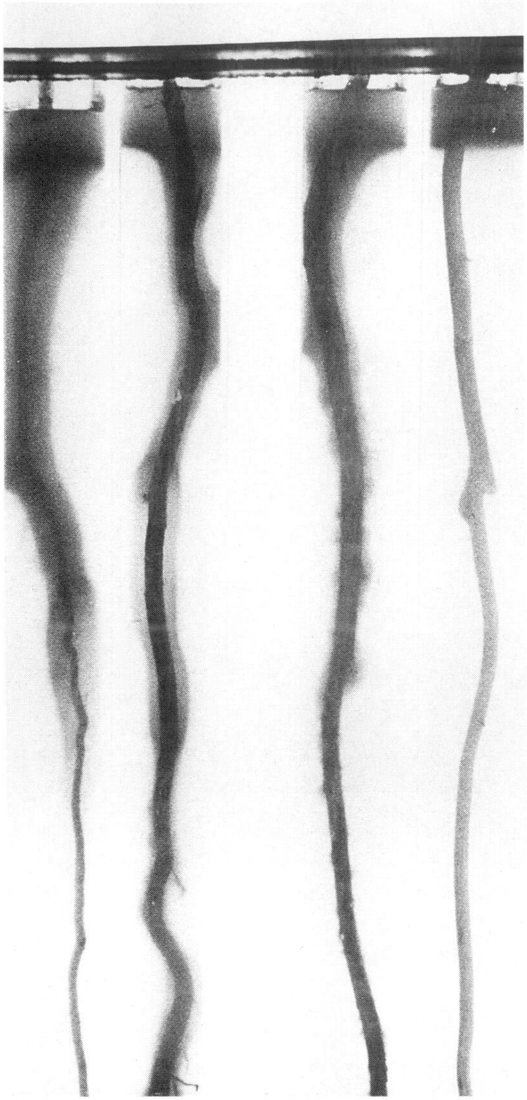

(a)   (b)   (c)   (d)

Fig. 5.13 Oxygen transport in tree roots of *Picea sitchensis* and *Pinus contorta* demonstrated by the oxidation of reduced indigo carmine solution to a coloured form by oxygen diffusing from roots. From left to right the roots are: (a) *Pinus contorta*, with tissues external to the xylem removed; (b) *Pinus contorta* bark intact; (c) *Picea sitchensis* bark intact; and (d) *Picea sitchensis*, tissues external to the xylem removed. The roots are submerged to a depth of 40 cm in the dye. (Photograph supplied by courtesy of M.P. Coutts and J.J. Philipson, Forestry Commission, Northern Research Station, Roslin, Scotland.)

Table 5.2 Survival time under inundation of some flood-tolerant trees. (For further details *see* Crawford, 1982b.)

Species	Flooding survival years
Quercus lyrata	3
Q. nuttali	3
Q. phellos	2
Q. nigra	2
Q. palustris	2
Q. macrocarpa	2
Acer saccharinum	2
A. rubrum	2
Diospyros virginiana	2
Fraxinus pennsylvanica	2
Gleditsia triacanthos	2
Populus deltoides	2
Carya aquatica	2
Salix interior	2
Cephalanthus occidentalis	2
Nyssa aquatica	2
Taxodium distichum	2
Celtis laevigata	2
Quercus falcata	1
Acer negundo	0.5
Crataegus mollis	0.5
Platanus occidentalis	0.5
Pinus contorta	0.3

1   Anoxia survival with shoot extension.
2   Anoxia survival without shoot extension.
3   Anoxic death in 7 days or less.

The most remarkable group of plants are those where anaerobically incubated rhizomes not only survive several weeks of complete oxygen deprivation but also manage to grow under these conditions. It has been said that the rice coleoptile is the only plant organ able to grow in the absence of oxygen (Pradet and Bomsel, 1978). Growth of rice under anoxia, however, is restricted to coleoptile extension and very limited protrusion of the seminal roots. With further incubation in an oxygen-free environment the coleoptile exhibits epinasty and the seedling dies. Barnyard grass (*Echinochloa crus-galli*) has also been reported to be capable of coleoptile extension but not root growth under anoxia (Rumpho and Kennedy, 1981). This species is a common weed of rice paddy fields and its great tolerance of anoxia in

Fig. 5.14 Riverside forest by the Rio Negro (Brazil) where an unusually extended period of flooding has caused the death of many trees. Normally these trees can withstand flooding for several months to a depth of 6–8 m (*see* Fig. 5.15). However some respite is needed for survival and flooding prolonged beyond this period results in forest death.

its early stages of development enables it to compete with rice. Coleoptile extension from germinating seeds of rice or barnyard grass is a short-term phenomenon in comparison with the shoot extension that can be observed under anoxia in rhizomatous species (Fig. 5.1). Rhizomes of *Scirpus maritimus* (Fig. 5.16) can survive 2 months under anoxia and even towards the end of this prolonged period of oxygen starvation are still capable of extending shoots from pre-formed buds. This ability to grow shoots and coleoptiles under anoxia is, as far as is known, merely an extension growth of pre-formed cells. There is no firm evidence as yet that any higher plants have evolved the capacity to break the general rule that, with the exception of some yeasts and related species, eukaryotic organisms require oxygen for mitosis.

The second class of plants observed in relation to this artificial testing of the limits of tolerance of anoxia comprises those species that remain alive and are capable of resuming normal growth when the oxygen supply is restored to the tissues. These species include the aquatic grasses such as *Phragmites australis* and *Phalaris arundinacea* as well as sedges and many rhizome-forming monocotyledons. The third group of wetland species comprises those wetland plants that are unable to tolerate even anaerobic incubation for 1 week. These species are the stress avoiders whose adaptations have already been considered above. In *Glyceria maxima* death normally ensues after 5–7 days anoxia but in spring, before carbohydrate supplies are depleted by shoot renewal, as long as 15 days anaerobic incubation can be necessary for a complete kill (*see* p.108).

5.3 Causes of anoxic and post-anoxic injury

Depriving higher plant tissues of oxygen will in all cases eventually lead to death. The length of time that higher plants can survive without oxygen varies with species and the conditions under which the anoxia is imposed. In cotton and soybean roots death can take place in as little as 3 and 5 hours respectively with a proportion of the roots being dead within 30 minutes (Huck, 1970). By contrast the rhizomes of some monocotyledonous species can be kept alive in an anaerobic incubator under strict anoxia for over 2 months (Crawford, 1982b). In

Fig. 5.15 A freshwater sponge growing on a tree on the banks of the Rio Negro near Manaus. The sponges survive as tree epiphytes as in the flood-plain forests of this region of the Amazon basin the trees are inundated for over 6 months in the year.

Fig. 5.16 Stimulation of shoot growth under anoxia: left, rhizome segment with terminal bud of *Scirpus maritimus* that has been incubated under anoxia for 7 days; right, aerobic control. Rhizomes of this species have the capacity to survive over 2 months incubation in the absence of oxygen and still produce new shoots.

some anoxia-sensitive species the damage caused by periods of anaerobiosis may not be immediately apparent, especially if the anoxic treatment is brief and the tissues are restored to air before any symptoms of injury are evident. Thus in pea seedlings a period of anaerobiosis does not immediately kill the seedling but when subsequently restored to air is seen later as damage to the continued growth of the shoot (Fig 5.17). Many tissues including rhizomes of *Iris germanica* and *Pteridium aquilinum* (not a higher plant) emerge from several days of anaerobic incubation with apparently healthy rhizomes which then begin to degenerate within hours of being exposed to air. After a few hours in air these post-anoxic rhizomes show evidence of peroxidative damage to their cell membranes by the production of malondialdehyde (Hunter, Hetherington and Crawford, 1983) and also of ethane. Both these products are considered to indicate peroxidative destruction of fatty acid chains.

Experimentally it is therefore possible to define two types of anaerobic injury, (1) anoxic and (2) post-anoxic, depending on when the damage is caused to the plant tissues. The time at which the injury is observed, however, does not necessarily determine when the injury was caused. The death of the shoot in pea seedlings, although seen only on the extension of the shoot during the post-anoxic phase, may have taken place during the period of anaerobiosis. However, the peroxidative damage that has been reported

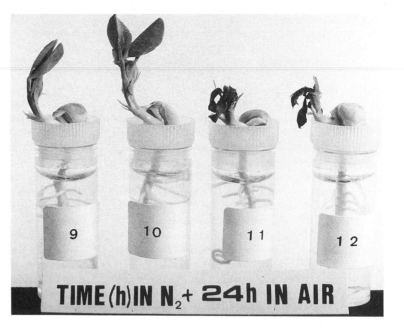

Fig. 5.17 Post-anoxic injury in pea seedlings. The seedlings have been exposed to total anoxia for 7–14 hours. The damage visible as a result of 11 hours or more of anoxia only became apparent after a 24 post-anoxic period in air.

in rhizomes on being restored to air after a period of anoxia (Hunter, Hetherington and Crawford, 1983) is clearly a post-anoxic injury as it can take place only when there is access to a source of oxygen.

The lethal targets for both anoxic and post-anoxic injury are as yet unclear and some of the controversy that surrounds the possible causes of anaerobic death and flooding injury in plant tissues may be due to the the fact that plants vary in the nature of the lethal target (Crawford, 1987; Vartapetian *et al.*, 1987). Death that is caused by a few hours anoxia as in soybean and cotton roots (Huck, 1970), may take place for entirely different reasons than those which are eventually responsible for the death of tissues that are capable of surviving several days or even weeks under anoxia.

5.3.1 RAPID CELLULAR MALFUNCTION UNDER ANOXIA

The absence of oxygen impedes a number of aspects of cell function. The supply of energy to the plant cell will be drastically reduced with the cessation of ATP generation by the electron transport system. Consequently the quantity of ATP available for essential cell processes falls immediately. This is seen in the effect of anoxia on the energy charge of the plant cell. Energy charge (EC) is defined as:

$$EC = \frac{(ATP) + 0.5\,(ADP)}{(ATP) + (ADP) + (AMP)}$$

Pradet and Raymond (1983) have shown for a large number of species that within minutes of their being deprived of oxygen the normal value for energy charge falls from about 0.8 to 0.4 or less (Fig. 5.18).

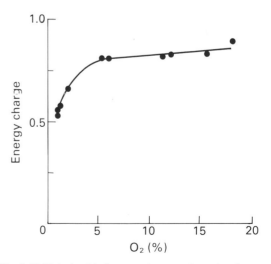

Fig. 5.18 Relationship between 'energy charge' and duration of anoxia in a number of bullrush rhizomes. (Data by courtesy of Dr. R. Braendle.)

In most aerobic species energy charge will remain low throughout the period of anoxia and it must therefore be assumed that there will be a shortage of the chemical energy necessary for the metabolic activity that is required to maintain the essential integrity of the cell especially protein (enzyme) biosynthesis. Membrane and enzyme damage could thus result for no other reason than that there is not sufficient metabolic energy available to maintain cell ultrastructure and the integrity of cell components. In species which are able to withstand anoxia a recovery of energy charge values is often observed as the anaerobic treatment is prolonged. To achieve this re-establishment of normal energy charge values under anoxia would require a considerable acceleration of glycolysis as under anoxia one molecule of glucose will yield only two molecules of ATP compared with the 38 molecules that are normally available for each molecule of glucose respired aerobically. There is no evidence to suggest that plant metabolism ever accelerates to this extent under anoxia. The restoration of the energy charge levels that does take place under anoxia could however be achieved if there was a diminution in energy expenditure brought about by a reduction in metabolic rate in anoxiatolerant organs. Evidence on this point awaits further research.

The rapid cessation of oxidative phosphorylation inhibits ATP requiring processes such as H^+ transport from cytoplasm to vacuole (Bennett and Spanswick, 1984). The removal of oxygen will also inhibit fatty acid desaturation and sterol biosynthesis. Thus alterations in energy metabolism and membrane biosynthesis may well be important factors in the pathology of anoxia and will be affected immediately when the oxygen supply is removed. Changes due to anaerobic metabolism will come about as a consequence of anoxia and will therefore take place more gradually as conditions in the cell change, due to prolongation of anaerobic metabolism. Thus, reduction in cytoplasmic pH (glycosidic acidosis) is a consequence of hypoxia, and tissue acidification can begin to take place within 2 minutes of the onset of hypoxia leading to root tip death in 12–24 hours (Roberts *et al.*, 1984)

The accumulation of ethanol (Crawford and Zochowski, 1984) has also been suggested as a possible mechanism whereby the continuation of anoxia will adversely affect the cell and could contribute to anaerobic injury. Ethanol accumulation is progressive and depends on the balance between the rate of production and removal by diffusion (Monk *et al.*, 1984) and it is unlikely therefore to be the cause of sudden anaerobic death that is seen in some sensitive roots. The time scale for the accumulation of potentially dangerous ethanol concentrations would suggest that it may play a role when anaerobic injury takes place over a matter of days or even weeks (Barclay and Crawford, 1981; 1982b).

5.3.2 ETHANOL AND CARBON DIOXIDE TOXICITY

Considerable debate has taken place on the role of ethanol in anaerobic injury. The ability of aerated plants to tolerate high concentrations of ethanol (Fig. 5.19) has raised doubts as to whether the amounts accumulated in plant tissues during periods

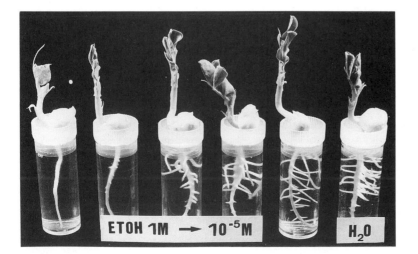

Fig. 5.19 Effect of increasing ethanol concentration on root development in pea seedlings. The ethanol concentration is 1 M in the left-hand tube and is reduced by a factor of 10 in each tube to the right, with the water control on the extreme right. Although ethanol at high concentrations reduces the development of lateral roots there is no evidence in this experiment where the shoots have access to air that it causes any damage to either roots or shoots.

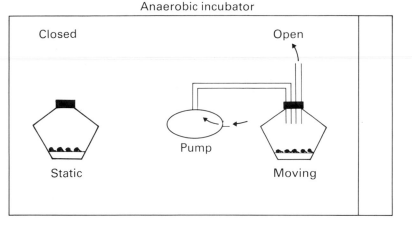

Fig. 5.20 Diagram of arrangement in anaerobic incubator for maintaining a static and moving anaerobic atmosphere while still retaining rigorous anoxia. The atmosphere in the anaerobic incubator contains 90 per cent nitrogen and 10 per cent hydrogen. Circulation of the atmosphere over a palladium catalyst ensures removal of traces of oxygen (*see* Crawford and Zochowski, 1984).

of flooding or experimental anoxia are sufficient to have a directly toxic effect on plant cells (Jackson, Herman and Goodenough, 1982). Recently it has been demonstrated that a circulating as opposed to a static anaerobic atmosphere increases the anaerobic life of chickpea seedlings (Crawford and Zochowski, 1984). The experimental arrangement for these experiments is shown in Fig. 5.20 where both the static and moving anaerobic environments are maintained in an anaerobic incubator in order to ensure the same rigorous exclusion of oxygen.

Fig. 5.21 shows the effect of circulating the anaerobic atmosphere around germinating seedlings on anoxic survival for a number of crop species. It can be seen that in lettuce, chickpea, turnip, flax, and wheat a circulating atmosphere produced significant prolongation of anoxic life. In chickpea seedlings (Fig. 5.22) (Crawford and Zochowski, 1984) it was observed that circulating the anaerobic atmosphere reduced the ethanol content of the anoxic tissues to one-thirteenth of that found in the static environment. These experiments therefore concluded that the removal of the volatile compounds generated under anoxia (for example carbon dioxide and ethanol) increased the survival of seedlings in the post-anoxic phase. Analysis of the composition of the anaerobic head space shows that the presence of high concentrations of carbon dioxide on an anoxia-sensitive species such as chickpea results in a marked increase in ethanol production (Fig. 5.23). The above and the recently reported effects of glycosidic acidosis on root viability under anoxia (Roberts *et al.*, 1984) suggest that there is a need for attention to be paid to the possibility of interactive effects between carbon dioxide concentration and ethanol accumulation on anoxic survival.

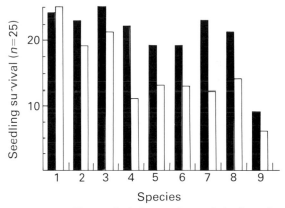

Fig. 5.21 Seedling survival after varying periods of anoxia in moving and static anaerobic environments. The anoxic treatment varied from 24 to 96 hours depending on the sensitivity of the species. Species (1) maize 96 hours; (2) soybean 96 hours; (3) chickpea 96 hours; (4) wheat 96 hours; (5) flax 48 hours; (6) cabbage var. harbinger 48 hours; (7) turnip var. Milan purple top 24 hours; (8) lettuce var. Webb's wonderful 24 hours; (9) rice 96 hours. (■ Moving atmosphere; □ static atmosphere.)

5.3.3 CYANOGENESIS

Some of the plants most intolerant of flooding are those that contain cyanogenic glycosides in their roots. The bark of apricot, plum and peach (*Prunus armeniaca, P. domestica, P. persica*) in both stem and roots is rich in cyanogenic glycosides and these species can be killed by as little as 24 hours flooding (Rowe and Beardsell, 1973). Anaerobiosis causes a hydrolysis of the cyanogenic glycosides in their roots, cyanide evolution takes place and tissue death

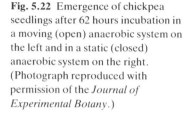

Fig. 5.22 Emergence of chickpea seedlings after 62 hours incubation in a moving (open) anaerobic system on the left and in a static (closed) anaerobic system on the right. (Photograph reproduced with permission of the *Journal of Experimental Botany*.)

rapidly takes place. In these trees flooding-sensitivity can be directly related to glycoside content of the roots. In species which have both cyanogenic and non-cyanogenic forms, moderately wet conditions usually favour cyanogenesis as this gives protection against grazing by the higher densities of snails that occur in such sites (Crawford-Sidebotham, 1972). However, in excessively wet areas this predator–defence mechanism exposes the possessor to the risk of death by flooding. Thus *Lotus corniculatus* which occurs on a variety of soils has both cyanogenic and non-cyanogenic forms while its close relative in British marshes *L. uliginosus* is never cyanogenic.

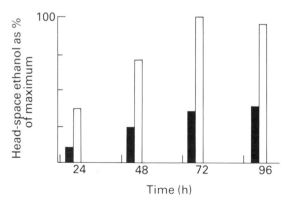

Fig. 5.23 Effect of addition of 25 per cent carbon dioxide to a static anaerobic atmosphere on the accumulation of ethanol in the head space above chickpea seedlings. The results are expressed as a percentage of the maximum concentration obtained over a 96-hour anaerobic incubation. (■ Anoxia with no carbon dioxide; □ anoxia with 25 per cent carbon dioxide.)

5.3.4 DEATH BY ANAEROBIC STARVATION

As shown above, it is the plants of the most extreme anaerobic habitats that have the greatest ability to survive under anaerobic conditions. Anaerobic metabolism is costly in terms of carbohydrate consumption as compared with normal respiration: those species that can survive extended periods of anoxia and live in the most anaerobic environments also have large carbohydrate reserves. These carbohydrate reserves show considerable fluctuations throughout the year with a typical maximum in autumn and a minimum in early summer after shoot extension (Steinmann and Braendle, 1984a) (Fig. 5.24). In both *Typha latifolia* and *T. angustifolia* the highest dry weight values for the rhizomes are found in late autumn (Fiala, 1978). Similarly tracer experiments with $^{14}CO_2$ on *Spartina alterniflora* show that the rhizome is the major carbohydrate sink in the growing season (Lyttle and Hull, 1980). Over winter, a sharp fall in the carbohydrate content has been observed in the rhizomes of *Typha latifolia* from 45 to 27 per cent of their total dry weight (Kausch *et al.*, 1981). In *Spartina alterniflora* the high winter levels of carbohydrate reserves are most severely depleted in spring in zones of rapid development (Gallagher *et al.*, 1984). A similar picture is observed in *Schoenoplectus lacustris* (Steinmann and Braendle, 1984b). These observations show that rhizomes that live in anaerobic mud typically have high concentrations of carbohydrates in early autumn and minimal levels in early summer. These strongly-marked fluctuations are evidence of the high metabolic cost to the plant of resuming growth in spring from rhizomes that are deeply submerged in an anaerobic habitat. In *A.*

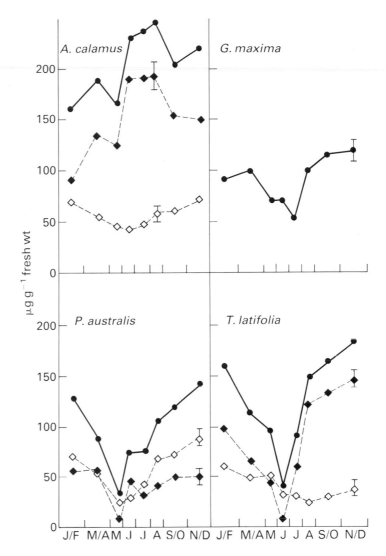

Fig. 5.24 Annual course of total carbohydrate ●, soluble sugars ◇ and starch ◆ content in rhizomes of *Acorus calamus*, *Glyceria maxima*, *Phragmites australis* and *Typha latifolia*. (Data by courtesy of Dr. R. Braendle, *see also* Braendle, 1985.)

calamus it is noticeable that the carbohydrate reserves never sink in early summer to as low a level as that found in those of the deeper burried species and that these fluctuations are not so pronounced. (Haldemann and Braendle, 1986). It may be that surface-living rhizomes make less demands on their reserves by not having to extend shoots as far and by prolonging the photosynthetic season through the possession of green rhizome tissues in a way that is not possible in the species with buried rhizomes.

There also appears to be a difference between anoxia-tolerant and anoxia-intolerant species in their ability to conserve carbohydrate supplies when experimentally deprived of oxygen. As carbohydrate consumption by plant tissues is highly temperature dependent, it is striking that under the conditions of experimental anoxia (22°C) *Schoenoplectus lacustris* showed only a 2 per cent reduction in total non-structural carbohydrate reserves. By contrast under the same conditions of experimental anoxia *Glyceria maxima* (which is intolerant of prolonged periods of anaerobiosis) showed a 46 per cent reduction in total non-structural carbohydrate (Barclay and Crawford, 1983). The soluble sugars also gave a similar pattern of behaviour with a marked reduction over 4 days in *Glyceria maxima* and little change in *Schoenoplectus lacustris*. Thus, even although *Glyceria maxima* rhizomes may normally possess considerable

carbohydrate reserves they are likely to suffer rapid
depletion when exposed to low oxygen stress. Such
behaviour is maladaptive in habitats where the oxygen
supplies may be limited. There appears therefore to
be an ecological relationship between carbohydrate
conservation and the degree of anoxia that is found
in the preferred habitats of these species.

5.3.5 POTENTIALLY HARMFUL POST-ANOXIC METABOLITES

The sensitivity to peroxidative injury that comes
about as a result of anaerobic incubation may be a
consequence of the inability of the plant to protect
its tissues against oxygen damage on return to air as
they may lack some of the necessary enzymes, for
example catalase, peroxidase or superoxide dismutase. During anoxia the enzymes that protect
tissues against the toxic effects of oxygen are sometimes reduced in activity thus rendering the plant
sensitive to peroxidative damage on return to air.
Alternatively, on returning to air, the oxidation of
anaerobically accumulated metabolites such as
ethanol may generate potentially damaging products
such as acetaldehyde, by the peroxidatic activity of
catalase (Oshino *et al.*, 1973):

$$\text{catalase} + H_2O \rightarrow \text{catalase} - H_2O_2 \text{ (compound I)}$$
$$\text{catalase} - H_2O_2 + CH_3CH_2OH \rightarrow \text{catalase} + 2H_2O + CH_3CHO$$

Hydrogen peroxide generation occurs through the
action of free ferrous iron and oxygen radicals
(Hendry and Brocklebank, 1985):

$$Fe^{2+} + O_2 \rightarrow Fe^{3+} O_2^{\cdot-}$$
$$\text{superoxide dismutase}$$
$$O_2^{\cdot-} + O_2^- + 2H^+ \rightarrow H_2O_2$$

Within 30 minutes of return to air there can be
observed in *Glyceria maxima* rhizomes a five-fold
increase in acetaldehyde concentrations as compared
with the anoxic phase. These persistent high levels
of acetaldehyde are likely to prove considerably
more damaging to plant cells than the higher concentrations of ethanol from which they were produced.
Therefore the prevention of ethanol accumulation
during the anoxic phase will reduce the generation
of potentially harmful acetaldehyde concentrations
when the plants are returned to air.

Buried plant organs have two possibilities for
controlling ethanol accumulation. Either they limit
its production as in the case of the rhizomes of the
bullrush (*Schoenoplectus lacustris*) which show only

very low glycolytic rate when incubated for prolonged
periods of anoxia, or else they facilitate the removal
of ethanol. This latter strategy is adopted by the
sweet flag (*Acorus calamus*) which has porous rhizomes which lie near the surface of submerged soils
and where constant bathing by free water avoids the
accumulation of any considerable concentration of
ethanol (Monk *et al.*, 1984).

A curious feature of wetland plants that has long
attracted attention is the highly developed aerenchyma
spaces that are frequently observed. The obvious
function of such spaces is usually considered to be
the facilitation of the downward passage of oxygen
from shoots to roots. However the excessive size of
such spaces (Fig. 5.6) has prompted the question
that the air space may be bigger than is necessary for
just facilitating the downward diffusion of air (Williams
and Barber, 1961). In order to explain the evolutionary forces which could have brought about
such an extensive development of aerenchyma in
some species it is necessary to postulate some functional significance for large air-spaces. The suggestion
that they might serve as oxygen reservoirs is not
tenable. Fig. 5.25 shows that even in species with the
largest amount of root air space the oxygen demand
of aerobic roots would only be sustained for 120
minutes. If however the important feature is the
dilution of potentially toxic volatile compounds then
the larger the air space the greater its efficacy. It
may be that the functional role of aerenchyma should
be considered not just from the point to view of
oxygen moving downwards but also of toxic compounds moving upwards. Another common name of
the water lily (*Nuphar lutea*) is the brandy bottle, on
account of the smell that emanates from its aerial
shoots which suggests that this plant is well adapted
for the upward movement of fermentation products
from its submerged organs. It appears that this
process is facilitated actively by solar radiation (Dacey
and Klug, 1982). These authors suggest that as the
petioles of young water lily leaves have a higher
resistance to air diffusion than the petioles of older
leaves then the expansion of air as the leaves warm
up will lead to a net down-flow in the young petioles
and an up-flow in the older tissues (Fig. 5.26). Solar
energy will thus cause a forced ventilation of the
submerged rhizome and remove volatile products of
anaerobic respiration. As the moving and static anaerobic environment experiments described above
have shown, any system which removes the volatile
products of anaerobiosis will be likely to increase
anaerobic survival. Not only will direct toxic effects

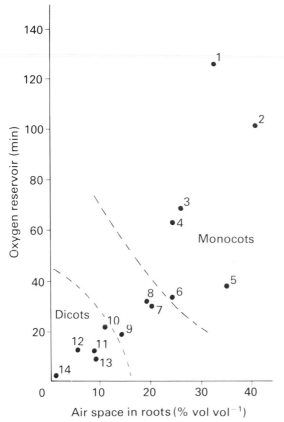

Fig. 5.25 Relationship between the capacity of the oxygen reservoir in roots to maintain aerobic respiration at the normal rate in air and the percentage air space in the roots of a number of wetland species.

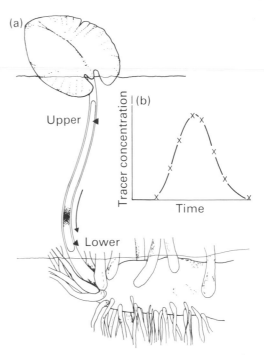

Fig. 5.26 Tracer movement of ethane used to follow the influx of air into the young emergent petioles of a water lily leaf. (a) Tracer represented by stippling was injected at the upper end of the petiole and sampled at the lower end. (b) The plot of tracer concentration as a function of time at the sampling point took the form of a typical elution curve seen in gas chromatography. The volume rate of gas flow was calculated and the elution curve integrated to show that virtually all the injected tracer passed the sampling point. The direction of gas flow was reversed in the older emergent (efflux) petioles. (Reproduced with permission from Dacey, 1980.)

be avoided but the potential danger of post-anoxic injury will be reduced as the precursors of toxic oxidative products will also be removed. If ethanol accumulation is avoided during anoxia then acetaldehyde production in the immediate post-anoxic phase is diminished.

5.3.6 SUPEROXIDE DISMUTASE AND THE REDUCTION OF POST-ANOXIC INJURY

Oxidative damage in biological systems may be brought about by various oxygen species and the superoxide radical (O_2^-) represents one potential source of toxicity. There are a number of cellular reactions which proceed by a single electron transfer from the substrate onto each molecule of oxygen used, producing superoxide. Halliwell (1984) has

cited the enzymes nitropropane dioxygenase, galactose oxidase and xanthine oxidase as being among those that produce superoxide in plant tissues. Studies with micro-organisms (aerobes, aerotolerant anaerobes, and strict anaerobes) have shown that SOD is essential for survival in the presence of oxygen and illustrate the central role of this enzyme in protection against oxygen toxicity (Fridovich, 1974). The damage to heart muscle as a result of heart attacks is in part a post-anoxic phenomenon. Post-ischemic injury in animals is comparable with post-anoxic injury in plants in that superoxide radicals are considered a major source of tissue damage and increased superoxide dismutase activity which destroys such radicals significantly reduces the extent of post-ischemic injury (McCord, 1985). It is therefore very significant that in the yellow flag iris (*Iris pseudacorus*) estimations of SOD activity showed a 13-fold increase in the activity of this enzyme during a 28-day anoxic incubation (Fig. 5.27).

This remarkable increase in SOD activity in *I. pseudacorus* distinguishes it clearly from the anoxia-intolerant species *I. germanica* and *Glyceria maxima* that were also examined. These latter species showed little change in SOD activity as a result of anaerobic incubation and 14 days anaerobic incubation is usually sufficient to cause post-anoxic failure in both these species (*see* p.108). The increase in SOD activity in *Iris pseudacorus* appears to be a *de novo* synthesis as it can be prevented by treating the rhizome with a protein synthesis inhibitor cyclohexamide. There are very few cases of polypeptide synthesis under anoxia with the best-known example being alcohol dehydrogenase. Thus this activity of SOD may well be an adaptive feature worthy of further examination, especially when it is realized that it represents an interesting parallel with mammalian tissue. Heart muscle is reported to be better able to recover from post-ischemic injury if SOD levels are high as a serious source of damage after heart attacks is the damage caused to muscle by superoxide radicles.

5.4 Soil toxins

Oxygen starvation is not the only environmental hazard that plants which use the *anaerobic retreat* have to endure. Reducing conditions in the soil give rise to the production of reduced ions many of which are potentially toxic to plants. Fig. 5.28 illustrates the sequential production of potentially toxic ions in the soil in relation to length of flooding. The reduction of organic matter can also give rise to short chain organic acids which are phytotoxic in their undissociated form (Lynch, 1977) at pH 5 or below. In warm climates hydrogen sulphide production in anaerobic soils can also prove a limiting factor in flooded soils which are rich in organic matter.

5.4.1 MANGANOUS AND FERROUS IONS

In aerated soils manganese occurs as the manganic ion Mn^{4+} which has limited solubility and the more soluble divalent ion. In this manner manganese differs from iron which only occurs as the divalent ion under reducing conditions. The reducing conditions of flooded soils however bring about a rapid increase in the highly soluble and phytotoxic manganous ion Mn^{2+} and many flood-sensitive species are adversely affected by manganous ions. The toxic effects of manganous ions are attributed to two progressive conditions. In the first, high manganese concentrations induce iron chlorosis. This deficiency can be relieved by giving the plants high concentrations of soluble iron salts. The antagonism between ions for uptake by roots causes the absorption of manganous ions to be hindered by the presence of ferrous iron. The second condition of manganese toxicity is marked by the appearance of brown necrotic spots which cannot be prevented by additional supplies of iron. The balance of other ions also

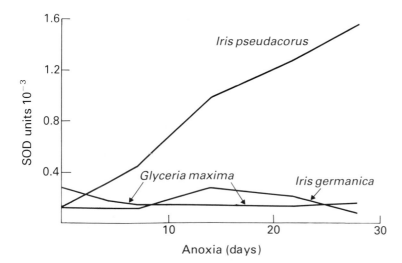

Fig. 5.27 Time course of superoxide dismutase activity in rhizomes of three species which vary in their tolerance of anoxia. In the most tolerant species *Iris pseudacorus* there was a 13-fold increase in the activity of this enzyme over the course of 1 month's anaerobic incubation. Use of protein synthesis inhibitors suggests that this increase is due to a *de novo* synthesis of the enzyme during anoxia.

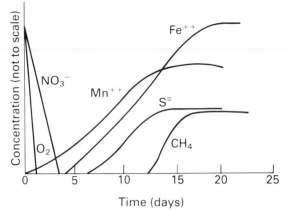

Fig. 5.28 The general sequence of reduction of redox systems in an oxygen depleted soil. (Reproduced from Patrick, 1978.)

Table 5.3 Tolerance of certain woodland species to ferrous iron in water culture (Martin, 1968)

	ppm Fe^{2+} (observations on root system)	
	Survival	Death
Mercurialis perennis	2	4
Endymion non-scripta	5	10
Brachypodium sylvaticum	–	15*
Geum urbanum	10	10–20
Circaea lutetiana	10	15
Primula vulgaris	10–20	20+
P. elatior	20	30
Carex sylvatica	30	30–40
Deschampsia cespitosa	50	80–100

* Not tested below 15 ppm Fe^{2+}

affects the degree to which manganese can produce toxic effects (Osawa and Ikeda, 1976). An increase in calcium decreases manganese uptake and its toxicity is also reduced when phosphorus supplies are adequate. Flooded soils are frequently phosphate deficient and this can predispose the plants to the dangers of manganese toxicity. The heath species *Erica cinerea* shows marked increases in leaf manganese content in flooded soils (Jones and Etherington, 1970) while other wetland species are able to restrict their uptake of iron from waterlogged soils. Differences can even be found between waterlogging-tolerant and non-tolerant populations of several grass species in their ability to restrict both the uptake of reduced manganese and iron (Etherington and Thomas, 1986).

Iron unlike manganese only becomes readily soluble once reducing conditions have become established in the soil. However the higher concentrations that are frequently present in flooded mineral soils can produce concentrations that are highly toxic. Higher plants that are able to withstand flooding usually have a greater ability to withstand iron poisoning than non-flood-tolerant plants. In as study on the effect of flooding on some woodland plants Martin (1968) was able to show that the possession of aerenchyma is correlated not only with a tolerance of wet soils but in particular with resistance to iron toxicity. The ranges of iron affecting the survival of these woodland plants is shown in Table 5.3. It is generally thought that the diffusion of oxygen downwards through the root and then escaping into the rhizosphere protects the plant against the uptake of

potentially harmful concentrations of iron by oxidizing the ferrous ions to the ferric form. This can take place either in a core of soil around the root or on the root surface as well as in the cortical tissues (including the aerenchyma) exterior to the stele. However the quantities of iron that are sometimes present in flood waters can be sufficient to overcome the detoxifying capacity of roots even of such flood-tolerant plants as rice (Howeler, 1973).

The ability of some species to tolerate iron concentrations that are toxic to others, as well as being due to the presence of aerenchyma, may be related to the varying mechanisms that plants adopt to obtain adequate supplies of iron in different soil conditions. Plants which live in soils where the available iron concentration is low and yet succeed in obtaining adequate supplies are termed *iron efficient* (Ambler et al., 1971). Such plants are able to excrete reductants which reduce the relatively insoluble ferric iron to the soluble ferrous form. It is inevitable therefore that plants which produce a reducing environment at their root surface will be totally unadapted to prevent the uptake of excess ferrous iron by oxidation in flooded soils. Consequently plants of wetland habitats that oxidize their rhizosphere will be inefficient iron absorbers in drier soils. Iron thus provides an interesting example of specialization to a particular habitat placing a limit on the tolerance range of a species. The adaptation that has evolved as an adaptive response to a particular stress proves a disadvantage when that stress no longer exists.

The dangers of iron toxicity have been particularly studied in rice and have led to the identification of two properties which assist resistance namely:

1 A capacity to reduce the soluble iron concentration in the medium around the root by oxidation or by changes in pH.
2 The prevention of absorbed iron ions in the root cortex form entering the stele and shoot.

This last property is particularly important as the toxic effects of iron on plant growth become evident when iron ions are translocated to the shoot causing an increase in the iron concentration in leaves. It is possible that the aerenchyma tissue that is so well developed in iron-resistant plants serves as a filter for the retention of precipitated iron in the outer layers of the root. This role for aerenchyma could be another cause of the evolution of apparently excessively large aerenchyma tissue from the point of view of oxygen diffusion that occurs in some species (*see* p.110). It has been argued (Williams and Barber, 1961) that the extent of development of aerenchyma in some plants is greater than can be accounted for by any selection pressure that was merely operating through increasing the diffusion of oxygen from shoot to root. As discussed above (p.125) the very large aerenchyma development found in many wetland plants may be a means of dilution of the volatile toxic products of glycolysis but this would not prevent it also serving as a filter for the removal of iron. The greater the radius of the root, the greater the surface area reoxidized for the deposition of iron on the root surface. Depending on the type of aerenchyma produced there may also be an improved interface area between air space and cell surface for the deposition of oxidized iron exterior to the stele on cell walls. The greater this total surface area the longer the root will remain serviceable for the uptake of nutrients and water, before it is clogged with an iron deposit.

5.4.2 SULPHIDE TOXICITY

In most temperate soils the conditions such as waterlogging that give rise to high sulphide concentrations also reduce iron to the ferrous state. Thus any sulphide formed anaerobically by bacterial action will be precipitated as insoluble ferrous sulphide and be less likely to be phytotoxic. In normal soils the presence of soluble ferrous iron is considered to be sufficient to keep the hydrogen sulphide level below 1×10^{-8}M (Ponnamperuma, 1972). Low pH conditions could increase the amount of dissolved hydrogen sulphide that could exist in equilibrium with a particular concentration of ferrous iron but as low pH also increases the solubility of ferrous ions the soil sulphide content is likely to remain low. It is mainly in soils of hot climates with large quantities of organic matter that sulphide poisoning is likely to be a hazard to plants. In rice, cloudy conditions produce a reduction in vitality that has been attributed to high hydrogen sulphide concentration which the plant is unable to detoxify during periods of inclement weather.

5.4.3 HORMONAL IMBALANCE

The root apex is the major site for the synthesis of gibberellins and of cytokinins and when subjected to anaerobic conditions their synthesis and translocation will be reduced. Many of the symptoms of flooding injury that are observed in crop plants that are suddenly inundated are characteristic of hormone imbalance. These symptoms are observed in the shoots and include wilting, chlorosis, and epinasty. In some cases the avoidance of shoot damage can be achieved merely by removing the roots before standing the shoots in water. Also spraying the leaves with gibberellin and cytokinin solutions can reduce flooding injury symptoms (Jackson and Campbell, 1979). Abscisic acid increases within hours of flooding in a number of crop plants and can lead to stomatal closure and reduced growth. Similarly, increased concentrations of ethylene have been reported in a number of crop plants susceptible to flooding injury. This ethylene appears to be synthesized in the aerial parts of tomato plants in response to the translocation from the root to the shoot of an ethylene precursor that is produced in roots of waterlogged plants (Jackson *et al.*, 1978). This precursor has been identified as 1−amino cyclopropane−1−carboxylic acid (Bradford and Young, 1980), the same compound that is thought to be converted to ethylene in the induction of aerenchyma formation in flooded roots (*see* p.112).

The hormonal effects observed in flooding injury are usually the consequence of a sudden change in root aeration in plants that have not evolved any means for surviving flooding. Even plants that are tolerant of flooding will show these symptoms if suddenly flooded. The leaves of alder wilt when the tree is suddenly flooded. The old leaves are shed and a new set of foliage which is unaffected by the flooding of the roots grows in their place. This is then followed by the breaking of dormancy of basal shoot buds and the flooded alder then grows with

Fig. 5.29 Typical bush form of alder (*Alnus glutinosa*) growing on a wet carr site. Alders growing on dry sites preserve the pole form.

the characteristic bush form that is found in natural wetland sites such as alder carr (Fig. 5.29). When alder is grown in cultivation in dry land sites free from flooding then the tree grows with the normal single trunk form.

The disruption of normal air supplies to the root will automatically cause a dysfunciton of all oxygen-requiring processes and among these a hormone imbalance and its effects on shoot growth will be an inevitable consequence. Thus although hormone imbalance points to a maladaption to flooding it does not necessarily indicate any prime cause for the death of plants when flooded.

5.5 Conclusions

The ecological exploitation of the *anaerobic retreat* is strictly open only to those plants that are able to tolerate a deprivation of oxygen or else provide through morphological adaptations an adequate supply of oxygen to their submerged organs. These are not necessarily exclusive strategies. Plants such as the bullrush (*Schoeneplectus lacustris*) can both obtain oxygen when it has an emergent shoot and survive prolonged periods of anoxia when totally submerged. This dual adaptation allows it to colonize areas denied to non-adapted plants. The ability to survive and even grow at low oxygen concentrations is also found in a number of seeds. The seed itself is a micro-anaerobic retreat for the germinating embryo. Seeds that can germinate with minimal oxygen supplies can germinate in areas where oxygen supplies are limiting. In the tropics there are also examples of trees which can survive prolonged inundation and yet remain alive. The remaining wetland plants that do not possess this metabolic resistance to anoxia live on the fringes of the anaerobic habitat. These species are the stress avoiders. In evolutionary terms they may possibly be the more advanced plants as they avoid the stress rather than modify their growth and metabolism to tolerate the limitations of the habitat. Ecologically avoidance and tolerance either on their own or combined at different seasons of the year increase the diversity of wetland plants and offer a wider range of habitat for their survival.

Fig. 6.1 Salt-drenched cliffs at Marwick Head, Orkney. The vegetation of this exposed site is particularly rich in varied ecotypes of *Plantago maritima*, *P. coronopus*, and *P. lanceolata* as well as having a colourful halophytic flora with *Scilla verna* and *Armeria maritima*.

6

Survival in coastal habitats

6.1 Hazards of the maritime environment

The study of coastal vegetation presents in a microcosm the effects of wind, drought, salt, erosion, burial and pH change interacting on closely adjacent plant communities. For the physiological ecologist therefore, some of the major stresses that limit plant growth, both in natural ecosystems and in cultivation, can be observed at varying intensities of operation on a gradient from exposed foreshore to sheltered hinterland. The nature of this gradient will depend on the topography of the coastline. Where shores are rocky with extensive cliffs then the changes in vegetation on moving inland can be abrupt and depend mainly on the height of the cliffs and their exposure to onshore gales and salt deposition (Fig. 6.1). The salt deposition on cliff vegetation can be very high and produces a distinctive community on the cliff face and the ground immediately behind the cliff top. In areas where the coastal vegetation itself determines the position of the high water mark by anchoring sand and silt, there is usually a more graded interaction between environment and vegetation.

Coastal sites provide habitats for pioneer plant communities which by their nature have to establish themselves on young skeletal soils. Thus the plants that live on dunes and dune slacks are in habitats where the soil has little power of retention of water or nutrients. Such communities are therefore susceptible to climatic stress, in that these young soils cannot offer any buffering action against drought or nutrient shortage. The sparse nature of the vegetation cover also contributes little to protecting the plant and soil against climatic extremes. Consequently the gradient effects imposed by distance from the sea are frequently very marked. Aerial views of dune systems that have escaped agricultural improvement or transformation into golf courses can show very pronounced variation over a few hundred metres (Fig. 6.2).

The following sections in this chapter examine the degree of specialization that is found in plants from a number of different maritime habitats and the relevance of the varying survival mechanisms which they have evolved. Of all terrestrial plant communities the coastal sand dune system is probably the one that lends itself best to the study of its development in terms of vegetation succession. Dune systems can be unfailingly ordered by age in relation to their proximity to the sea. This orderly progression from young to old in moving in one direction (inland) is not something that can be readily found in other supposedly seral plant communities. The aquatic communities bordering on lakes and river edges can suffer reversal of successional changes by flood movement of soils and vegetation. Although blow-outs can cause destruction of sand dunes they do not usually obscure the pattern of development that can be observed on moving inland.

6.2 Succession and adaptation in sand dunes

6.2.1 FORESHORE AND EMBRYO DUNE DEVELOPMENT

The parent material of the foreshore is in constant movement whether it be shingle, sand, or silt. The most mobile foreshores are those where the deposited material is sand. The colonizing species, apart from having to solve the problem of anchoring themselves in an unstable medium, need to avoid the hazards of burial if they are to survive. For annual species this is achieved with rapidly completed life cycles that can exploit the period of relative calm that exists in high summer away from the hazards of the equinoctal gales. The strand-line community contains a mixture of annual and perennial species the relative proportions of which vary with the exposure of the foreshore. Table 6.1 lists some typical species from western European foreshores. When the foreshore is very exposed and the growing season short the annual species can disappear entirely.

Behind the strand-line vegetation there can arise small embryo dunes where tidal currents and prevailing wind direction permit. The existence of an embryo dune system (Fig. 6.3) is an indication that the shore-line is actively accreting sand and probably advancing seawards. Any obstacle to sand movement can start an embryo dune and they even begin to form around annual species. However if

Fig. 6.2 Oblique aerial view looking north over the rapidly accreting dune and slack system at Tentsmuir, Fife. Moving landwards from the sea there can be seen a zone of mobile dunes followed by a dry slack, an alder zone, a wet slack and a grey or fixed dune. (Photograph: J.K.S. St Joseph, Crown Copyright Reserved.)

Table 6.1 Annual and perennial species of European sandy foreshore and early sand dune communities.

Annuals	Perennials
Atriplex glabriuscula	*Honkenya peploides*
A. littoralis	*Elymus farctus*
A. patula	*Leymus arenarius*
A. triangularis	*Ammophila arenaria*
A. prostrata	*Tussilago farfara*
Cakile maritima	*Rumex crispus*
Cerastium atrovirens	*Cirsium arvense*
Salsola kali	*Euphorbia paralias*
Polygonum oxyspermum ssp. *raii*	*E. portlandica**
Cerastium semidecandrum	*Carex arenaria*
Senecio vulgaris	*Senecio jacobea*
Phleum arenarium	*Cynoglossum officinale**
Myosotis ramosissima	*Viola tricolor*
Erophila verna	*Cirsium arvense*
Anagallis arvensis	*C. vulgare*

* Exists as a biennial or perennial

Fig. 6.3 Vertical aerial view of the most rapidly accreting part of the Tentsmuir dune and slack system showing the cluster of embryo dunes that are developing on the accreting sand bank. (Photograph: J.K.S. St Joseph, Crown Copyright Reserved.)

the rate of sand accumulation is in any way substantial, annuals are soon buried and further development of the embryo dune requires a perennial species with greater powers of emergence through deposited sand. Two principal methods of emergence are possible; growing out horizontally or growing up vertically. The colonizers of the embryo dunes are principally species that rely on growing out horizontally to achieve survival. The sand couch grass (*Elymus farctus*) on germination can extend a 7 cm long root down to the zone of permanently humid sand within 10 days. The initial rosette of tillers once established can send out oblique stolons up to distances of between 5 and 30 cm. This can be repeated for two seasons and as the plant becomes further established even greater lengths of horizontal extension are possible (Gimingham, 1964). If the stolons have their ends killed by being continually buried as they extend laterally, then other buds will grow upwards and can reach the surface if the plant is not buried too deeply. In studies of burial and emergence in Scottish sand dunes 23 cm has been found to be the limit for this type of emergence in *Elymus farctus* (Gimingham, 1964). The maximum vertical growth that can be achieved by sand couch grass is usually in the order of 30 cm. Live buds can be

found at greater depths of up to 60 cm (Gimingham, 1964) and although this is an important property allowing the plant to resume growth after periods of erosion, it does not contribute to the augmentation of dune height.

The sea lyme grass (*Leymus arenarius*) which is largely a northern species found from the Arctic to the shores of Brittany (northern France) is a remarkable plant for its capacity to grow horizontally. When the two species are together (*Elymus farctus* and *Leymus arenarius*) and where there is a sufficient supply of onshore sand deposits, very high rates of coastal accretion can be observed. Fig. 6.4a and b show the steady rate of coastal accretion that has been achieved by these two species together at Tentsmuir (east Scotland) in recent years. In places the rate of accretion has been between 7 and 14 m per annum. This is greater than the lateral extension growth of these species. However in favourable seasons fragments of rhizomes that have been dispersed beneath embryo dunes eroded by winter storms can suddenly spring up and establish a continuous plant cover over areas that before were largely devoid of plants. This rapid lateral growth of the dunes fixes a large portion of the available sand so there is little vertical development of the

(a)

Fig. 6.4 (a) Oblique aerial view looking south-east in 1957 over the accreting dune and slack system at Tentsmuir, Fife. Note the line of anti-tank blocks erected on the high water mark in 1940 (*see* arrow). (b) Similar view to (a) but taken in 1973. Note the consolidation of embryo dune area in the top left-hand area of the photograph. (Photograph J.K.S. St Joseph, Crown Copyright Reserved).

(b)

Fig. 6.5 A blow-out from a newly accreted mobile dune with isolated fragments of colonization containing sea-lyme grass (*Leymus arenarius*) and sand couch grass (*Elymus farctus*).

dune systems. Consequently after severe winter storms these embryo-dune systems are frequently eroded (Fig. 6.5) but due to the dispersal of the rhizomes throughout the sand there is usually a rapid recovery in the following season. The capacity of plants to anchor sand and build out new coastlines is dramatically illustrated in the steady rates of coastal accretion that have been observed at Tentsmuir in eastern Scotland. The recent accretion rates noted above have probably been maintained over several thousand years. It is possible to identify the post ice-age coastline at Tentsmuir and its present distance from the sea demonstrates that there has been an average coastal accretion rate of 2 m per annum for the last 12 000 years. The actual rate must have been greater as at various times in the past 12 000 years the sea has readvanced over the land as in the Flandrian succession.

6.2.2 MOBILE AND FIXED DUNES

The dunes that arise behind the embryo dunes are usually referred to as mobile or yellow dunes. The sand on these dunes is not yet fixed by the vegetation and they are therefore prone to movement under the direction of the prevailing wind. As the sand is not covered by vegetation they have a typically yellow colour as compared with later stages in dune

colonization referred to as fixed or grey dunes. The latter name comes from the change in colour due to the extensive development of ground cover by mosses and lichens. Two vicarious species of marram grass replace one another as the dominant yellow dune vegetation in the northern hemisphere on moving from the Old World to the New. In North America *Ammophila breviligulata* is the characteristic dune marram and this is represented in Europe by *Ammophila arenaria*. In both North America and Europe, *Ammophila* is the species which is responsible for the greatest fixation of sand in the yellow dune systems. It is capable of both extensive lateral and vertical growth. Both the American and European species have similar maximum capacities for vertical growth of about 1 m per annum. In fact the continuous vigorous growth of *Ammophila* requires a fresh deposition of sand. The older root system rapidly senesces and the plant requires a fresh deposition of sand around the shoot to encourage new root growth from the stem. *Ammophila* is not unique in this respect as a similar phenomenon is seen in another sand dune grass *Corynephorus canescens*. This latter grass being similar in behaviour to *Ammophila* in its requirement for fresh sand deposits for continued vigour has been used for studies on root longevity (Marshall, 1965) as it is easier to grow under cultivation. Careful examination of the root systems showed that cortical disintegration took place with-in 2 to 9 months resulting in the roots being devoid of cortex after 12 months, thus necessitating the need for fresh burial to encourage new adventitious root development.

The skeletal nature of young soils under pioneering communities shows marked changes as distance from the sea increases. This is particularly the case when the sand is low in shell fragments. Once away from the immediate zone of salt-spray deposition non-calcareous sand is quickly leached of its sodium chloride content with the result that there is a rapid change in pH on moving inland. Fig. 6.6 shows a profile through a sand dune system that has been rapidly accreting at Tentsmuir (east Scotland) to-gether with the pH profile of the sand at a depth of 10 cm. It can be seen that there is a very abrupt transition zone where pH falls rapidly. Furthermore, the zone where the pH falls rapidly moved seawards by 150 m in 5 years. This rapid change in pH does not take place where the sand dune system has a significant incorporation of shell sand. It requires upwards of 300 years to remove the free carbonate from the top 10 cm of sand even in humid climates

(Ranwell, 1959). In the Lake Michigan sand dunes a thousand years may be necessary to remove the carbonate out of the top metre of sand (Olson, 1958). Time scales for dune development will vary therefore with both the type of sand and its rate of deposition. Nevertheless, all dune systems once fixed will show a change in vegetation cover in relation to the time that has elapsed since their formation.

Fig 6.7 shows a map of changing coastlines at Tentsmuir (east Scotland) that has been obtained from aerial photographs covering the past 45 years and for the preceding century from Admiralty charts, accurately plotted for navigation into the nearby Tay Estuary. Here, as already pointed out, the sand is low in carbonate and there is a rapid zonation of the vegetation on moving inland. The complete change from young embryo dune to dune-heath encompassed by this region takes place in less than 100 years. The rate of change is so rapid in places that it is possible to measure an annual advance of creeping willow, birch, and alder across slacks that have only recently become free from the danger of sea flooding and salt inundation. At the same time there is a retreat of the salt-tolerant species such as

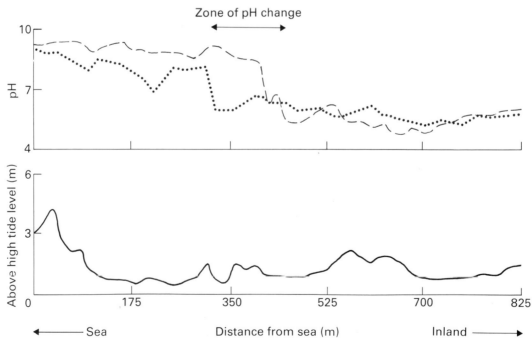

Fig. 6.6 Change in pH levels over 5 years (1970–75) along a transect at Tentsmuir, Fife running inland from the newly accreted zone shown in Fig. 6.4b. Note the movement seawards of the boundary marking the change from high to low pH (— pH 1970; . . . . . pH 1975).

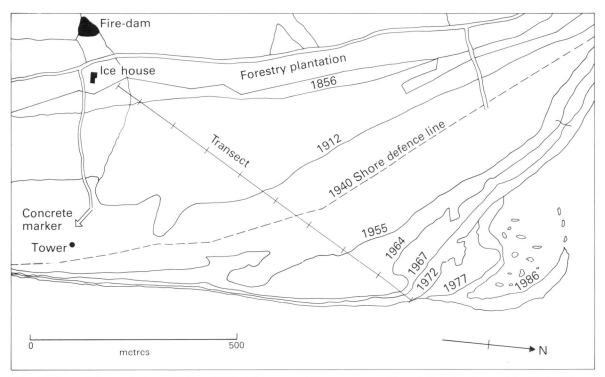

Fig. 6.7 Coastal changes between 1856 and 1986 at Tentsmuir, Fife reconstructed from Admiralty charts of the Firth of Tay and aerial photographs.

Juncus gerardii, *Centaurium erythraea*, and *Glaux maritima* along with the fall in pH. The Tentsmuir study is of interest for two reasons. First, it shows the remarkable capacity that plants have for accreting fresh sand given suitable conditions. Many coastal sites where erosion is taking place as a result of recreational pressure would have their long-term survival better ensured if local authorities had more faith in the growth capacity of plants and placed less reliance on wood, iron, and concrete. Secondly, it allows a real assessment of vegetation succession in relation to time. In many successional studies the times scale is only relative and all that can be said is that certain communities are younger than others. At Tentsmuir however it is possible to quantify the time factor.

Sandy soils have a low capacity for nutrient retention which means that apart from providing only minimal mineral nutrition they do not accumulate large quantities of salt. Thus in the humid climates, that are found on most temperate coasts, rainfall is sufficient to keep the soil profile relatively free from high salt concentrations. This does not mean the coastal dune vegetation does not have to be salt resistant. The foliage of all plants on exposed coasts, and particularly those of fore dunes, will be regularly exposed to heavy salt depositions after storms. However, the root systems of dune plants do not live in permanently saline soils as do the plants of salt marshes and some dune slacks. The hazards that beset dune plants are largely those of burial, drought, heat, and erosion together with exposure to salt spray and wind. Dune and fore dune vegetation, although subject to sea spray, is therefore not saline (Salisbury, 1952; Ranwell, 1972). The foliage does not accumulate salt and in this it is quite distinct from the plants that grow in salt marshes and areas that are regularly inundated with salt water.

6.2.3 DROUGHT AND HEAT TOLERANCE IN SAND DUNE VEGETATION

The low retention of water by sand inevitably means that dune vegetation will be prone to water shortages during periods of drought. Although marram grass grows up rapidly through sand, the root system as discussed above, has only a limited active life and it is the new roots produced in the upper layers of the dune that are mainly responsible for the uptake of water and nutrients. Dune plants are not phreatophytes, i.e. they do not belong to that class of plants that survive drought by having a root system that penetrates to great depths to obtain a reliable water source from a permanent water-table. In sand dunes even where the sand is fine (30—50 microns in diameter) the capillary rise in water-level is not more than 40 cm. As the active rooting zone for most dune plants, including marram grass, does not extend to a depth of more than 1 m this means that in dunes that are 3—4 m in height, the plants are effectively isolated from the water-table. Dune vegetation is therefore highly dependent on the moisture in the upper regions of the sand dune. In fixed dunes *Trifolium pratense* will use up with evapotranspiration all the available moisture in its rooting zone in less than 4 days (Salisbury, 1952).

Despite this apparent vulnerability to drought, dune vegetation does not wilt as rapidly as might be expected during periods when there is no rainfall. The proximity of dunes to the sea, combined with the porous nature of the sand, leads to an internal condensation of dew which has been estimated at up to 0.9 ml per 100 ml of sand per night. The source of this dew has been the subject of varying interpretations. Salisbury (1952) considered that the dew was a direct result of moist sea air moving into the upper layers of the sandy soils where internal condensation would preserve the distillate from re-evaporation in the morning and thus retain the moisture gained over night. Alternatively, as the upper layers of the sand cool at night this will cause an upward distillation of moisture from the warmer, moister layers of sand below. It is an old debate as to whether dew distills upwards or downwards. The argument has been considered pointless by some who maintain that dew is nowhere until it is formed (Simpson, 1929). Whatever the origins of dew there is at least agreement that when it is internally condensed on a large surface area then it can provide a considerable input to the water resources of the vegetation.

The ability of plants to obtain water from dew condensed within a porous medium has been known since classical times. In the Negev desert there are stone mounds called *teleilat el einab*, literally the hillocks for grapevives, which are believed to have been constructed for the culture of vines or other small plants by Nabatean communities in the first and second centuries AD. (Glueck, 1959; *see also* Monteith, 1963). Under favourable conditions these mounds can condense up to 0.5 l of dew per hour giving a total output for one night of 2—3 l which would be sufficient to support some limited plant growth. Even larger structures up to 10 m high and 24—30 m in diameter are to be found at Theodosia in the Crimea (Hitier, 1925). These may be expected to condense up to 300 l of water in one night as dew (Monteith, 1963). The higher figure of 12 000 per day that has been claimed for these mounds, and for which underground channels leading into subterranean cisterns were constructed, are beyond the capacity for dew formation and would require periods of continuous fog cover.

The ecological importance of fog cannot be overestimated. In coastal regions in particular sea fogs can make a very significant contribution to precipitation. The coastal redwood forests of California are one example and the pine forests of the Canary Islands are another (*see* Chapter 3 p.78). However the most spectacular demonstration of a fog-supported vegetation is in the Atacama desert in Chile and Peru. The desert here is one of the driest in the world. At Antofagasta measureable rainfall can sometimes be detected in less than 6 years out of 20 and then the precipitation does not exceed 4—6 mm per annum. However from May to November a dense cold mist the *garua* (Peru) or *camancha* (Chile) rolls in with the sea breeze from the cold Humboldt current. It is particularly dense at night but is sufficiently thick by day to obscure the tropical sun for days at a time. In the month of August some areas exposed to this mist have only 36 hours of sunshine and an average temperature of 13°C while less than 800 m above the temperature rises sharply to 24°C. This mist supports a lichen-dominated vegetation over extensive areas of the fog-shrouded hills facing the sea. However in certain favoured spots there exists a unique forest vegetation, the 'lomas'. The precipitation under the trees can be eight times that which is condensed in the open. Typical tree species are *Carica candens*, various species of *Eugenia*, *Caesalpinia tinctoria*, and near Lachay, *Acacia macrocantha* (Hueck, 1966). The lomas have a high proportion of endemic species, both trees and ground flora, and were probably much more widespread in the past. Fog-dependent ecosystems are very fragile. The damage that has been done by cutting for fuel and grazing by goats has reduced this vegetation to a few isolated pockets, but some success has been achieved by planting imported *Eucalyptus* and *Casuarina* which has led to a partial recovery of the local flora.

Despite the ability of plants in sand dunes to

supplement their water resources with internally condensed dew they frequently suffer from water deficits. In summer such deficits are readily observed in *Ammophila arenaria*, *Cynoglossum officinale*, *Potentilla anserina*, and *Senecio jacobea*. The stomata of *Ammophila arenaria* close in the morning and those of *C. officinale* and *S. jacobea* at midday. Some of the above species were also noted to wilt temporarily during dry periods (Willis and Jeffries, 1963). Similarly it has been observed that some American sand dune species can survive periods of daily wilting during dry spells. Wilting is damaging to plants not just for the loss in turgor and prolonged period of stomatal closure but for the danger of overheating of the leaves due to the loss of cooling power from transpiration. In mesophytes up to 50 per cent of heat loss can be accounted for by transpiration (Gates, 1968).

The sand dune habitat will aggravate the dangers of heat injury as the surface layers of dune soils can reach very high temperatures. Surface temperatures as high as 60°C have been recorded and the temperature of the air above the sand can rise by up to 10°C (Salisbury, 1952). This will further contribute to leaf heating particularly in species such as *Senecio jacobea* which have leaf rosettes near the sand surface.

The annual habit is one which opts out of the problem of an accumulating stress by accomplishing the life cycle before conditions become too adverse. Thus in foreshore and dune communities there are usually numerous annual species, except in those areas where the growing season is either too short or the exposure too extreme. Where conditions are sheltered and establishment during the summer is not at the mercy of severe wave action, then the annual foreshore community will advance seawards and can significantly alter the profile of the foreshore. Also in the dune communities many annuals develop over winter, flower in early summer and then survive the heat and drought as seeds. A feature common to many dune annuals is that they do not appear to have a viable seed bank. In *Vulpia membranacea* and *Cerastium atrovirens* the entire viable annual seed production germinates usually in late summer or early autumn (Table 6.2) (Harper, 1977). This lack of input into a seed bank has presumably evolved as a result of low survival of buried seed in these habitats. Apart from the hazards of survival in moving or accreting dune systems, sand does not provide a suitable habitat for the development of prolonged dormancy in seeds. Large numbers of viable seeds are found in soils where natural com-

paction results in poor aeration which can be further aggravated by flooding so as to reduce the concentration of oxygen and increase that of carbon dioxide. Carbon dioxide is well known to act as a powerful inhibitor of seed germination. Lack of oxygen reduces the peroxidation damage to membrane lipids and it is this gradual damage to membranes in aerobic conditions that is thought to be a major factor in determining the viability of seeds stored under areobic conditions. Such conditions do not readily exist in sand dunes where the soils are well aerated and where soil respiration contributes less to the build up of high concentrations of carbon dioxide.

6.2.4 STRESS METABOLITES IN SAND DUNE PLANTS

The stress metabolites (proline and glycine betaine) and the polyols (mannitol and sorbitol) (Fig. 6.8) are so named because they accumulate in a wide variety of plant tissues in response to various climatic stresses ranging from salt exposure to freezing and drought. It has recently been shown that in sand dune plants these same metabolites will increase the heat stability of certain enzymes (Smirnoff and Stewart, 1985). In *Ammophila arenaria*, proline, betaine, sorbitol, and mannitol increase the heat stability of glutamine synthetase and glutamate: oxaloacetate aminotransferase. With glutamine synthetase the effect increased with solute concentration (Figs. 6.9 and 6.10). Increased heat resistance will have an important survival value for plants under water stress as leaf water deficits inhibit the synthesis of most proteins (Bradford and Hsiao, 1982). If leaves heat up after stomatal closure, as they are likely to do in a sand dune habitat (*see above*), there will be a shift in the balance between enzyme synthesis and thermal inactivaton. Any small increase in enzyme heat stability caused by the accumulation

Table 6.2 Annual species characteristic of fixed sand dunes and having no buried seed bank (Harper, 1977).

Aira praecox
A. caryophyllea
Erophila verna
Filago minima
Arenaria serpylifolia
Cerastrium semidecandrum
C. atrovirens
Erodium cicutarium
Mibora minima

Quaternary ammonium compounds

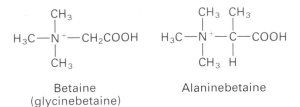

Betaine
(glycinebetaine)   Alaninebetaine

Amino acids

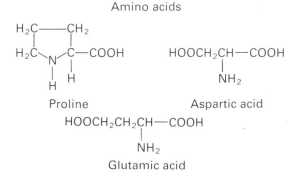

Proline   Aspartic acid

Glutamic acid

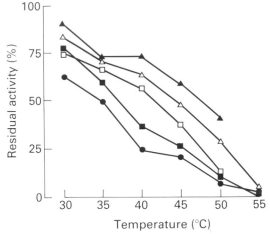

Fig. 6.9 Effect of various solutes on the heat stability of glutamine synthetase from *Ammophila arenaria* leaves. The enzyme was heated for 4 minutes: ● control; ■ 1 M proline, □ 1 M betaine, ▲ 1 M mannitol, △ 1 M sorbitol. (Redrawn with permission from Smirnoff and Stewart, 1985.)

Polyols

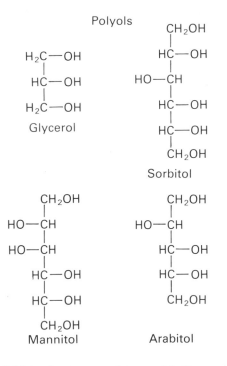

Glycerol

Sorbitol

Mannitol   Arabitol

Fig. 6.8 Molecular structure of stress metabolites commonly found in higher plants.

of stress metabolites could lead to a shorter lag period in the resumption of full metabolic activity on the return of favourable conditions (Smirnoff and Stewart, 1985).

In some foreshore communities there can be found species which progressively change from being C_3 plants to CAM (Crassulacean Acid Metabolism) plants as the growing season advances. This is the case with a number of the Mesembryanthemaceae which can grow either on the foreshore or else on cliffs of coasts from southern Britain through Europe to the Mediterranean and South Africa. Typical of such species are *Mesembryanthemum crystallinum*, *M. nudiflorum*, *Aptenia cordifolia*, and the hottentot fig, *Carpobrotus edulis* (Fig. 6.11). During the mild Mediterranean winter they are C_3 plants but as the summer advances and conditions become progressively drier they change to a C_4 metabolism (Winter *et al.*, 1978; Osmond, 1978). Experimental studies of the induction of Crassulacean acid metabolism in the Mesembryanthemaceae have shown that it can be induced by water stress (Winter, 1974), salt stress (Winter and von Willert, 1972), length of photoperiod (Queiroz, 1974), and thermoperiod (Neales, 1973). The Mesembryanthemaceae are mainly halophytic species in that in coastal habitats they accumulate salt even when growing on the foreshore. In this

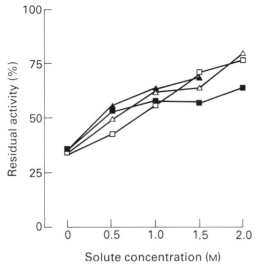

Fig. 6.10 The effect of solute concentrations on the heat stability of glutamine synthetase from *Ammophila arenaria* leaves. The enzyme was heated for 4 minutes at 40°C (*see* Fig. 6.9 for solutes). (Redrawn with permission from Smirnoff and Stewart, 1985.)

Fig. 6.11 The hottentot fig (*Carpobrotus edulis* [Mesembryanthemaceae]) a facultative CAM plant growing on a foreshore in northern Portugal.

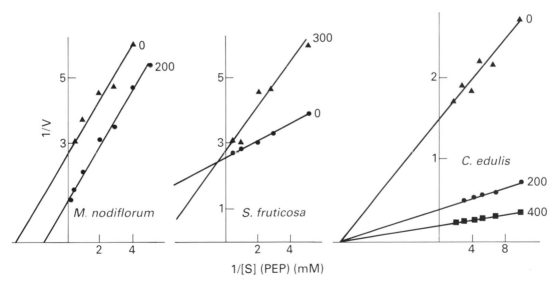

Fig. 6.12 Double reciprocal plots of enzyme activity with increasing phosphoenol pyruvate concentrations using PEP carboxylase isolated from halophytes grown under different saline conditions. No sodium chloride was added during the assay. NaCl concentrations of growing solutions (0–400 mM) are noted for each curve. Note that species differ in the manner in which enzyme activity is increased by growing in saline conditions. In *Mesembryanthemum nodiflorum* both V_{max} is increased and K_m reduced while in the other two species only V_{max} or K_m are significantly altered (*see also* Table 6.4). (Reproduced with permission from Treichel *et al.*, 1974.)

they are exceptional by comparison with the majority of foreshore plants. However in common with other foreshore species they are xerophytic and this distinguishes them from the halophytic species of salt marshes which transpire freely making use of a large throughput of water to alleviate their exposure to high salt concentrations.

The change from C_3 photosynthesis to CAM in the Mesembryanthemaceae is marked by a large increase in the specific activity of PEP carboxylase together with changes in enzyme kinetics (Fig. 6.12) (Treichel *et al.*, 1974). This increase is seen when extracts are made of this enzyme after growing the plants on increasing concentrations of salt (Table 6.3). It is important to note that this enzyme in common with all those tested for halophytic species is just as sensitive to salt as those from glycophytic species when extracted and tested *in vitro* (*see* Fig. 6.22). When the plants have adopted the Crassulacean acid metabolism pattern of stomatal opening at night for carbon dioxide uptake and closure during the day, the consequent reduction of transpiration can result in tissue temperatures rising 10−15°C above air temperatures (Nobel, 1978). In their adaptation to the combined stresses of heat, drought, and salt the Mesembryanthemaceae combine a number of adaptive features. These include succulence (*see below*) which when it develops in xerophytic plants acts as a water storage mechanism reducing the dangers of tissue dehydration and excessive salt concentrations. In the Mesembryanthemaceae this succulence is combined with the possession of a dense covering of leaf bladder cells. The walls of these cells are highly permeable and provide a water and salt reservoir that is readily exhangeable with the subcuticular cells of the leaf (Steudle *et al.*, 1975). They thus provide the leaf with an additional water and salt buffer with which to protect its photosynthetic tissues (Fig. 6.13). The coast-dwelling species of the Mesembryanthemaceae are a unique group of plants which employ succulence, extra-epidermal solute storage together with high salt uptake and CAM to enable them to remain metabolically active under the combined stress of heat, drought, and high salt exposure.

6.3 Dune slack communities

Behind the front line of dunes can usually be found the first in a series of habitats of level ground referred to as dune slacks or winter lochs (Fig. 6.14). The word 'slack' comes from the Old Norse *slakke* to make wet and is also used in English in the expression

Plant	Specific activity Ea min^{-1} mg Protein^{-1}	K_m (mM)	$V_{max} \times 10^{-3}$
Aster tripolium			
400 mM NaCl	38	0.18	0.11
Plantago maritima			
Control	48	0.21	0.13
200 mM NaCl	44	0.32	0.13
Salicornia fruticosa			
Control	69	0.10	0.39
300 mM NaCl	62	0.24	0.36
Honkenya peploides			
Natural site	143	0.10	0.71
Mesembryanthemum nodiflorum			
Control	45	0.32	0.39
200 mM NaCl	142	0.71	0.83
Mesembryanthemum crystallinum			
Control	49	0.08	0.29
300 mM NaCl	272	0.08	1.82
Carpobrotus edulis			
Control	116	0.08	0.69
200 mM NaCl	420	0.08	2.63
400 mM NaCl	960	0.08	5.55

Table 6.3 Specific activity of PEP carboxylase isolated from halophytes grown under different saline conditions (From Treichel *et al.*, 1974.)

Fig. 6.13 Leaf bladder cells on the surface of a leaf of *Mesembryanthemum nodiflorum*. The walls of the cells are highly permeable and provide a water and salt reservoir that is readily exchangeable with the subcuticular cells of the leaf (*see* Steudle *et al.*, 1975). (Photograph supplied by U. Lüttge.)

to slake one's thirst. The slack habitat is not widespread and occurs only where dune systems have developed in series. They therefore require a considerable undisturbed hinterland behind the front line of foreshore dunes. Probably the main claim to fame for dune slacks and their greatest economic importance in certain parts of the world is that they are the habitat which saw the invention of the game of golf. The ancient game of gold appears to have been invented in St Andrews shortly after the founding of the University in 1411.

Already by 1457 the game had become sufficiently popular for the Scottish parliament to have to pass laws discouraging this sport saying it should be 'utterly cryit doune and nocht usyt' (Young, 1969). The first site where the game developed was the dune slacks that are particularly well developed at St Andrews where the original site must have closely resembled the undisturbed dune and slack system shown in Fig. 6.2, except that the level ground which provided more succulent pasture than the adjoining dunes

would be regularly grazed and thus facilitate the search for golf balls. Thus in mediaeval times the slacks provided the natural fairway and the dunes the rough. Slacks although not permanently wet tend to flood in the winter hence the name 'winter lochs' that is sometimes given to these slacks in Scotland. As sand dunes impede drainage to the sea the water-table rises more rapidly in the landward slacks (Fig. 6.14). In winter when evapotranspiration is at a minimum this gradient increases with the result that the rear slacks flood first and stay under water for the longest time. The dune slacks in their development from seaward to more inland sites therefore represent a series of habitats where plants are flooded for varying lengths of time. They also differ however in their relative dampness in summer.

The most usual classification of slacks is to classify them as dry slacks and wet slacks depending on whether the water-table in summer is more or less than 1 m from the soil surface. A distance of less than 1 m from the soil surface for the water-table

(a)

(b)

Fig. 6.14 A landward dune slack at Tentsmuir, Fife; (a) flooded in winter and (b) late spring when the flooding has subsided. The landward slacks are the first to flood with water draining from inland areas and impeded by the dunes from reaching the sea. The flooding is therefore always from freshwater in these slacks.

will maintain a permanently humid soil profile within the rooting zone of most plants throughout the year. A depth of more than 1 m will probably produce in most years a period of relative water shortage. There is also a third gradient running through the slack system, and this relates to the nutrient status of the soil water. Most slacks are inundated by freshwater. Only in the very first slacks behind the first line of dunes is there a danger of salt flooding. As freshwater is less dense than salt water the upper part of the phreatic zone is usually fresh. In areas where the surface of the sand is only a little distance above the high water mark then at times of high tide this can raise the salt water level within the rooting zone of the plants in the front slack. However in most cases it is the freshwater layer alone which rises within the rooting zone. On moving inland the danger of periods of salt immersion to roots vanishes and the nutrient status of the soil water increases as drainage from more fertile adjoining land feeds into the slack. It is therefore possible to detect three gradients running through any series of slack communities in moving from the front duneline inwards, namely:

1 Increasing winter flooding.
2 Changes in depth of the summer water-table.
3 Increasing nutrient status of the soil water.

Table 6.4 shows species combinations from a number of different dune slacks shown in the aerial photographs (Figs. 6.2–6.4). The most rapid changes have taken place in the most seaward of the dune slacks. Within 10 years a community that was dominated by the halophytic *Glaux maritima* along with *Armeria maritima* and *Juncus gerardii* has been replaced with first creeping willow (*Salix repens*) and *Juncus balticus* with the latter disappearing as seed-

lings of birch and alder became established. Such can be the speed of change that in places birch forest now exists where only 20 years ago there was just a dwarf vegetation of salt-tolerant slack plants. It is interesting to note that the sex ratio of the creeping willow (a dioecious species with separate male and female plants) shows a predominance of female plants (60 : 40 – female : male) only in older sites. The same phenomenon is also found in the sea buckthorn (*Hippophae rhamnoides*) where again it is the female plants that appear to have the greatest capacity for survival especially as competition develops on the older sites.

The alder zone is a particularly conspicuous feature of this dune slack series. The alders become established at the edges of the first line of freshwater slacks. The fruits wash up here and establish themselves as a floodline association. During the period of establishment the young tree has a reasonable probability of not being flooded. Once established the alder has a greater resistance to flooding and will survive periods of inundation which would be damaging at the seedling stage.

Ultimately on moving inland the special characteristics of the dune slacks with their dwarf, flood-tolerant vegetation are lost and the low-lying land between the ancient dune ridges is occupied by typical marshland vegetation often dominated by species such as *Glyceria maxima* and *Filipendula ulmaria*.

6.4 Salt marshes

Plants that live in salt marshes (Figs. 6.15 and 6.16) are exposed with great regularity to saline conditions when their rooting zone and even the shoot itself may be covered with sea water. Consequently all plants in a salt marsh have to have the capacity to

Table 6.4 Typical species combinations found in wet and dry slacks on the east coast of Scotland. (Data adapted from Crawford and Wishart, 1966.)

Salt slacks	Nutrient poor dry slacks pH>5	Nutrient poor dry slacks pH<5	Alder flood-line slacks	Nutrient rich wet slacks
Glaux maritima	*Lotus corniculatus*	*Erica tetralix*	*Alnus glutinosa*	*Juncus effusus*
Festuca rubra	*Salix repens*	*Galium palustre*	*Angelica sylvestris*	*Glyceria maxima*
Rhinanthus minor	*Festuca rubra*	*Salix repens*	*Primula veris*	*Potentilla palustris*
Agrostis stolonifera	*Juncus balticus*	*Holcus lanatus*	*Agrostis stolonifera*	*Filipendula ulmaria*
Juncus gerardii	*Parnassia palustris*	*Carex flacca*	*Corallorhiza trifida*	*Hydrocotyle vulgaris*
Centaureum erythraea	*Carex arenaria*	*Ammophila arenaria*		
Plantago maritima	*Betula verrucosa*	*Betula verrucosa*		

Fig. 6.15 Oblique aerial view of the Eden Estuary, north-east Fife showing the location of salt marsh in relation to extensive mud-flats. This mud-flat was planted with *Spartina anglica* in 1948 in an attempt to reduce the erosion of the salt marsh. Due to the northern location of this estuary flowering occurs too late for the *S. anglica* to set seed and the spread of the species is slow and limited to vegetative reproduction. (Photograph: J.K.S. St Joseph, Crown Copyright Reserved).

Fig. 6.16 Close up view of the *Spartina anglica* colonies that have resulted from the introduction of this species in 1948 to the Eden Estuary, north-east Fife (*see* Fig. 6.15).

survive salt water inundation. For those species that inhabit the most seaward end of a salt marsh this may take place twice daily. For plants in the more inland situations this may happen only at periods of spring-tides. There therefore exists a double stress in salt marshes to which the plants have to be adapted namely, salinity and flooding. Numerous investigations have been made to determine whether the marked zonation that is found in salt marshes is due to varying degrees of salt stress or differences in the flooding frequency of the marshes. In the Baltic, detailed study of water-table levels and salinity (Tyler,

1971) suggests that here drainage rather than salinity is the main factor controlling the distribution of plants. However the Baltic is less saline than the open sea and there is good reason to implicate both salinity and aeration in the causal factors of salt marsh zonation in other areas. In a factorial study of the effects of drainage and salinity on the growth of salt marsh species Cooper (1982) was able to show that there were differences in response between upper and lower marsh species. In the upper marsh species such as *Festuca rubra* and *Juncus gerardii* growth was limited by salinity and waterlogging. In the lower marshes *Puccinellia maritima* preferred waterlogged soils but only *Salicornia europea* had a higher growth rate under saline conditions (Ranwell, 1972; Waisel, 1972; Cooper, 1982).

Tidal flooding leads to the development of strong reducing conditions which are inimical to the survival of a number of species which are capable of withstanding high salt concentrations in well-drained soils. The varying ecology of the two American species of cord grass *Spartina patens* and *S. alterniflora* (Fig. 6.17 and *see* Fig. 2.5) appears to be due to the inability of *S. patens* to adequately ventilate its below ground organs thus making it sensitive to the

reducing conditions in the lower reaches of the salt marsh (Gleason and Zieman, 1981).

In the early days of physiological ecology (Schimper, 1898) it was thought that halophytic species were physiological xerophytes in that they would react to the salinity of sea water by restricting their transpiration rates. However the transpiration rates of true halophytes (plants of salt marshes) are just as great as those of glycophytes growing on freshwater. Salt marsh species here differ from plants of foreshore and dune habitats in that they have very few xerophytic adaptations. They have thin cuticles and the stomata are usually frequent and unsunken. The cuticles of halophytes are often so thin that they can absorb useful amounts of water through their stems and reduce the water demands of the shoot. The misunderstanding of Schimper arose from the observation that glycophytes, which were used as the experimental material by the earlier physiologists, did restrict their transpiration rates when flooded and it was incorrectly assumed that halophytes would respond in the same manner. Whether or not plants restrict their water uptake in areas prone to salt exposure depends in most instances on whether they are plants of the foreshore and frontal dune systems

Fig. 6.17 Oblique aerial view of the upper region of the Mississippi delta where brackish water and less severely flooded sites are the preferred habitat of *Spartina patens* (*see also* Fig. 2.5 for *S. alterniflora*).

where salt has to be endured from deposition on shoots and water is limited, or whether they are plants of salt marshes where there is no shortage of water although it is saline (Delf, 1912).

An interesting exception to the absence of drought adaptation in salt marsh plants of potentially great economic importance has been noted in relation to the active research that took place in the 1970s into C_4 plants. Certain salt marsh species and most noticeably *Spartina anglica* (Long and Woolhouse, 1978) possess an active C_4 metabolism. This effectively increases their stomatal resistance to water vapour loss while increasing their uptake of carbon dioxide. Plants which transpire less will also burden themselves less with the problems of disposing of unwanted minerals. Whether or not this is related to salt tolerance is not clear. There is however a striking similarity between the xerophytism of certain mire species which live with their roots exposed to high concentrations of reduced ions and the habitat of the C_4 *Spartina anglica*. The greater efficiency in fixing carbon dioxide in relation to water loss may be related to the ability of this plant to survive with its roots in mud-flat soils that are very rich in ferrous sulphide.

In a survey of 30 species of higher plants collected from coastal sand dunes and salt marshes three were found to have the typical *Kranz* leaf anatomy of the C_4 plants (Long *et al.*, 1975). That is, the chloroplast containing cells are concentrated in a wreath (*Kranz* = German for wreath) around the vascular bundles of the leaf. These included *Salsola kali* and *Atriplex laciniata* which occur as foreshore plants. Other species to join this list now are listed in Table 6.5 It is open to question as to why some salt-tolerant plants have a C_4 metabolism. As well as increasing the water efficiency of the plant it has been suggested (Long *et al.*, 1975) that the lower carbon dioxide compensation point of C_4 plants enables them to

Table 6.5 Salt marsh plants in Britain which have been reported as showing C_4 carbon dioxide fixation. (From Long *et al.*, 1975; Smith, personal communication.)

Salsola kali
Atriplex laciniata
Cynodon dactylon
Cyperus longus
Spartina townsendii
S. anglica
S. maritima

increase their photosynthetic yield in a cool climate. The leaf photosynthetic rate of the *Spartina townsendii* at 25°C is about 50 per cent higher than the maximum rates reported for the major herbage grasses in Britain. Salt marshes in general are among the highest yielding areas in terms of total herbage produced. Whether this is due to C_4 metabolism directly however is not clear. For the true salt-adapted halophyte there is no shortage of water in the salt marsh and this together with the high nutrient status of these marshes is probably the major factor in producing the high yields.

Whatever the evolutionary causes of this combination of C_4 metabolism with halophytism it is a promising field for research into finding plants that can survive irrigation, without demanding too much water and thus causing an unacceptable degree of salt accumulation to take place in the upper regions of the soil. The inevitable accumulation of salt in freely-irrigated soils in arid climates is one of the most limiting factors to increasing food production in both developed and developing countries.

6.5 Surviving salt exposure

The means whereby plants survive exposure to high concentrations of salt vary with the manner in which this salt enters the plant tissues. When the salt stress is due to deposition on foliage as takes place in foreshore habitats, cliffs and forward dune systems then most plants attempt to resist the uptake of excessive amounts by means of various exclusion, and secretion processes. However in the truly saline habitat where the plants grow with their roots permanently in salt water such regulation mechanisms fail and the plants exist by coming quickly into equilibrium with the salt concentrations in the soil. Salt-tolerant vegetation has been divided physiologically into osmoregulators and osmoconformers (Wyn Jones and Gorham, 1983) making a parallel between plants and animals with regard to osmoregulation. This physiological division can be used to classify plants in relation to their tissue water potentials. Plants growing in saline soils can have tissue water potentials in healthy unwilted leaves in the range -2 to -3 MPa (Jeffries *et al.*, 1979). In crop plants leaf water potentials of -1.2 to -1.6 MPa are found only in leaves of wilted plants. Well-watered leaves generally have water potentials of -0.2 MPa or more.

The water relations of the plant can be represented in terms of water potential by:

$$\psi = \psi_p + \psi_m + \psi_s$$

where:

ψ = the water potential of the cell, tissue or plant

ψ_p = the pressue (turgor) or hydrostatic potential of the cell

ψ_m = the matrix potential, or potential of structurally bound water

ψ_s = the solute or osmotic potential of the cell contents

In ecological situations not all species will conform neatly to any physiological division. However if -2 MPa is taken as the dividing line then most species that live in salt marshes are able to attain water potentials as low as this without wilting and can therefore be classified as osmoconformers. Species such as *Atriplex littoralis* are different in that they can maintain more or less constant leaf ψ_s up to an external ψ of -0.9 MPa which would classify this species as an osmoregulator (Stewart and Ahmad, 1983). *Limonium vulgare* and *Plantago maritima* behave in an intermediate fashion between osmoconformers and osmoregulators. Both the above species are intermediate in their ecology as well as their physiology. *Atriplex littoralis* occurs in foreshore sites which is consistent with its osmoregulating properties but can also be found in salt-rich sites due to its ability to accumulate NaCl as an active osmoticum. For the purposes of producing some rationalization into this account of salt tolerance, plants from unflooded sites will be considered apart from those of the more frequently flooded salt marshes. With one or two exceptions (for example the Mesembryanthemaceae) the foreshore and dune plants belong to the class of osmoregulators as discussed above and the salt marsh to the osmoconformers. It is important to note however that even the highly salt-tolerant osmoconformers of the salt marshes have an upper limit to their tolerance of salt and they too practise some of the various means of salt regulation described below with reference to foreshore plants.

6.5.1 SURVIVING SALT SPRAY IN FORESHORE AND DUNE HABITATS

Dune and foreshore plants as discussed above are not plants of saline habitats in that their roots are not immersed in salt water or salt-saturated soils. Also, their shoot tissues do not accumulate salt, which as we shall see below is a common feature of halophytic species. However plants on dunes do have to be resistant to salt spray and for this they

need some form of short-term avoidance of salt injury. In the foreshore and dune habitat water is not so freely available as in the salt marsh. Thus they differ from salt marsh plants in having a greater development of xerophytic characteristics both morphologically and physiologicaly. Foreshore and dune plants also differ from true halophytes in that their growth is usually reduced when the salt concentration around their roots reaches 1/32 of the sea water. A similar inhibition of growth in halophytes is found only when the salt concentration reaches 1/8 to 1/2 of that of sea water.

Tolerance to moderate salt exposure as in the plants of the foreshore and dune system appears to rest on prevention of salt accumulation in the shoot. Comparison of different species and varieties in relation to their salt tolerance has largely been carried out with crop plants. In comparisons of salt-sensitive and salt-tolerant barley (Greenway, 1962) it was found that the salt tolerant varieties were able to maintain lower levels of Cl^- in their leaves than the sensitive ones. In the sensitive barleys the high level of Cl^- in the leaf was due to rapid uptake in the root coupled with rapid transport to the shoot. In similar studies with salt-sensitive and salt-tolerant soybeans the tolerant varieties do not transport large quantities of Cl^- to their leaves but retain it in the older portions of the root. A similar pattern of behaviour appears to operate for foreshore and dune plants. The means whereby these wild species restrict their salt uptake into leaves varies between species and as most species use more than one method it is not practical to classify plants either as salt excluders or salt excretors. For purposes of description however it is convenient to consider the control of salt accumulation under the following headings:

1 exclusion
2 excretion
3 succulence
4 passive removal

6.5.2 THE CONTROL OF TISSUE SALT CONTENT BY EXCLUSION, SECRETION, SUCCULENCE AND PASSIVE REMOVAL

Some grasses of salt-exposed habitats have low concentrations of both sodium and chloride in their leaves. In salt-tolerant species of *Puccinellia* a double endodermis develops which retards the passage of salt into the xylem. Apart from such physical means of exclusion it appears that the xylem parenchyma cells can function as pumps which actively pump sodium out of the xylem. Potassium on the other hand appears to be pumped in actively. This arrangement will allow the exclusion of sodium and avoid the potassium deficiency which is common in glycophytes subjected to saline conditions.

The active removal of salt from plant tissues by specialized organs has attracted scientific investigation since the nineteenth century. Although originally studied only in a few families, for example Tamaricaceae and Plumbaginaceae (Fig. 6.18) the occurrence of salt-excreting glands is now known to occur in the Gramineae, Primulaceae, Convolvulaceae and others, as well as in the mangrove families Avicenniaceae and Acanthaceae. Plants with salt glands are particularly common in salt marshes where the problems posed by high concentrations of salt are in part alleviated by an abundant supply of water. The presence of salt-excreting glands is linked with a higher tolerance of salt than in those species which are without them (Munns *et al.*, 1983). The structure of salt glands varies from the two-celled hair gland found in certain grasses (for example *Spartina anglica*) and the genus *Atriplex* to the more complex structure of 16 cells found in the Plumbaginaceae (for example *Armeria maritima*). The cells of the salt glands have certain features which distinguish them from other leaf cells. They lack a vacuole and have a high concentration mitochondria. Although they may have high concentrations of salts as compared with surrounding cells they are not accumulating organs but are designed for the active movement of ions outwards.

The gland cells like the endodermis also have cutinized and suberized cell walls which ensure that ion transport can only take place through the symplast. These impervious layers also prevent re-entry of secreted salt by leakage back into the leaf. Salt glands differ in whether they are sunk into the endodermis or not. Thus the glands of *Spartina* spp. which are sunk into the leaf are essentially the same as the leaf hairs found in *Atriplex* spp. (Fig. 6.19). The mealy appearance of *Atriplex* leaves is due to their extensive development of leaf hairs for the secretion of salt. These hairs (trichomes) have developed a bladder-like function and have received considerable attention in relation to their role in salt excretion. The trichomes have a large highly-vacuolate bladder cell attached to a stalk, which is in turn attached to an epidermal cell. The emission of salt from these hairs is by rupturing and collapse of the bladder cells. A number of swollen and ruptured

Fig. 6.18 Thrift (*Armeria maritima* [Plumbaginaceae]) a species which controls the salt levels in its tissues by actively excreting salt through specialized multicellular glands on the leaf surface.

cells can be seen in Fig. 6.19a. These hairs are not the most effective of salt-excreting organs as they cease to function when ruptured. However they can remove considerable quantities of salt from the leaf. In a study of *Atriplex hymenlytra* in Death Valley in California it was found that over the growing season one-third to one-quarter of the salt content of the leaf was removed by these hair glands (Bennert and Schmidt, 1983). Sodium was the main cation excreted but chloride was removed in smaller quantities. The cation excess in the excreted salts was balanced by large quantities of the organic oxalate. The presence of large quantities of sodium oxalate crystals could be seen on the leaf surface.

In all salt glands the metabolic work is carried out by a basal cell or stalk gland which pumps salt into the upper cell from which it passes passively out of the leaf, or else accumulates to rupturing point in the leaf hairs. Secretion by salt glands can be pre-

(a)

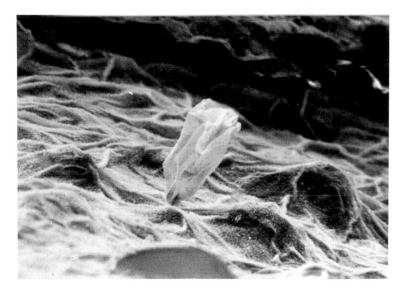

Fig. 6.19 Scanning electron micrograph of the leaf surface of *Atriplex* showing (a) numerous swollen leaf hairs (trichomes) which serve as salt excreting glands and (b) close up of a leaf hair which has a large highly vacuolated bladder cell attached by a stalk cell to the leaf. (Photograph A. Edwards.)

(b)

vented by either lowering the oxygen supply or the temperature of the leaf. The concentration of salt in the brine excreted by the glands is greater than that in the water round the roots and shows that the glands are capable of transporting salt against a concentration gradient. The mechanism of gland action has been extensively studied using strips of *Atriplex spongiosa* leaves (Lüttge and Osmond, 1970). In these glands chloride excretion into the bladder is actively enhanced by light. The amount of photosynthetic activity that takes place in the gland stalk and bladder cytoplasm is too small to account for the rate of chloride movement that takes place. There must therefore be a biochemical linkage between the site of energy production and utilization.

The development of thick, succulent leaves at

high salinities is found frequently in dicotyledons but seldom in monocotyledonous species. The succulence that develops in response to sodium chloride and other salts is due to an increase in the volume of the cells of the spongy mesophyll. It is usually accompanied by other morphological changes. The mesophyll tissue typically has fewer intercellular spaces and fewer chloroplasts than non-succulent leaves. There are also changes in the chloroplasts which in *Atriplex halimus* become much shorter and thicker. The thickening of the leaf increases its volume and at the same time reduces the surface area to volume ratio. The extent of succulence development is always greatest in those species that do not possess salt-excreting glands. It would appear that species which can control their salt accumulation by excretion do not develop succulence so readily.

There is an ecological distinction which reveals two different types of succulent species. In the more xerophytic species succulence develops from germination and provides a water reservoir for the leaf during times of drought. The bladder cells which develop on the leaves of *Mesembryanthemum* (Steudle *et al.*, 1975) are an example of a succulent species adapted to arid environments (Fig. 6.13). The Mesembryanthemaceae thus combine succulence and CAM metabolism as specific adaptations against drought. In more strictly halophytic species which inhabit salt marshes such as *Glaux maritima* succulence develops in response to salt concentration and in the field can be seen to increase as the growing season advances. In some cases it is possible to demonstrate that this change in volume (Fig. 6.20) is sufficient to reduce the sodium chloride concentration in the leaves when expressed in terms of leaf volume. As a large part of the leaf volume is taken up by vacuole space this will reduce the vacuolar concentration of salt.

The problem of the progressive accumulation of salt is particularly acute for perennial species. In some, the salt is removed from the plant by shedding of the maturer leaves. The death of these leaves may in fact be hastened by the amount of salt they absorb. This form of passive removal of salt is a constant feature of *Juncus maritimus* and *J. gerardii*. In these species leaf shedding presents no great disadvantage as the aerial portions of the plant die down in the autumn and result in a general salt cleansing of the plant. In evergreen species there is a danger that the perennating leaves will accumulate harmful quantities of salt. In such cases passive removal of the salt also takes place by leaching from

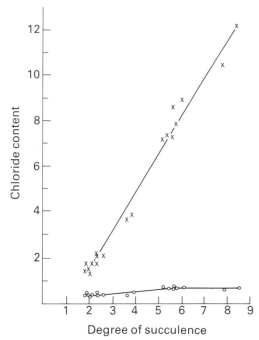

Fig. 6.20 The effect of succulence in reducing tissue chloride in leaves of the mangrove *Laguncularia racemosa* is illustrated in the manner in which the leaf content of salt is expressed: x = mmol δm^{-2} of leaf surface and shows as chloride content rises succulence increases; o in mmol g^{-1} of sap which remains constant due to the increased volume of the leaf with increasing succulence. (Redrawn with permission from Kinzel, 1982.)

the leaves with rain water. Winter rainfall is often relatively high, while potential transpiration is lower than summer. Thus salt can be removed at a rate greater than that at which it accumulates. In spite of this reduction in salt content, the water potential of the leaves does not rise. In winter the leaves of such evergreen species replace their fall in osmotically-active salts with an increase in the soluble carbohydrate content of their leaves. In winter the leaves of *Halimione portulacoides* and *Limonium vulgare* both contain more glucose and fructose than is summer. Similarly in *Cochlearia anglica* there is an increase in raffinose as well as in the above sugars (Kappen, 1969). This increase in soluble sugars will also increase the frost hardiness of these species as well as protect them from potentially damaging exposure to salt in winter storms. This type of limited control of salt accumulation is typical of the foreshore plants. From time to time they will be exposed to drenchings of sea water but they do not have to exist with their roots permanently in a soil which is permanently or regularly inundated with sea water as in the case of salt marsh plants.

Osmoregulation as the sole means of surviving in saline habitats is an adaptation of only limited value as when exposure to high salt is prolonged the exclusion mechanisms eventually fail. It is therefore not surprising that these plants differ from true halophytes in that their growth is reduced by prolonged exposure to sea water. Thus the foreshore annual species *Atriplex halimus* can have its growth reduced by as little as 120 mM NaCl in its culture solution (Gale and Poljakoff-Mayber, 1970). For some species the gradual accumulation of salt that takes place inexorably during the growing season presents insuperable problems. Thus in these species senescence appears to be inevitable and this may account for the selection of annual forms that are found in many foreshore species. In other species the plants shed their older leaves as their salt accumulations increase. Although wasteful in that leaves are shed before they have reached the end of what might have been a longer photosynthetically active life it is the price that these plants pay for being able to survive exposure to salt.

6.5.3 ADJUSTMENT TO OSMOTIC STRESS IN SALINE SOILS

From equation (1) (p.149) it can be seen that a change in leaf water potential can arise in a variety of ways. In most studies on salt tolerance in higher plants it is considered that ψ_s is the component which most affects leaf ψ. The results of Stewart and co-workers (Fig. 6.21) show that changes in leaf ψ are accompanied by very similar changes in ψ_s. In *Limonium vulgare* and ψ_s decrease in parallel over a change in external ψ from near zero to -2.7 MPa. The other major component ψ_p remains more or less constant up to -1.8 MPa.

Alterations to solute or osmotic potential can occur either by an accumulation of solutes taken up from the external medium or by the synthesis of organic solutes within the plant. A third possibility however is the reduction of cell water content by dehydration of the stressed organ. The latter however is not strictly a compensation mechanism. It does not occur in those stress conditions where the plant enters a dormant state as in the resistance to frost desiccation in overwintering shoots (*see* Chapter 3). In actively growing plants it is only the first two

possibilities that contribute to the adjustment of osmotic potential that is necessary for survival in a salt marsh. In most halophytic higher plants this is achieved by uptake of NaCl (Flowers *et al.*, 1978). The results in Table 6.6 taken from studies by Stewart and Ahmad (1983) show that in species such as *Atriplex littoralis*, *Plantago maritima* and *Suaeda maritima* over 70 per cent of ψ_s can be accounted for by Na^+ and Cl^-. In other species, such as *Cochlearia danica* Na^+ and Cl^- accounts for only 46–50 per cent of ψ_s.

Although it is obvious that halophytic species can continue their metabolic activities at low internal water potentials and with high leaf contents of salt they do not appear to have enzyme systems that are in anyway more tolerant of salt than glycophytic species. Fig. 6.22 shows the effect of increasing salt concentration in the assay mixture for phosphoenol pyruvate carboxylase in four halophytic species. In common with other enzymes such as malate dehydrogenase glucose–6–phosphate dehydrogenase, isocitrate dehydrogenase and RuBP carboxylase from the salt-tolerant plants *Atriplex spongiosa* (Greenway and Osmond, 1972) and *Suaeda maritima* (Flowers, 1972) there is an inhibition by NaCl concentrations that is equivalent to those found in leaves of halophytes. The inhibition appears to be non-specific

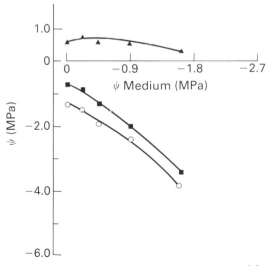

Fig. 6.21 Leaf water potential and component potentials in sea scurvy grass (*Cochlearia officinalis*) grown at different salinities: Leaf $\psi = \blacksquare$; $\psi_s = \bigcirc$; $\psi_p = \blacktriangle$. (Reproduced by courtesy of Professor G.R. Stewart.)

Table 6.6 The contribution of NaCl to leaf osmotic potentials. (From Stewart and Ahmad, 1983.)

Species	Leaf ψ_s MPa	ψ Na$^+$ + Cl$^-$ MPa	% of leaf ψ_s
Atriplex littoralis	−2.82	−2.0	70
Avicennia nitida	−3.29	−1.88	57
Cochleria danica	−2.28	−1.05	46
Cochleria officinalis	−2.42	−1.24	50
Limonium vulgare	−2.70	−1.56	57
Plantago maritima	−2.04	−1.38	68
Suaeda maritima	−2.63	−1.93	73

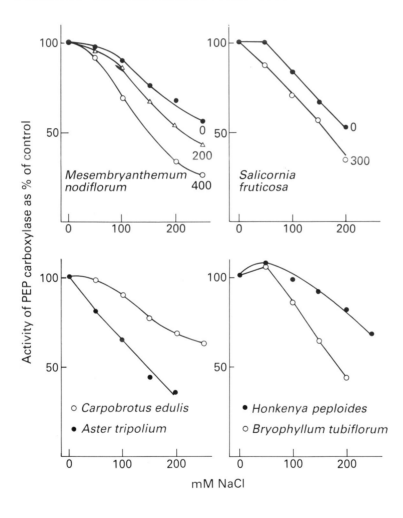

Fig. 6.22 The effect of increasing concentrations of NaCl on phosphoenol pyruvate carboxylase activity when added to the assay mixture for a number of halophytic species. In *Mesembryanthemum nodiflorum* and *Salicornia fruticosa* note that irrespective of whether or not the plants were grown under saline conditions salt in the assay mixture reduces enzyme activity. Compare this data with that in Table 6.3 and Fig. 6.12 where growth in saline conditions is shown to increase specific activity. (Reproduced with permission from Treichel *et al.*, 1974.)

resulting in a reduction in V_{max} and an increase in K_m. It is thought that the salt affects the ability of enzyme molecules to take part in the conformational changes which are essential for catalysis. The effect of sodium chloride in inhibiting *in vitro* enzymes assays must not be confused with the ability of plants when grown in the presence of sodium chloride to increase the specific activity of an enzyme, for example PEP carboxylase (*see* p.141).

Given therefore that halophytes and glycophytes do not differ at a molecular level in their sensitivity to salt their ability to function with high tissue contents of NaCl must be dependent on some degree of compartmentation within the cell. In vacuolated cells up to 95 per cent of the cell volume may be taken up by a large central vacuole. A number of studies have given reasonable evidence that the bulk of the leaf content of sodium chloride is located within the vacuole. Technically this presents problems since large losses of ions can take place during tissue preparation. Data from leaf tissue that was prepared by freeze substitution and examined by analytical electron microscopy in order to have a non-aqueous system so far gives contradictory results that may be overestimating the amount of Cl^- and Na^+ in organelles such as chloroplasts (Harvey *et al.*, 1981).

The compartmentation of NaCl into the vacuole requires an equivalent lowering of cytoplasmic water potential. It is here that the amino acid, proline (Fig. 6.23) and the quaternary ammonium compound, glycine betaine (Storey and Wyn Jones, 1975) appear to function as compatible cytoplasmic solutes. Many other methylated onium compounds apart from glycine betaine have been shown to occur in the leaves of halophytic species. In the Plumbaginaceae, β-alanine is found and related sulphonium compounds are present in *Spartina anglica* (Larher *et al.*, 1977). Similarly β-alanine betaine is found in *Limonium vulgare*. It is now even possible (Stewart and Ahmad, 1983) to distinguish different groups of halophytic species in terms of the compounds they synthesize in

Fig. 6.24 Sea arrow grass (*Triglochin maritima* [Juncaginaceae]) an example of a halophytic species which is common in salt marshes and where proline is the major compound synthesized in response to salt exposure.

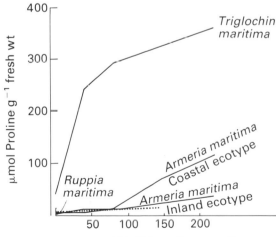

Fig. 6.23 Effect of increasing salt concentration in the growth medium on proline accumulation in a number of halophytic species. Note that in the coastal ecotype of *Armeria maritima* there is a greater response than in the inland ecotype. (Redrawn from Stewart and Lee, 1974.)

response to salt stress. In one group proline is the major compound (for example *Triglochin maritima*) (Fig. 6.24), while in a second it is glycine betaine, and in a third methylated onium compounds together with proline. There is a fourth distinct group characterized by the genus *Plantago* where sorbitol increases in response to salt exposure. *Plantago maritima* shows a very close relationship between leaf content of NaCl and sorbitol (Ahmad *et al.*, 1979). Other metabolites can also be found which increase in response to salt. Thus in *Puccinellia maritima* the amides, asparagine, glutamine, serine, and glycine increase in response to salt stress. Wild asparagus which lends its name to the amide asparagine is after all a plant of cliff top vegetation where there is a constant risk of salt exposure. In the mangrove *Avicennia nitida* as well as an accumulation of betaine the ureides allantoin, and allantoic acid also increase with salt exposure. When enzyme activity is examined in relation to the accumulation of proline and glycine betaine there is little or no inhibitory effect up to concentrations of 1 M (Flowers and Hall, 1978). In one instance their accumulation has even been shown to reduce the inhibitory effects of NaCl (Pollard and Wyn Jones, 1979).

The mode of action of these protective substances is still the subject of some debate. It is possible that apart from their function as osmotic regulators they also act as modifiers of water structure and thus enhance protein hydration at low water potentials (Schobert and Tschesche, 1978). They are not exclusively accumulated in the cytoplasm. In *Puccinellia maritima* when it is grown on 600 mM NaCl there is an increase of approximately 240 mM in organic solutes and Na^+ and Cl^- by 800 mM. With the cytoplasm representing only 5–10 per cent of the cell volume some of the organic solutes must be in the vacuole otherwise the cytoplasm would have a lower osmotic potential than the vacuole (Ahmad *et al.*, 1981) (Fig. 6.25). Cells will differ in their accumulation of solutes just as they differ in the extent of their vacuolation. In non-vacuolated meristematic cells higher concentrations of organic solutes are to be expected. Thus in the stem apices and in expanding leaves of water-stressed wheat very large amounts of proline of up to −0.6 MPa have been recorded (Munns *et al.*, 1979).

6.6 Conclusions

An ecological examination of the zonation of coastal vegetation shows salt to be the one environmental stress that impinges at some time in the life cycle of maritime vegetation. The great diversity of maritime

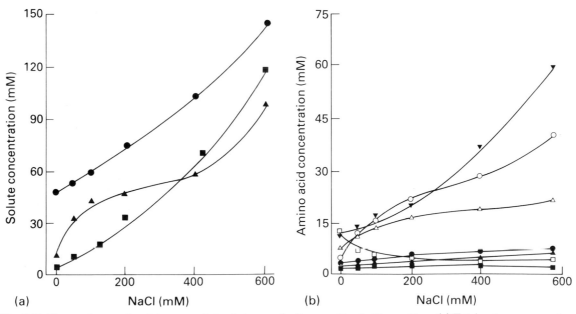

Fig. 6.25 Changes in organic solute accumulation in leaves of salt grown *Puccinellia maritima*: (a) Total amino compounds ●; proline ■; soluble carbohydrates ▲. (b) Changes in amino compounds: glutamine ▼; △′-acetlyornithine ○; asparagine △; glutamate □; asparatae ■; serine ●; alanine ▲. (Reproduced with permission from Ahmad *et al.*, 1981.)

plant communities is in part a reflection of the different means that they employ to avoid the harmful effects of salt. There is no evidence that any plant enzymes have evolved a resistance to salt inhibition and that the resistance mechanisms that have developed operate by keeping the active metabolic machinery of the cytoplasm free from too much sodium chloride. This common result is however achieved in various ways with some plants allowing salt to accumulate in certain organs (for example old leaves) and even within the cell vacuole. All plants need to exercise some degree of regulation. When the habitats are unflooded then it is even possible for exclusion mechanisms to operate. However as salt increases in the soil water these methods eventually fail. The increasing use of irrigation in arid climates and the dangers this imposes on crop production shows that the different strategies of plants in response to salt exposure needs to be matched in agriculture just as it is in nature. A possible lesson from natural evolution is the high productivity that is achieved when C_4 metabolism is combined with salt tolerance as is found in certain salt marsh grasses. For most salt marsh species the maintenance of active metabolism and photosynthesis does however rely on a plentiful supply of water. This however is not without problems as flood irrigation will produce even higher concentrations of salt in the upper regions of the soil. In agriculture as in ecology the answer will probably lie in some compromise with a sub-optimal solution that will give an acceptable chance of survival without over-exposure to environmental stress or the deposition of too much salt in the upper regions of the soil.

7
Survival on the forest floor

7.1 The forest environment

Plants that live on the forest floor enjoy a habitat that is buffered from extremes of temperature, exposure to wind, and desiccation. In comparison with grass and heathland sites the forest provides a measure of protection against trampling and heavy grazing (Fig. 7.1). The forest soil, depending on the type of dominant tree and the age of the forest, frequently provides the ground vegetation with a more even supply of water throughout the year. The tree canopy by increasing evapotranspiration reduces the tendency of soils in humid areas to become waterlogged in winter and its shade in summer lessens the danger of drought. In addition soil respiration can provide the plants that live near the ground with an enriched source of carbon dioxide, a factor that is generally limiting for plant growth. Fig. 7.2 shows the carbon dioxide concentration at several heights in a forest during a temperature inversion. The slope of the curve at night gives an indication of forest respiration at each height. The highest levels of carbon dioxide are found in the zone nearest the soil indicating the considerable potential of soil derived carbon dioxide for light fixation. The degree to which soil respiration contributes to the photosynthetic carbon fixation pool has been investigated using the isotope discrimination ratio $\delta^{13}C$ (Medina and Klinge, 1983). Carbon dioxide from soil respiration has a more negative value than average atmospheric carbon dioxide. Thus analysis of the $\delta^{13}C$ values for foliage at different heights above the soil shows the greater negative value of $\delta^{13}C$ for the lower canopy leaves as compared with the upper canopy and thus demonstrates the importance of soil respiration as a source of carbon dioxide for photosynthesis in the lower regions of the forest. It follows that the importance of soil respiration as an enriched source of photosynthetic carbon dioxide will be even greater for the plants actually that live on the forest floor (Table 7.1).

To be able to benefit from these advantages,

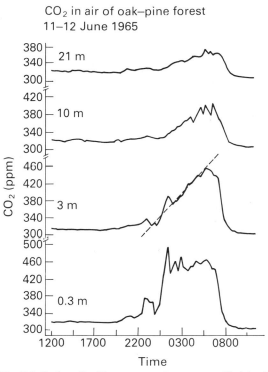

Fig. 7.2 Carbon dioxide concentration at several heights in a forest during a temperature inversion. The slope of the curve showing the increase in carbon dioxide concentration during the inversion is an index of the rate of respiration of the forest at night. Note the higher level of carbon dioxide accumulation at the level nearest the ground due to the additional component from soil respiration. (Redrawn from Woodwell and Bodkin, 1970.)

plants of the forest floor have to be able either to make a net annual photosynthetic gain with a light intensity that may be as little as 5 per cent of that which falls on the forest canopy or else endeavour to capture sufficient light during intermittent periods of greater intensity of illumination. This can be either seasonal as in the vernal flora of deciduous wood-

Fig. 7.1 (*opposite*) *Linnaea borealis* a shade-loving species found in woods and in the shade of rocks. It is particularly characteristic of Scottish and Scandinavian pine forests. Note the well spaced non-overlapping leaves.

Table 7.1 Height ditribution of $\delta^{13}C$ values in a tropical rain forest in San Carlos de Rio Negro. (Reproduced from Medina and Klinge, 1983.)

	$\delta^{13}C\permil$
Upper canopy leaves (>20 m)	−29.5±1.8
Lower canopy leaves (2−10 m)	−33.7±1.5
Undergrowth leaves	−35.2±1.2
Seedlings	−28.7
Composite leaf litter	−29.4

lands, or on an occasional basis when trees fall. Sunflecks have also been shown to contribute significantly to the photosynthetically usable radiation in woodland sites (Anderson, 1964). Depending on the type of wood and the time of day the radiation reaching the ground in a sunfleck can be sufficient to raise the temperature of a leaf of *Impatiens parviflora* in a few minutes by as much as 9° above air temperature (Rackham, 1975). High-temperature damage from sunfleck radiation is also thought to be the the explanation of leaf necrosis that can be seen in some years in Cambridgeshire woods in *Mercurialis perennis* (Fig. 7.3) (Rackham, 1975). Apart from shade, the other major limiting factor in woodland habitats is mineral nutrition. In some forests and in particular tropical rain forests the large biomass of the tree vegetation holds the bulk of the nutrients with little left in the soil. In temperate zones the accumulation of needle litter and podzolization of the soil under coniferous forests can also produce a mineral deficient soil.

Despite the disadvantages of low-light regimes and low-mineral status the ameliorating effect of tree cover on climatic extremes enables many plants to extend the limits of their geographical distribution beyond what is possible in treeless habitats. Thus *Adoxa moschatellina* which in Britain is common in open habitats in the western and more oceanic regions but is restricted to woodland sites in the east. The common primrose (*Primula vulgaris*) will grow in open grassy habitats in the west of Britain and is even successful as a cliff top species on exposed Atlantic coasts, even in the far north in Orkney and Shetland. However, elsewhere it is typically a woodland species. When the woods are felled it will flourish for a while but is eventually replaced by other species. The primrose needs the woodland or hedgerow habitat to protect it against aggressive competitors, except in areas where the climate alone is sufficiently rigorous to provide protection.

The plant communities of the forest floor provide examples of the varying ecological strategies as defined by Grime (1979) as:

1 *Competitors*, usually perennial plants with a well-developed capacity for resource capture and high levels of input of resources into new organs.
2 *Stress toleraters*, again perennials but with a conservative utilization of captured resources and able to survive for long periods with little growth or reproduction.
3 *Ruderals*, plants with rapid growth rates and

Fig. 7.3 *Mercurialis perennis* (dog's mercury) a typical forest floor species capable of growing at low-light intensities and prone to damage when exposed to full solar radiation from sunflecks in summer (*see* text).

short life-spans and often capable of prolonged seed dormancy and able to exploit irregular occurrences of suitably disturbed habitats.

As in other environments where resources are severely limited the imposition of a stress on the plants is more in the nature of an entrance examination than a constant feature of daily existence. The plants that survive in the habitat do so because their adaptations allow them to live in a manner that avoids stress. Entry to the habitat of the forest floor implies a life-style that avoids the stresses of low light and possibly low nutrient availability. Among the competitors are those species that are able to complete their life cycle in the shade of the forest canopy. *Impatiens parviflora* has been a particularly successful invader of woodlands in northern and central Europe since its introduction from central Asia in 1837 (Coombe, 1966). Apart from physiological adaptation many shade-tolerant species possess morphological characters which aid survival in low light environments. *Milium effusum* and *Zerna ramosa* are examples of the broad-leaved morphology which is a common adaptive trait in shaded habitats. The leaves of this last species can even survive 45 days in the dark (Coombe, 1966). The avoidance of overlapping leaves is also characteristic. Figs. 7.4 and 7.5 show two species where leaves are well spaced out and avoid mutual shading, *Chrysosplenium oppositifolium* and *Trientalis europaea*. The latter although frequently now found in open upland sites is usually indicative of the presence of former forest cover.

Stress tolerators include species which germinate in the shaded habitat and remain there in a suppressed condition. Illumination is often below the compensation point (the amount of light necessary for photosynthesis to balance the needs of respiration). Some tree saplings can survive for years in light conditions that are just about the compensation point until a suitable opening appears in the tree canopy when they will grow rapidly and fill the gap. Typical examples include beech (*Fagus sylvatica*) and yew (*Taxus baccata*) and the eastern hemlock (*Tsuga canadensis*). When the age structures of forests are examined it is often possible to trace back the establishment of certain cohorts to specific times when openings were created in the canopy. In a study of the black cherry (*Prunus serotina*) growing in oak forest in Wisconsin, it was found that 58 per cent of the trees were 20–30 years old and their establishment dated back to a period in the 1930s when the farmers cut the foliage for cattle fodder (Auclair and Cottam, 1971). This opening of the

Fig. 7.4 *Chrysosplenium oppositifolium* an extreme shade-tolerant species of stream sides, springs, wet rocks and wet ground in woods. The shade tolerance of this species also allows it to inhabit the entrance to caves.

canopy not only allowed the suppressed saplings to fill the gaps but also promoted extensive germination.

Periodic catastrophes are part of the natural biology of forests and many species have evolved mechanisms which allow them to take advantage of these events. In the forests of the southern USA the long-leaved pine (*Pinus palustris*) stays in a genetically determined suppressed state called the 'grass stage' until there has been a forest fire. Without fire there is no vertical extension of the growing bud which remains sessile on the ground surrounded by a rosette of long needles (Fig 7.6). The bud is well protected to stand

Fig. 7.5 *Trientalis europea* (chickweed wintergreen) a shade-tolerant species of pine forests, now often found under heather in former forest sites.

Fig. 7.6 The so-called 'grass stage' of the long-leaved pine of the southern USA (*Pinus palustris*) which remains in this genetically controlled dwarf stage until exposed to the heat of a forest fire.

heat and only after the thermal shock does the terminal bud extend and the dwarf sapling take on a tree form.

The vernal flora consists of plants such as *Anemone nemorosa* and the bluebell *Hyacinthoides non scripta*. The density of bluebells in deciduous English woodlands has been shown to be governed by the amount of illumination that reaches the plant in the high-light phase of spring rather than that in the low-light phase when the leaves have fully expanded (Blackman and Rutter, 1946). Such species are obvious stress avoiders in that they complete the bulk of their growth and reproduction before the closing of the canopy. Sweet cicely (*Myrrhis odorata*) (Fig. 7.7) although probably an introduced plant to most of Britain, is a widespread feature of the vernal flora of many woodlands, particularly where there has been human disturbance. The plant was used in mediaevel times for struing before church doors at certain festivals and this has probably contributed to its present widespread occurrence throughout the British Isles. It also grows early in the year which made it

Fig. 7.7 *Myrrhis odorata* (sweet cicely) a widespread feature of the vernal flora of many woodlands. This species was probably introduced into Britain as its early growth and aromatic properties made it a useful plant for struing at Church festivals between Easter and Whitsun.

particularly useful for events in the Church calendar from Easter to Whitsun. This combination of a vernal growth and sweet scent is also found in the much rarer holy grass (*Hierochloe odorata*) which was introduced to Scotland by monks from Prussia for this same purpose. Both species are able to grow in deciduous woodlands and complete the bulk of their growth before the summer canopy closes.

The ruderal species include a few annuals and a number of specialized perennials which are able to invade forest clearings. The foxglove (*Digitalis purpurea*) typifies a reproductive pattern that is particularly successful in this situation. This species is often referred to as a biennial in that it flowers in the second year of its life. However foxglove, like so many of the so-called biennials, outside cultivation should really be considered as perennial as it can survive for 3, 4 or more years before dying (Silvertown, 1984). As pointed out by Silvertown more precise terms are semelparous (single reproducing) and iteroparous (multiple reproducing) perennials. True obligate biennials appear to be rare (Klemow and Raynal, 1981). Semelparous perennials are also not common. Only 1.4 per cent of the North American flora belong to this class compared with 21.3 per cent of annuals (Hart, 1977). However in certain taxonomic groups semelparous perennials are very frequent. In the Umbelliferae 30 per cent of the European species show this behaviour (Silvertown, 1984). The characteristic feature of the foxglove and other species which have the capacity to invade forest clearings, for example fireweed (*Chamaenerion angustifolium*), is the ability to produce large quantities of small seeds which can be widely dispersed. The small seed has the disadvantage that there is an initial lag phase in the development of the seedling and the plant cannot normally flower in the first year of growth. However the delay in reproduction is compensated for by the large number of seeds that are eventually produced. By contrast the annual species invest more resources in their seeds in terms of dry matter per seed and therefore do not have this lag phase. Their dispersal however is often not so efficient due to the greater seed size. Annuals therefore need habitats that provide space that is available regularly in the same locality. Semelparous perennials such as foxglove can fill intermittent gaps and when squeezed out eventually by iteroparous perennials, can establish new populations due to the efficient dispersal of very large numbers of small seeds. The semelparous perennial frequently has to reach a critical size before it can flower. This ability to queue for reproduction by size rather than by age is likely to be a superior strategy in deteriorating environments such as would arise when shade is re-established over a forest clearing. In comparison with annuals, biennials appear to be at a disadvantage due to the delay in reproducing. However this disadvantage also disappears (Harper, 1977) in declining successional populations as biennials die more slowly than annuals.

7.2 **Establishment on the forest floor**

The first problem for any plant that has to survive on the forest floor is that of initial establishment whether by seed or by the spread of vegetative organs. Seed germination is most readily seen in forest clearings or areas where the soil has been disturbed by rooting animals. In the large natural forests of southern beech (*Nothofagus* spp.) regeneration is best in those clearings where wild pigs have rooted through the soil. Many seeds develop a light requirement for germination when buried and thus lie dormant until brought to the surface and exposed to light (Wesson and Wareing, 1969). The requirement for light for germination is particularly a phenomenon of small seeds, these include not only many herbaceous species but also the seeds of pioneer tree species which are usually the first to germinate in extensive clearings (for review *see* Smith, 1982). Large seeded species such as climax trees in rain forest can germinate readily in the dark and are more able to sustain growth on their own reserves until they reach an adequate source of light.

It is not necessary for the seeds to be buried for germination to be inhibited. The leaf canopy alone is capable of preventing germination in a number of small seeded species (Fig. 7.8) (Taylorson and Borthwick, 1969; King, 1975; Silvertown, 1980). The effectiveness of the leaf canopy in preventing germination may be due to a reduction in the ratio of red to far-red (R : FR) radiation. Light-requiring seeds of *Cecropia obtusifolia* and *Piper auritum* cannot germinate when placed on the forest floor in a dense rain forest but are capable of germinating when the far-red radiation is removed by a narrow-band red filter placed immediately above the seeds (Vazquez-Yanes, 1980). These seeds require a red light illumination for germination. The amount of red light under the tree canopy is small but sufficient for germination if the inhibiting far-red component is removed. Dense shading by green leaves not only reduces the quantity of light but also changes its quality. The light which acts on the red : far-red reversible photoreceptor phytochrome is conveniently described by the ratio of red (660 nm) : far-red (730 nm). In sunlight the 660 : 730 (energy) ratio is about 1 : 1.15 but is reduced to lower levels in shaded habitats depending on the species composition of the canopy and the sky conditions. In shaded habitats under clear skies the red : far red ratio is lower than when the sky is overcast or hazy. Fig. 7.9 summarizes the range of values found under different canopies using data compiled by Morgan and Smith (1981).

There are two separate phenomena illustrated by the above example. The first is dark dormancy and the second a light requirement for germination. At

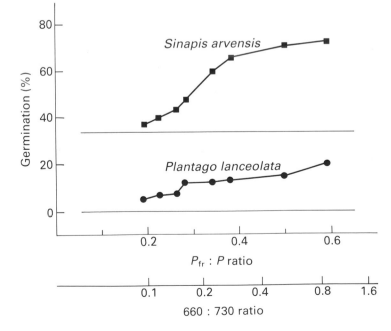

Fig. 7.8 Effect of only 1 hour of daylight filtered through leaves of mustard plants grown at various densities on the germination of seeds of mustard (*Sinapis arvensis*) and plantain (*Plantago lanceolata*). The light treatment was given 10 hours after sowing the seeds in the dark. (Reproduced with permission from Frankland, 1981.)

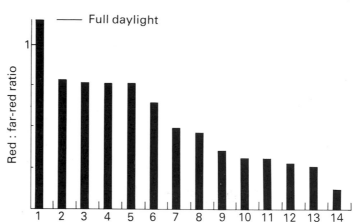

Fig. 7.9 Effect of the canopy of different species on the ratio of red: far-red photon fluence rates in the resulting shade-light: 1 (control), full sunlight; 2, birch; 3, oak; 4, red pine; 5, jack pine; 6, beech; 7, white pine; 8, spruce; 9, alder; 10, montane tropical rain forest; 11, hawthorn; 12, sugar maple; 13, lowland tropical forest; 14, sweet chestnut. (Data adapted and condensed from Morgan and Smith, 1981.)

first these may seem just different aspects of the same response but they represent different stages in the control of seed germination. Many small seeds although they may not have a light requirement for germination when first shed from the mother plant rapidly develop a dark dormancy when buried in the soil (Wesson and Wareing, 1969). The development of dark dormancy has been correlated with the retention of chlorophyll in the seed-bearing structures of the mother plants (Cresswell and Grime, 1981) (Figs. 7.10 and 7.11). When seed embryos mature

inside green tissues the ratio of R : FR radiation will be low and this condition appears to impose a light requirement for germination. Thus the potential for development of dark dormancy depends on conditions during the ripening of the seed. The inhibition of germination under the forest canopy on the other hand depends on the red : far-red regime experienced by the imbibed seed.

The need for adequate energy supplies for establishment in forest conditions means that most shade-tolerant species have large seeds. There is a direct relationship between seed size and shade tolerance and the height to which seedlings can grow in reduced light conditions. The small seeded plants of woodland habitats are typically the opportunist plants of clearings such as foxglove (*Digitalis purpurea*) and fireweed (*Chamerion angustifolium*). The requirement for large seeds in shade-tolerant plants has resulted in these species being the most poorly represented in the British Flora in comparison with the species richness that exists in forest vegetation on the southern side of the English Channel. Heavier-seeded plants are not so readily dispersed as the pioneer grasses and herbs of more open communities. Therefore in the plant migration northwards and westwards at the end of the last ice-age many of the forest herbs arrived at the southern side of the English Channel too late to cross into Britain. The species poverty in Britain's forest plant communities is not due to the lower temperature regimes north of the English Channel but is caused by the barrier to migration that this comparatively narrow stretch of water produced when the North Sea was flooded during the Atlantic period.

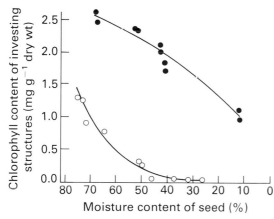

Fig. 7.10 Relationship between moisture content of seeds and chlorophyll concentration of the investing structures at various stages of inflorescence development in *Tragopogon pratensis* (●) and *Hordeum murinum* (○). (Reproduced with permission from Grime, 1979.)

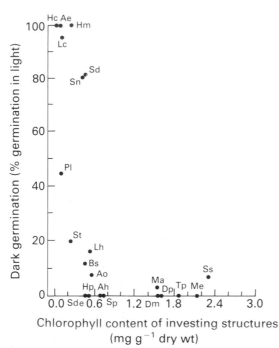

Fig. 7.11 Relationship between chlorophyll concentration of the investing structures and the dark germination of mature seeds in various species of flowering plants. As chlorophyll content varies with moisture content (*see* Fig. 7.10) values for chlorophyll are taken at the mid-point in the range of seed moisture content measured during development. Ae, *Arrhenatherum elatius*; Ah, *Arabis hirsuta*; Ao, *Anthoxanthum odoratum*; Bs, *Brachypodium sylvaticum*; Dm, *Draba muralis*; Dp, *Digitalis purpurea*; Hc, *Helianthemum chamaecistus*; Hm, *Hordeum murinum*; Hp, *Hypericum perforatum*; Lc, *Lotus corniculatus*; Lh, *Leontodon hispidus*; Ma, *Myosotis arvensis*; Me, *Milium effusum*; Pl, *Plantago lanceolata*; Sd, *Silene dioica*; Sde, *Sieglingia decumbens*; Sn, *Silene nutans*; Sp, *Succisa pratensis*; Ss, *Senecio squalidus*; St, *Serratula tinctoria*; Tp, *Tragopogon pratensis*. (Reproduced with permission from Grime, 1979).

7.3 Growth and survival in shaded habitats

With the exception of vernal and pioneering herbs and colonizing trees such as birch and ash, all forest plants, from the dominant trees that form the top canopy to the ferns of the ground layer, are able to survive with 20 per cent or less of full sunlight throughout the major part of the growing season. In the mixed forest of *Quercus robur* and *Carpinus*

betulus originally studied by Salisbury (1916) illumination during the low-light phase varied between 0.2 and 5 per cent of full sunlight. The shade-light of the forest floor also differs in quality from sunlight and as will be seen below plants respond to both changes in quantity and spectral composition of light. In the shaded habitat there is a diminution in intensity of all wavelengths but selective absorption by chlorophyll of blue and red light causes a relative increase in the proportion of green and far-red radiation. The morphogenic effects (effects on plant form) of changes in light quantity and qualify act together in nature but for simplicity in understanding the differing adaptability of plants to shade it is more convenient to discuss them separately.

7.3.1 ADAPTATIONS TO LOW FLUENCE RATES

Light intensity or fluence rate (quantum flux density) can differ in contrasting terrestrial habitats by at least two orders of magnitude in the daily quantum flux available for photosynthesis (Björkman, 1981). The characteristic differences between sun and shade plants are seen in the studies of Björkman and co-workers on the sun plants *Nerium oleander* and *Encelia californica* as compared with the shade plant *Cordyline rubra* (Fig. 7.12). The shade plant due to low levels of dark respiration quickly reaches the light compensation point where the oxygen evolved through photosynthesis balances the uptake of aerobic respiration. As has long been known shade plants cannot attain the higher rates of photosynthesis that are found in sun plants and a plateau is reached in *Nerium oleander* at a light intensity that matches only 5 per cent of maximum daylight. In shade-tolerant plants prolonged exposure to high-light intensities can even lead to a photoinhibition of photosynthetic activity due to photo-oxidation (Foyer and Hall, 1980).

Most shade-tolerant plants are perennials and in these species it is not so much the length of time that their leaves are kept above or below compensation point that is important as the net energy balance for the year. Thus provided the dark respiration rate is not high, then even the short periods of active photosynthesis that can take place in sunflecks or early in the year before the canopy closes may be enough to provide the plants with a net energy surplus for the year. Plants that can survive and flower in shaded habitats or even just to sustain years of suppression have selected a mode of existence which illustrates once again that ecological fitness is conferred by the operation of the Montgomery effect (Montgomery,

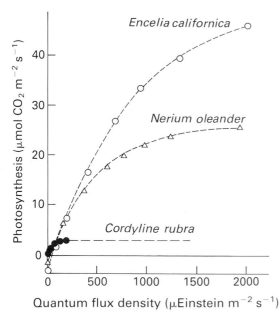

Fig. 7.12 Rate of net photosynthesis as a function of incident quantum flux density (400–700 nm) for the sun species *Encelia californica* and *Nerium oleander* grown under natural daylight (≈ 40 Einstein m^{-2} day^{-1} and the shade species *Cordyline rubra* grown in its native rain forest habitat (≈ 0.3 Einstein m^{-2} day^{-1}. (Reproduced with permission from Björkmann, 1981.)

1912), i.e. where ecological advantage is conferred by low growth rates in areas of low environmental potential. Thus not only are most shade-tolerant plants slow growing with low dark respiration rates, they are also less sensitive to temperature than plants intolerant of shading (Grime, 1966). The survival value of being insensitive to temperature in terms of metabolic rate will avoid the danger of overactive catabolic activity in the heat of summer when shade is dense and the likelihood of carbohydrate depletion greatest. Maintaining a low and steady metabolic rate is yet another factor which ensures that after a warm summer on the dim forest floor the shade-tolerant plant will have an energy surplus.

7.3.2 SHADE AND CARBON DIOXIDE FIXATION

At a cellular level the effect of shade on intolerant plants of barley (*Hordeum vulgare*) is to reduce greatly the amount of fraction I protein in the leaf. The carbon dioxide-fixing enzyme RuBP carboxylase is contained in this fraction and it appears that under shaded conditions barley leaves are unable to main-

tain the enzyme complement necessary for maximum photosynthesis (Blenkinsop and Dale, 1974). This becomes evident when the plants are returned to full sunlight as the leaves are unable to resume carbon dioxide fixation at their normal rate. The decline in fraction I protein is much greater than any other fraction of leaf protein. Prolonged shade therefore not only causes a degradation of fraction I protein but reduces the level of carbon dioxide fixation which further diminishes the balance between protein synthesis and breakdown. The result is an acceleration of the decay of the carbon dioxide fixing mechanism. In a comparison of the leaf protein of 10 shade species and 13 sun species from a tropical rain forest it was found that the shade plants had only 15–25 per cent of the protein content of sun plant leaves (Goodchild *et al.*, 1972). The major part of this saving in protein content was due to a lower content of RuBP carboxylase. Similarly in clones of *Solidago virgaurea* taken from open and shaded habitats the plants from the shaded habitats always had a lower RuBP carboxylase activity (Björkman, 1968). The turnover rate of leaf proteins places a high maintenance cost to the plant. The high quantities of RuBP carboxylase that have to be maintained in sun plants in order to be competitive with each other do not justify their maintenance costs in the shaded environment.

The need for efficiency of energy utilization also accounts for the scarcity of C_4 plants from shaded habitats.

The efficiency of light usage by plants is usually expressed in terms of quanta. The quantum yield which is based on the utilization of incident light is termed ϕ. The efficiency with which this absorbed light is then used is defined as $\phi_i = a\phi_a$ where ϕ_a is the efficiency of utilization of absorbed quanta. If plants use the same pathways for the generation of reducing power and for carbon dioxide fixation, it is to be expected that ϕ_a should be similar at low-light levels (low incident quantum flux densities). Numerous studies (*see* Björkman, 1981) have shown that for sun and shade leaves of the same species and for sun and shade grown plants of the sun-ecotype of *Solidago virgaurea* that this expectation is confirmed. When the metabolic pathway for carbon dioxide fixation differs as in C_3 and C_4 plants then differences in ϕ_a become apparent. If the oxygenase activity of ribulose bisphosphate (RuBP) carboxylase is reduced by measuring photosynthetic activity at low oxygen concentrations (Fig. 7.13) it can be seen that C_4 plants have a ϕ_a that is about one-third lower of that measured for C_3 plants (Ehleringer and Björkman,

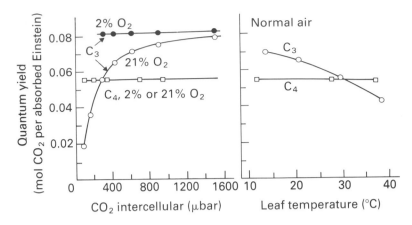

Fig. 7.13 Effect of intercellular carbon dioxide pressure and leaf temperature on the quantum yield of photosynthetic carbon dioxide uptake in C_3 and C_4 plants. The carbon dioxide and temperature dependence of ϕ_a was determined at 30°C and 325 μbar CO_2 and 210 μbar O_2 respectively. (Reproduced with permission from Björkman, 1981.)

1977). Under normal oxygen pressures due to the temperature dependence of the oxygenase activity of RuBP carboxylase the quantum yield of C_3 is highly temperature dependent. Consequently C_4 plants only achieve a superior quantum yield to C_3 plants at high temperatures.

This basic inefficiency of C_4 plants at low temperatures may account for their absence from the Arctic (*see* Chapter 3, p.58) and the northern limit to their distribution in some temperate habitats. The expansion of cord grass (*Spartina anglica*) through the estuaries of Britain has been noted as an outstanding ecological success for a C_4 species. However, it is rarely found north of 57°N and Fig. 6.16 shows its second most northerly station in the British Isles at 56° 22′N where it grows too slowly to set seed successfully. This same inefficiency in resource utilization when light is in short supply may also explain the scarcity of C_4 plants in shaded habitats. One outstanding exception is found in the tropical rain forest of Oahu, Hawaii where an arborescent spurge *Euphorbia forbesii* shows marked C_4 metabolism yet grows under the shade of other C_3 tree species (Fig. 7.14) (Pearcy and Calkin, 1983). The constant warm temperatures of this habitat may account for this anomalous case.

When light is not limiting, the capacity of plants to photosynthesize, i.e. the level of light-saturated photosynthesis (P_s) is controlled by the mechanisms of carbon dioxide fixation and not by the efficiency of light absorption. The relationship between RuBP carboxylase activity and the light-saturated photosynthetic rate (P_s) has been summarized for a range of sun and shade leaves of a number of C_3 plants (Björkman, 1968). Figs. 7.15 and 7.16 show a graphical representation of this data and illustrate the very

high degree of correlation between light-saturated photosynthesis (P_s) and RuBP carboxylase activity. Note however there is no similar correlation between chlorophyll content and photosynthetic activity when compared on a surface area basis. High-light intensities nevertheless require an increased capacity for photosynthetic electron transport. The limiting component in the photosynthetic electron transport chain appears to be cytochrome f. When sun plants are grown in high-light conditions there is a substantial increase in the ratio of cytochrome to chlorophyll (Wild, 1979). Shade plants appear to lack this response.

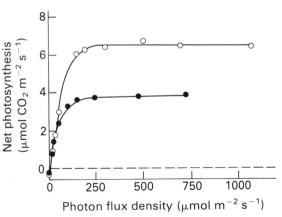

Fig. 7.14 Response of net photosynthetic Co$_2$ uptake to incident photon flux density in *Euphorbia forbesii* (o) and *Claoxylon sandwicense* (●). The plants were growing in the shaded understorey of a mesic evergreen forest in Hawaii. The *Euphorbia* sp. despite growing in the shade of C_3 tree species showed marked C_4 metabolism (Pearcy and Calkin, 1983).

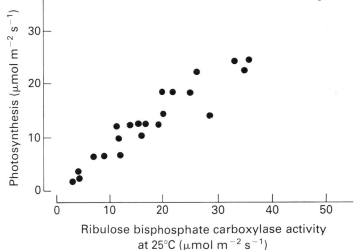

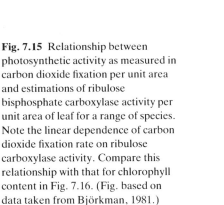

Fig. 7.15 Relationship between photosynthetic activity as measured in carbon dioxide fixation per unit area and estimations of ribulose bisphosphate carboxylase activity per unit area of leaf for a range of species. Note the linear dependence of carbon dioxide fixation rate on ribulose carboxylase activity. Compare this relationship with that for chlorophyll content in Fig. 7.16. (Fig. based on data taken from Björkman, 1981.)

7.3.3 CHLOROPHYLL CONTENT IN SHADE LEAVES

A common feature of plants that live in deep shade is the dark green colour of their foliage (Fig. 7.17). Experiments to estimate the quantum yield of photosynthesis using leaves with differing chlorophyll contents show that over a wide range there is no appreciable effect on the amount of carbon dioxide fixed per Einstein of light absorbed (Fig. 7.18). Nevertheless although sun and shade species do not differ in the efficiency with which they use absorbed quanta they do differ in their ability to absorb incident light. The amount of light needed to attain the light-saturation rate will vary depending on the efficiency with which the photosynthetic pigments absorb light. Thus the higher the chlorophyll content the greater the amount of light absorbed by a leaf. The internal reflectance that takes place in leaves results in light absorption per unit of pigment being greater in leaves than in solutions as the light path for absorbance is increased. The high absorbance of chlorophyll in the red and blue regions of the spectrum however reduces the gain at these wave-

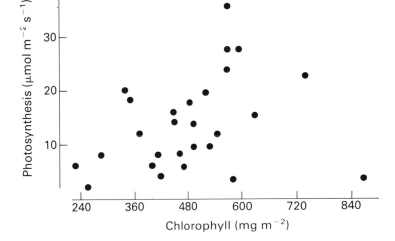

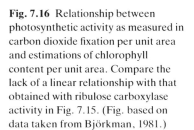

Fig. 7.16 Relationship between photosynthetic activity as measured in carbon dioxide fixation per unit area and estimations of chlorophyll content per unit area. Compare the lack of a linear relationship with that obtained with ribulose carboxylase activity in Fig. 7.15. (Fig. based on data taken from Björkman, 1981.)

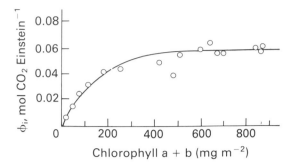

Fig. 7.18 Quantum yield of photosynthetic carbon dioxide uptake for leaves of differing chlorophyll contents. Each data point represents a different C_3 species or variety. (Redrawn from Björkman, 1981 using data of Gabrielsen.)

Fig. 7.17 *Allium ursinum* (wild garlic) a species of shaded woodland habitats with characteristic dark green foliage indicative of high chlorophyll content.

lengths with increased chlorophyll content. Thus a three-fold increase in chlorophyll content results in only a 2–3 per cent increase in absorption in these wave bands (Björkman, 1981). However in the far-red and green regions of the spectrum there is a much greater increase in absorption. Thus, although an increase in chlorophyll content results in a less than proportional increase in absorption of incident light this may still be of advantage in areas where the intensity of illumination is low.

Shade plants therefore do not economize in leaf chlorophyll content in the same way as they do in the carbon dioxide fixing enzymes. When grown under deep shade, obligate shade species such as *Cordyline rubra* (Björkman, 1981), contain just as much chlorophyll as sun plants grown at high-light intensities. The chloroplasts of shade plants can have particularly large stacks of grana with as many as 100 thylakoids per granum. The grana within the chloroplasts are not orientated in one particular direction as they are in sun plants presumably due to the non-directional nature of shade-light (Boardman *et al.*, 1975). This increase in the amount of light-trapping chlorophyll is also seen in the molecular

structure of the chlorophyll itself in that shade plants when grown in deep shade show an increase of chlorophyll b with a lower chlorophyll a to chlorophyll b ratio (Egle, 1960). As chlorophyll b is purely a light-harvesting pigment this increase means that there is a higher proportion of light-harvesting chlorophyll to total chlorophyll in the chloroplasts of shade-tolerant plants. It is considered (Butler, 1977) that the light-harvesting chlorophyll complex is mainly associated with photosystem II and it appears that shade plants compensate for the greater amount of far-red light relative to red by an increase in photosystem II relative to photosystem I. In this way the shade plants achieve a more balanced energy distribution between the two photosystems.

Apart from changes in pigment composition the chloroplasts of both higher and lower plants have the ability to respond to an imbalance in the quantal distribution of light between photosystems I and II (Barber *et al.*, 1981). Ecologically it is of interest that this response is only of importance when light conditions are limiting. This regulatory mechanism has been described as the 'carburettor effect' and has been invoked to explain the ability of plants to maintain a high quantum efficiency of photosynthesis over a wide spectral range (Myers, 1971). The mechanism appears to work through the phosphorylation and dephosphorylation mechanisms of photosystem II which affects the degree of spill-over of energy to photosystem I and that the mechanism is controlled by the redox state of the plastoquinone pool. As yet it is not clear how the redox state of plastoquinone controls kinase activity (Barber *et al.*, 1981). Thus both from the physical arrangement of chlorophyll molecules in the chloroplasts and the types of chlorophyll present, together with the control of energy utilization by the photoreactive centres, higher plants have a remarkable range of adaptations for abstracting the maximum energy from a wide spectral range when the intensity of illumination is low.

7.3.4 LEAF SIZE AND PUBESCENCE

The dominant factor controlling leaf size in plants of different habitats is not considered to be the capture of light but the optimizing of water-use efficiency. Parkhurst and Loucks (1972) made this assumption when testing mathematical models for leaf size and found that the predictions that this assumption gave fitted well with observed trends in diverse regions (tropical rain forest, desert, Arctic, etc.). Their model predicted that only in warm or hot environments with low radiation as in the forest floors of temperate and tropical regions would large leaves be advantageous. Although there is a great deal of variability in leaf size in the herb vegetation of the forest floor their conclusions are consistent with general trends in leaf size. The variability that does undoubtedly occur is thought to be due to the lack of environmental pressure on leaves on the forest floor. Carbon dioxide is usually not limiting due to soil respiration

and low wind speeds. Evaporation is low and considerable genetic drift can take place in relation to leaf size as it is not under continual environmental pressure (Gates, 1980).

The reflective properties of the leaf are extremely important in relation to the absorption of light. Pubescence can increase reflectance and the desert pubescent form of *Encelia* will absorb only 30 per cent of incident light compared with glabrous leaves of equal chlorophyll content which absorb 84 per cent. In these species the quantum yield for photosynthesis is directly proportional to incident light when radiation is limiting. However at high-light intensities the light-saturated rate of photosynthesis is the same in the glabrous and pubescent leaves (Ehleringer and Björkman, 1978).

7.4 Morphogenic effects of shade

7.4.1 LIGHT QUALITY AND EXTENSION GROWTH

The effects of light quality on plants grown in shade are best seen in those laboratory experiments where light intensity is kept constant and only the spectral composition is changed. The most satisfactory simulation of the morphogenic effects of natural shaded

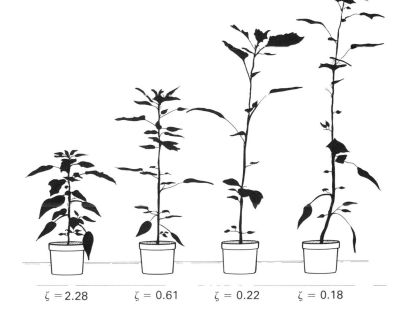

Fig. 7.19 Shadowgraphs of seedlings of *Chenopodium album* grown for 14 days in a range of simulated shade-light treatments in which the fluence rate of photosynthetically active radiation (400–700 nm) was kept constant but with additional far-red radiation being given to achieve different red: far-red ratios. R : FRζ = 2.28 for white flourescent light, R: FRζ = 0.18 simulated the ratio for deep shade. (Reproduced with permission from Morgan, 1981.)

ζ = 2.28   ζ = 0.61   ζ = 0.22   ζ = 0.18

conditions using constant quantum flux density are obtained with supplementary far-red illumination. Reducing the red : far-red ratio from R : FR = 3 down to 0.18 while keeping the intensity of illumination constant, induces a striking increase in internode elongation in seedlings of *Chenopodium album*. Far-red illumination also increases apical dominance. Thus the plant grown in simulated shade appears not only taller but also narrower due to the suppression of lateral bud expansion (Morgan, 1981) (Fig. 7.19). As with all red : far-red effects on plant form the photoreceptor is phytochrome. The ratio of red : far-red light although striking in its effect on plant growth is not in itself predictive when directly plotted against plant growth. If however the amount of phytochrome changed to the form absorbing light in the far-red (730 nm) i.e. P_{fr} is expressed as the phytochrome equilibrium (ratio of $P_{fr} : P_{total}$) then a linear relationship is obtained with stem extension (Fig. 7.20). Due to the absorption of red light by chlorophyll direct estimations of $P_{fr} : P_{total}$ are not possible in green leaves and use has to be made of data obtained from non-green tissues and mathematical models to deduce the conditions that prevail in the green leaf (Smith,

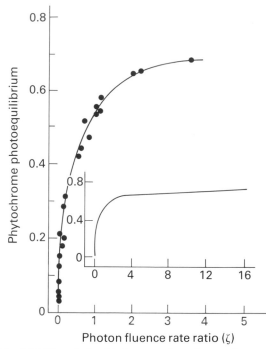

Fig. 7.21 The relationship between phytochrome photo-equilibrium, measured in etiolated pea epicotyl tissue, and the red : far-red ratio of the incident broad-band radiation. The photoequilibrium is most sensitive to the range of red : far-red found in terrestrial habitats (i.e. between 0 and 1). (Reproduced from Morgan, 1981.)

1982). Fig. 7.21 shows the relationship that is obtained from such modelling. The change brought about by the R : FR ratio in the shaded light of woods on the $P_{fr} : P_{total}$ equilibrium acts on a portion of the curve that shows very marked responses to changes in light quality. Thus phytochrome will make an excellent sensor of light quality in shaded habitats (Smith, 1982).

Plants also respond morphogenically and physiologically to blue light. Blue light is particularly effective in inducing stomatal opening. The opening of stomata at low-light intensities as at dawn can be brought about by low intensities of blue light (Zeiger *et al.*, 1981). In the shade-tolerant species *Impatiens parviflora* red light alone is insufficient to prevent etiolation at the seedling stage and it is only when blue and red light are given together that there is a reduction in this early stage of growth extension (Hughes, 1966). There appears to be a separate receptor for the morphogenic effects of blue light but as yet it has only been tentatively identified as a flavin.

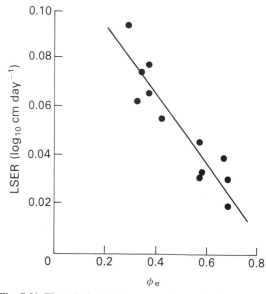

Fig. 7.20 The relationship between stem extension rate (LSER) of *Chenopodium album* grown for 9 days in a simulated shade-light (supplementary far-red, *see* Fig. 7.19) days and the phytochrome equilibrium (ϕ_e) that would be established in etiolated tissue by the light treatments. (Reproduced from Morgan, 1981.)

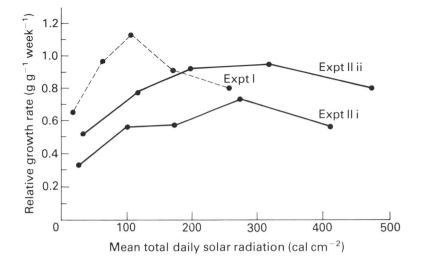

Fig. 7.22 The relative growth of *Impatiens parviflora* at varying light intensities showing the ability of this species to sustain growth rates even under intense shade. (Adapted from Hughes, 1966.)

7.4.2 SHADE AND TOTAL PLANT FORM

Both sun and shade plants change their overall form in relation to shade. The response although almost universal does however differ in extent between different species. The shade plants are distinct in their phenoplasticity in that they are able to maintain a constant relative growth rate (RGR, *see below*) over a range of light intensity from full sunlight to moderate shade (Fig. 7.22). This plastic response is due to the inverse relationship between dry matter

gained per unit of leaf area (ULR) and the leaf area per unit dry weight of leaf (LAR). Thus as the efficiency of the leaf declines in terms of relative productivity per unit of leaf weight, the area of the leaf is altered by the plastic response of the plant to low-light intensity and the productivity of the leaf as a whole is maintained. Research into the phenoplastic responses of plants to shade have been based mainly on the techniques of 'growth analysis' which defines the inter-relationships between the various types of assessment of growth that can be made by harvesting

Table 7.2 Terms and expressions used for growth analysis measurements

Term	Contraction	Expression for instant value	Units
Relative growth rate	RGR	$1/W \; dW/dT$	(weight weight^{-1} time^{-1}
Leaf weight ratio	LWR	L_w/W	dimensionless
Leaf area ratio	LAR	L_a/W	dimensionless
Specific leaf area	SLA	L_a/L_w	area weight^{-1}
Unit leaf rate (= net assimilation rate NAR)	ULR	$1/L_a \; dW/dT$	area^{-1} time^{-1}

where: L_a = total leaf area
L_w = total leaf dry weight
T = time
W = total dry weight

and
(1) RGR = ULR × SLA × LWR
(2) RGR = ULR × LAR

methods. Some of the terms used in growth analysis are defined in Table 7.2. The definitions are taken from Hunt (1978) where a fuller account is given of the units and methods of growth analysis.

The net effect of the changes in plant form is to increase the proportion of photosynthetically active leaf material. Thus in most species shading tends to increase the proportion of total dry matter that is present in the leaves. In the shade-tolerant meadow sweet (*Filipendula ulmaria*) and in yellow flag (*Iris pseudacorus*) as the level of illumination falls the amount of photosynthate that is transferred to the leaves increases (Whitehead, 1973). In fireweed (*Chamerion angustifolium*) reducing the light from 100 to 40 per cent daylight increased the leaf : weight ratio (Myerscough and Whitehead, 1966). This increase in leaf growth as a result of shading appears in most studies to be at the cost of root growth. In shaded habitats a reduction in root development may not be too harmful due to the buffered conditions of the forest floor referred to in the introduction to this chapter. Where nutrients are limiting however this change may prove unfavourable. The overall success of phenoplasticity in maintaining viable plants in shaded habitats is seen in *Impatiens parviflora* where the relative growth rate (RGR) in 10 per cent of full illumination was only reduced to 80 per cent of that found under the full light regime (*see* Björkman, 1981). The morphogenic effects of light thus enable the plant to compensate for the low intensity of photosynthetic energy and the relative growth rate is kept constant. The relationship between ULR and SLA for *Impatiens parviflora* grown under varying light intensities is shown in Fig. 7.23.

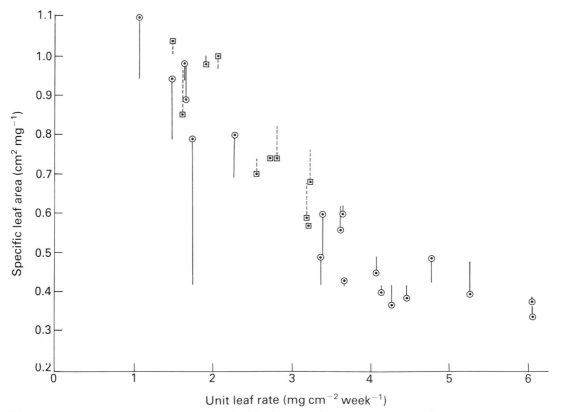

Fig. 7.23 Relationship between specific leaf area and unit leaf rate for plants of *Impatiens parviflora* grown under field (○) and controlled laboratory conditions (□) with different levels of shading. Lines join values of specific leaf area for the beginning and end of the growth period with the symbols marking the specific leaf area at the end of the experiment. (Redrawn from Hughes, 1966.)

7.5 **Mineral nutrition in shaded habitats**

Mineral deficiency is a common condition of certain forest soils notably the podzols of northern coniferous forests and the laterites of tropical rain forests. In the latter the mineral nutrients of the ecosystem are locked up in the large biomass of the tree vegetation of the rain forest. In the northern coniferous forests the low rate of mineral cycling particularly in pine forests together with the acidic nature of the pine litter causes an acidification of the soil profile which when combined with leaching, produces the typical nutrient deficient podzolic soil profile. In the temperate forests most tree species are infected with ectotrophic mycorrhizae. The ground flora also is rich in mycorrhizal associations. The ericaceous genera, *Calluna, Erica, Vaccinium, Arctostaphylos,* and *Rhododendron*, which are typical of forest floor communities from the northern coniferous forests to the tree covered slopes of the Himalayas, have a particular from of endomycorrhizal association which is characterized by the penetration of the root cortex by septate fungal hyphae forming intercellular hyphal complexes (Moser and Haselwandter, 1983).

In early studies on the benefits that accrue to the host plants by the development of mycorrhizal associations it was thought that the enhancement of nitrogen uptake played a major role. Although this may sometimes be the case as in ericaceous mycorrhizae (Read, 1978) it appears that the greatest influence that the association has on the host plant is to enhance phosphorus uptake from soils with low *P* availability. The increased uptake appears to come not from using a source of *P* that is unavailable to the host but from improved exploitation of greater soil volume by the association. In some cases however there has also been demonstrated a change in phosphatase activity by mycorrhizal infection and this could increase the amount of *P* that is available for uptake. The fungal associations accumulate and translocate the phosphate and deposit it in high polyphosphate concentrations in host cells. Up to 40 per cent of the *P* present in the fungal endophyte can be accounted for by this type of concentration.

7.6 **Conclusions**

For species that can adapt their energy expenditure to match low levels of light input the forest floor provides a habitat that is buffered against many of the commonest hazards of terrestrial life namely, heat, drought, and exposure to wind. The absence of these stresses allows leaves to develop in size and to have a minimal resistance to the entry of carbon dioxide. As most species are slow-growing perennials, the mineral deficiencies of some forest soils are not a serious limitation to plant growth. The universal limitation to plant growth, the low concentration of carbon dioxide in the atmosphere, is also alleviated by soil respiration in the sheltered environment of the forest. The growth form of many forest floor plants with their rosettes of leaves forming an umbrella over the soil will serve to trap carbon dioxide as well as light. The forest floor provides a predictable environment with many compensations for low light intensities and where once established long-term survival is reasonably assured.

8

Drought survival

8.1 Plant form and drought tolerance

The life-style of green plants, based as it is on exposing living surfaces to sun and air, makes desiccation injury and overheating constant hazards. It is therefore not surprising that practically every facet of plant life, from the morphology of leaves, stems, and roots to the timing of germination, vegetative growth, and flowering bears the imprint of the adaptations necessary to survive with the water supply of the chosen habitat. So successful are these adaptations that some species can survive in areas where it never rains and the only source of moisture is the dew that condenses from clouds or sea fogs. The trade wind cloud that blows onto the mountainsides of Tenerife (Canary Islands) can support a dense cover of broom (*Spartocytisus nubigenus*) in the crater of an ancient volcanic with negligible rainfall (Fig. 8.1). Although this is an extreme example it serves to typify the ability of plants to evolve adaptations which will fit them to the most extreme environmental conditions.

As with all aspects of environmental stress the plant can either avoid or tolerate the adversity. Avoidance of desiccation stress to the plant cell can be achieved either by buffering mechanisms in the morphological or physiological constitution of the plant or else by carefully timed development (phenology) so that sensitive tissues are not exposed at the period of greatest desiccation risk. Tolerance of the drought stress implies that the plant tissues are actually dehydrated. At a cellular level this depends on the protoplasm being in a condition where dehydration does no irreversible damage. Survival under these conditions usually requires the plant to enter a dormant state. In this, drought stress varies from the other limiting environments discussed in previous chapters, for example shade and salt, where tolerance mechanisms can keep the plant capable of growing while enduring the stress. The reliance on dormancy by plants to tolerate drought presents one of the greatest problems in improving the resistance

of crop plants to water stress. Agriculture in this instance cannot adopt nature's remedy for surviving drought through inactivity, as this would defeat the aims of production. Therefore for agriculture, avoidance rather than tolerance is more likely to be able to combat the effects of water stortage. One of the readiest remedies to ensure stable crop yields in areas of uncertain rainfall is to grow early maturing crops and thus minimize the exposure of the plants to water shortage (Jones *et al.*, 1981). However in years of adequate rainfall these crops will yield less than later maturing varieties.

The ability to withstand long periods of moisture shortage is not a prerogative of any one plant group or life form. In trees, shrubs, grasses, ferns, mosses, and lichens there are species which are drought tolerant and others which are restricted to areas of plentiful water. This evolution of drought-resistant species in all the major life forms of plants is sometimes overlooked as it easy to be deceived by the dramatic changes in vegetation that take place when passing from areas of high to low rainfall. In particular the reduction in tree cover which can make a very vivid impression on arriving in an arid zone can lead to an underestimation of the range of drought tolerance in tree species. The absence of trees in many dry areas is often as much due to man and his animals as to the reduction in precipitation. Rutter (1968) has shown that the water consumption of some forests is not very different from that of grassland. In *Eucalyptus* stands the difference is negligible while in spruce the difference from grassland is never more than 50 per cent.

Reduction in water supply produces a much greater effect on the species composition of a forest than on the existence of the tree form. In the cool rain forests of Chile and western Patagonia rainfall of 2500−3800 mm is needed to support the lush growth of the evergreen southern beech (*Nothofagus dombeyii*) (Figs. 8.2 and 8.3). Sixty kilometres to the east the rain-shadow of the Andes has reduced the rainfall to less than 350 mm yet in spite of this tree

Fig. 8.1 (*opposite*) Bushes of the broom *Spartocytisus nubigenus* growing in a volcanic crater in a zone of negligible rainfall at an altitude of 2240 m (8000 feet). The growth of the broom is dependent on moisture condensed from the cloud cover that regularly forms over the mountain.

Fig. 8.2 Evergreen forest of *Nothofagus dombeyii* growing in a zone of high rainfall (2500–3800 mm) in south-west Patagonia.

cover can still be found in areas which are not grazed. Fig. 8.4 shows such a dry habitat tree community consisting mainly of the southern cedar *Libocedrus chilensis*. The density of the tree cover is much reduced but this is a feature noted with all life forms in relation to drought. Walter (1939) showed in south-west Africa a linear relationship between precipitation and grassland productivity. The existence of this linear relationship means that per unit of rainfall the amount of transpiring plant material is constant.

A similar phenomenon has been observed (Woodell *et al.*, 1969) in the creosote bush (*Larrea divaricata*) in the Mohave and Sonoran deserts (Fig. 8.5). As can be seen in Fig. 8.6 there is a good correlation between rainfall and the density of the bushes. The conclusion from this study was that the spacing of the bushes was controlled by root competition for water. This observation mirrors Walter's view that the amount of water transpired per unit of plant material remains the same and the net effect of the water shortage is to reduce the amount of exposed plant tissue above ground. In this way the plants optimize their dispersal, probably by root competition

Fig. 8.3 Distant view of the evergreen southern beech *Nothofagus dombeyii* forest as seen from a distance of 30 miles east in the rain-shadow area of the Patagonian desert showing the strict limitation of beech to the montane high rainfall area.

so as to obtain the maximum amount of growth per unit of rainfall. Due to efficient root dispersal and penetration down to the water-table it is even possible to find trees in the Sahara desert which would be more plentiful if it were not that man and his animals use them more rapidly than they can regenerate. The genera most commonly found include *Acacia* and the phreatophytic *Tamarix* (Grenot, 1974). A phreatophyte is a plant that can absorb water from a permanent water-table (the phreatic zone).

Similarly with lichens, while some species are among the most xerophytic of plants known to exist, others such as *Buellia atrofusca* may require more than 2500 mm of rainfall per annum (Parker, 1968). Likewise in mosses, the genus *Sphagnum* is restricted to bogs where water shortages hardly ever exist, while other genera can survive years of desiccation and resume full metabolic activity on rehydration. In the pteridophytes a similar range of behaviour is found. Genera such as *Trichomanes* and *Hymenopyllum* live in wet climates exclusively, while species such as *Ceterach officinarum* are highly tolerant of drought. The latter species possesses large quantities of collenchyma and has vacuoles that solidify on drying thus preserving the structure of the plant under periods of severe drought. *Selaginella lepidophylla* the so-called resurrection plant of the southwest United States will survive long periods of

drought curled up in a ball and when exposed to water will open itself out and turn green within a few hours (Eickmeier, 1979).

Seeds also vary in their tolerance of drought. The possession of a seed coat is not in itself a guarantee of survival through long periods of desiccation. The longest records for seed longevity belong not to desert plants but to the imbibed seeds that are periodically unearthed from old pastures and peat bogs. In crop plants however drying the seed normally increases the length of time they remain viable. A useful rule of thumb is that a doubling of the life of the seed is achieved for each 1 per cent reduction in moisture content (Harrington, 1973). There is however a limit to which the drying process can be pursued even in seeds tolerant of desiccation. The longevity of seeds at 4−5 per cent moisture is usually less than that at 5−6 per cent. The damage caused by the extra dehydration has been attributed to oxidation of unsaturated lipids and the liberation of free radicals at the double bonds thus destroying membrane structure. In other species seeds are very sensitive to any reduction in their fresh weight and lose viability rapidly in dry air. Species in this category include genera of the large seeded hardwoods such as *Quercus*, *Fagus*, *Aesculus* and *Castanea*. The tolerance of a wide range of plant forms and structures to severe drought was recognized by Maximov

Fig. 8.4 The southern cedar *Libocedrus chilensis* growing in an area of the Patagonian desert remote from human habitation and having a rainfall of less than 350 mm. Despite the drought conditions tree growth is still possible provided the area is not subject to too much human and animal interference.

(1929) who after travelling in the deserts of Central Asia and North America became convinced that the true basis for xerophytism lay in the resistance of the plant protoplasm to withstand drying. Maximov's studies were largely restricted to prairies and steppes where there are relatively short and predictable periods of water stress. Species that inhabit these regions reduce their transpiration only slightly at the beginning of the dry season and thus develop large water deficits which their tissues are able to withstand (Walter and Stadelman, 1974). This type of behaviour however must not be thought of as universal. There are so many types of deserts and desert plants that some examination of differences in xerophyte responses to drought is necessary before any generalizations can be made on the methods of survival.

8.2 Plant reactions to water stress

The greatest division of plants in relation to their water supply is between those species which maintain an equilibrium between atmospheric humidity and the hydrature of their protoplasm (poikilohydric plants) and those in which the state of hydration of the protoplasm is independent of air humidity (homoiohydric plants). To the poikilohydric plants belong the terrestrial algae, fungi, lichens, and bryophytes as well as certain pteridophytes. In the higher plants most species buffer the effects of air humidity in relation to water content of the protoplasm and these are the homoiohydric species (Walter and Stadelman, 1974)(*see* Table 8.1). These authors further divide the homoiohydric plants on their method of buffering the effects of dry air on the protoplasm. The term xerophyte, irrespective of whether the species is poikilohydric or homoiohydric is usually restricted to those species that are non-succulent and non-halophytic and obtain their water supply from local precipitation and atmospheric moisture. The succulent xerophytes are placed in a group outside the poikilohydric/homoiohydric classification.

8.2.1 DROUGHT-AVOIDING SPECIES

Drought-avoiding species fall into two classes, therophytes and geophytes. The former survive the dry period as seeds while the latter survive as dormant, buried rhizomes, bulbs or corms. The therophytes can be further subdivided into summer and winter ephemerals. They both possess mechanisms whereby germination will not take place unless there has been a substantial fall of rain (Mulroy and Rundel, 1977). Many desert plants have seeds with elaborate methods for integrating the intensity and duration of precipitation in their habitat and subsequently triggering germination only when there has been a sufficient rainfall to carry their development through to flowering and fruiting. Winter ephemerals have typically a rosette form of growth and C_3 metabolism whereas the summer ephemerals which grow as a

Fig. 8.5 The Mohave Desert with a mixed sclerophyllous shrub vegetation showing in the foreground the creosote bush (*Larrea divaricata*).

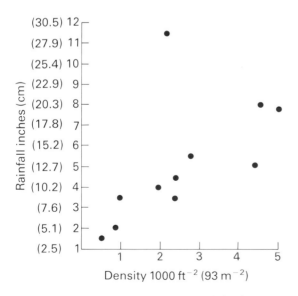

Fig. 8.6 Scatter diagram showing the correlation between rainfall and the density of the creosote bush *Larrea divaricata* in the Californian desert. (Reproduced with permission from Woodell *et al.*, 1969.)

result of summer rains usually have erect growth, and show the *Kranz* type anatomy of C_4 plants.

In addition to rapid development these ephemeral species characteristically manage to flower with a minimal development of vegetative structure. They have also a high degree of phenotypic plasticity. When rainfall is low vegetative growth is severely restricted and few seed and flowers are produced while, if rainfall is plentiful, they produce a greater abundance of vegetative growth, flowers and seeds (Mulroy and Rundel, 1977)

The phenological division of plants into drought avoiders and drought toleraters on the basis of whether or not they avoid the dry season as seeds or buried perennating organs, although convenient on a time scale, does not necessarily mean that they are physiologically different in their tolerance of tissue desiccation. It has been pointed out (Jones *et al.*, 1981) that the succulent plant with tight stomatal closure, an impermeable cuticle and suberization of its roots survives drought by avoiding tissue desiccation in just the same way as a fleshy rhizome of a buried geophyte. Conversely the seeds of drought escapers pass the season of water shortage with

8.2.3 HOMOIOHYDRIC XEROPHYTES

Malacophyllous species

Malacophyllous species or soft-leaved species (in Greek *malakos - soft, phyllos - leaf*) are characteristic of semi-desert conditions where winter or seasonal rains ensure the periodic alleviation of drought. Thus in contrast to the previous groups, these plants have at least the expectation that for a certain period of the year growth can take place without severe limitations due to water shortage. The leaves of such species wilt during drought, but are able to withstand the stress and recover after rain. Typical genera include *Cistus, Lavendula, Thymus, Rosmarinus, Artemisia*, and many more belonging to the Labiatae, Compositae, and Cistaceae. It was these species that suggested to Maximov that protoplasmic resistance was the essential factor in drought hardiness. When these malacophyllous species are exposed to extended periods of drought they shed their leaves and survive the drought by restricting further water loss to the smaller amounts that are transpired by dormant buds and stems.

Sclerophyllous species

Sclerophyllous species or hard-leaved species (in Greek *sclero* - hard) as well as describing plants with sclerophyllous leaves, is used to include the leafless species (aphyllous plants) which are able to maintain a favourable water balance by reducing their transpiration surface area during predictable drought periods. Sclerophyllous plants do not renew their leaves annually. Thus the cost of renewal in terms of water and other resources can be spread over a longer time period. In the parana pine (*Araucaria araucana*) the sclerophyllous leaves of this drought-tolerant tree can last for 25 years. The parana pine is the characteristics species of the drought zone that lies between 37 and 40°S in northern Argentina (Hueck, 1966). This type of xerophyte is a common constituent of Mediterranean climates and typical examples include the holm oak *Quercus ilex*, the umbrella pine *Pinus pinea* (Fig 8.11), the olive *Olea sativa* (Fig. 8.12) and in the aphyllous group *Spartium junceum* and *Genista* spp. *Spartium junceum* or by its common name Spanish broom, has spread rapidly since its introduction to South America and now

Fig. 8.11 The umbrella pine *Pinus pinea* in central Spain, an example of an xerophytic species with sclerophytic foliage.

Fig. 8.12 An ancient olive grove at Stari Bar on the Albanina—Yugoslav frontier. The olive is a typical example of a Mediterranean tree with sclerophytic foliage.

occupies large tracts of land in the Andean Highlands where the dry season at a high altitude produces an acute water stress. Other sclerophyllous species which grow alongside *S. junceum* in South America include various species of *Colletia* and *Berberis* together with *Schinus molle*. The very large number of species of *Colletia* to be found in the mountain deserts and scrubs of South America is of great interest when it is realized that this is another example of a genus outside the Leguminosae which bears nitrogen-fixing root nodules.

The sclerophyllous xerophytes as well as having xeromorphic shoots also have well-developed root systems. When these roots are sufficiently well developed to reach deep water-tables the species are usually described as phreatophytes (*see* p.179). The well-developed phreatic root of *Colletia spinosa*, for example, has a moist texture and appears to have considerable water reserves. The root is also rich in saponins. This diverse group of secondary metabolites is so named for its capacity to form a lather and it is for this property that the Indians of the Peruvian highlands gather *Colletia* roots for use as soap (Fig. 8.13).

8.2.4 SUCCULENT XEROPHYTES

The succulent water-storing species form a very distinct group ecologically. Walter and Stadelman (1974) in their review of the water relations of desert plants suggest that they should be considered apart from the true xerophytes as their tissues are never subjected to any significant water deficit. Succulent species like poikilohydric plants are able to survive long periods without any external water supply. The succulent species differ however in that they can remain metabolically active for a while during drought and do not lapse into a state of dormancy which

Fig. 8.13 Washing day in an Indian village in the central highlands of Peru. Soap in this community is a luxury and instead a lather is obtained from the natural saponins liberated when the clothes are rubbed with a piece of *Colletia* sp. root.

was already visible within 8 hours of wetting the soil (Walter and Stadelman, 1974).

Figs. 8.14 and 8.15 show a special feature found in some succulents where the entire plant is buried below the soil surface with only the flattened leaf tips exposed at the soil surface. This example is of *Lithops salicola* but other genera showing this habit include *Nananthus*, *Conophytum* (Mesembryanthemoideae), and certain soil cacti of the north Chilean desert (Walter and Stadelman, 1974). Succulent species commonly open their stomata by night and keep them closed by day. In this way they take up carbon dioxide during the period of minimum water stress and reduce water vapour loss. The mechanism of this dark fixation of carbon dioxide Crassulacean acid metabolism (CAM) is discussed below (p.198). During periods of extreme drought succulent species keep their stomata shut both day and night (*see* p.193 on CAM idling and CAM cycling) and no net gain of carbon dioxide takes place. During this period of suspended growth such plants are able to match the rate at which they lose water with their consumption of carbohydrate. In this way the reduction in dry matter keeps pace with the loss in moisture so that there is no appreciable change in the hydration of the protoplasm. Walter (1962) cites the example of *Opuntia phaeacantha-tooumeyi* which loses 60 per cent of its original weight during the dry period while the water content on a dry weight basis decreased only from 84.4 to 72.7 per cent. During prolonged periods of drought such species suffer from starvation rather than dehydration or loss of turgor. Thus xerophytic orchids after a period of prolonged drought turn yellow as carbohydrate consumption is followed by chlorophyll degradation. Despite a severe water stress and considerable loss of colour such orchids recover rapidly their green colour when water is available photosynthesis resumes, and the cell constituents that were degraded during the drought period are replenished.

normally overtakes other xerophytic types. During the rainy period the succulent species fill their tissues with water. They do not rely on deep tap roots and in some desert species of *Tillandsia* as mentioned in the opening chapter (*see* Fig. 1.2) they do not root at all but blow around the desert freely with the wind. It is therefore not surprising that many succulent species lend themselves to the epiphytic habit. In the rooted species the shallow roots die during the drought period but grow again quickly when the soil becomes wet. In *Opuntia puberula* root formation

8.3 Drought tolerance with tissue dehydration

8.3.1 THEORIES OF DROUGHT TOLERANCE

From early studies on poikilohydric plants and malacophyllous xerophytes it has long been known that the liquid contents of the vacuole can disappear entirely when resistant species are exposed to drought. The vacuole solidifies with gums and sugars, and provided the cell membranes are unruptured, will return to its normal condition on rewetting of the

Fig. 8.14 Close up view of the expanded leaf tips of the desert succulent *Lithops salicola*.

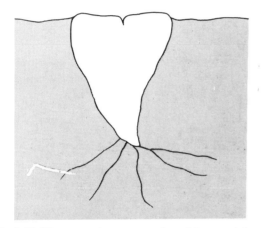

Fig. 8.15 Diagrammatic representation of the growth form of the desert succulent *Lithops salicola*.

tissue. On the basis of this and other observations Iljin (1957) suggested a mechanical theory of drought injury. Iljin's view of drought damage associated the death of the cell with the breaking of protoplast membranes and the plasmodesmata which attach the cytoplasm to the cell wall. If mechanical tensions could be avoided by slow drying and slow rehydration, he considered that the protoplasm of most plants was capable of surviving complete dehydration. Iljin's theory therefore considered that small cell size and the absence or reduction in size of

vacuoles was an advantage in drought-tolerant species as the geometry of such cells would have a greater area of cell wall in relation to the volume of cytoplasm. Thus there would be more cell wall elasticity to accommodate the shrinkage of the cell contents and consequently reduced mechanical strain on the plasmodesmata.

Although mechanical injury to the cell as envisaged in Iljin's theory will be part of the damage caused by drought the features he lists for tolerance such as small cell size and the absence of plasmodesmata are in reality questionable as their absence does not make a plant necessarily drought tolerant. Plasmodesmata have been found in drought-tolerant plants (Bewley and Pacey, 1978) and small cell size in species from drought-prone areas may be no more than a consequence of a scarcity of water during development and not a feature that confers any immunity to drought damage. Iljin based his analyses on whether or not tissues were killed by desiccation by testing for viability in terms of the cell's ability to retain dyes and deplasmolyse on rehydration. These features of cell activity are dependent only on an intact tonoplast and therefore do not distinguish between dead or living cells (for review, *see* Bewley and Krochko, 1982). Current opinion suggests that it is the physiological properties of the protoplasm and its ability to recover from dehydration rather than physical properties of stretching and breaking plasmodesmata that determines whether or not a cell can survive dehydration (Henckel and Pronina, 1973). According to Bewley and Krochko (1982) the survival of the cell is ultimately dependent not on avoiding damage to the protoplasm but on it being limited to a reparable level. This requires two conditions:

1   The maintenance of sufficient physiological integrity so that metabolism can be reactivated quickly on rehydration.
2   A repair mechanism that can be put into effect on rehydration and which can repair any damage caused to membranes and membrane-bound organelles.

The validity of this view is illustrated in the case histories discussed below.

8.3.2 DROUGHT TOLERANCE IN GROWING PLANTS

In *Myrothamnus flabellifolia* (Fig. 8.10) the most celebrated example of drought resistance in higher plants, the leaves, when in the the fully desiccated state, are so dry that they can be rubbed between finger and thumb into a fine dust with the minimum of effort. In these conditions using sensitive apparatus it is possible to show a continued, steady but small evolution of carbon dioxide by the folded leaves (Ziegler and Vieweg, 1970). Electron microscope studies of the leaves reveal a dense cytoplasm with the vacuole still present containing solidified material, probably of a proteinaceous nature. The membranes of the leaf organelles are only weakly visible with the thylakoid structure of the chloroplasts completely collapsed so that the grana appear as thick-packed lamellae. In contrast with mosses and liverworts the restoration of water to the tissues of poikilohydrous angiosperms does not give rise to an immediate resumption of photosynthesis. When twigs of *Myrothamnus* are placed in water-saturated air there is a steady rise in the evolution of carbon dioxide (Fig. 8.16) which continues until the respiration rate of the twigs is 100 times that of the tissues in the dried state. It would appear therefore that in the dry state the respiratory mechanisms of the leaf remain intact. This is further substantiated by the extraction and cstimation of aldolase activity together with NAD^+ specific triose-phosphate dehydrogenase and malic dehydrogenase. The soluble enzymes of photosynthesis, ribulose bisphosphate carboxylase, ribose 5 phosphate isomerase, phosphoribulo kinase and $NADP^+$ dependent triose phosphate isomerase are also fully active despite the inability of the tissues to fix carbon dioxide. The block in the photosynthetic

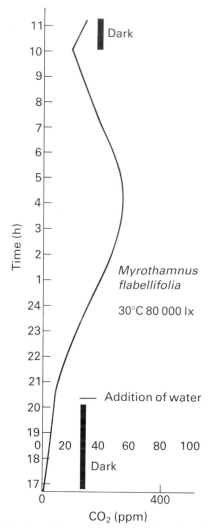

Fig. 8.16 Infra-red absorption curve of the carbon dioxide gas exchange of a branch of the resurrection plant *Myrothamnus flabellifolia* as it passes initially from the desiccated state through exposure to damp air in the dark and then into light with access to liquid water. Carbon dioxide production initially increases and then is reduced to the compensation point as photosynthetic activity is restored. (Reproduced with permission from Viehweg and Ziegler, 1969.)

mechanism causing the failure of oxygen evolution appears to lie in the electron transport system with a complete failure of the photophosphorylation mechanism. Thus the soluble enzymes in the chloroplast do not appear to be affected by the desiccation of the tissues, whereas the membrane-bound systems are all inactive. It is during the prolonged reactivation stage which can last up to 12 days in poikilohydric angiosperms that the reconstitution of the membrane-bound photosynthetic apparatus takes place. Inhibitors of protein synthesis such as chloramphenicol and actinomycin D can all prevent the reconstitution of the photosynthetic mechanism. As with freezing injury it is the membrane-associated enzymes that are denatured (Ziegler and Vieweg, 1970).

The ultimate point at which the cytoplasm is damaged beyond repair is reached when the enzymes that are necessary for the continuation of life are fully inactivated. In a study of the resistance of bacteria and viruses to desiccation Webb (1965) showed that cell survival does not depend on the continued activity of the usual enzymes associated with carbohydrate metabolism, i.e. glycolysis and the Krebs cycle. When desiccation injury does occur in micro-organisms it involves that portion of the protoplasm involved in the synthesis of new enzymes, namely the RNA-DNA complex.

8.3.3 DROUGHT TOLERANCE IN GERMINATING SEEDS

Many desert species require a period of 'after ripening' for the successful germination of their seeds. Such seeds although they lose water on the parent plant before they are shed will not germinate even when there is sufficient moisture in the soil. They remain instead in a period of dormancy while they complete a period of after ripening. Thus in winter annuals when the seeds are shed in early or mid-summer there is no germination until the seeds have dried further and then, in the autumn when soil moisture rises again, germination will take place rapidly.

The water absorption process in germinating seeds can be conveniently divided into three phases (Koller and Hadas, 1982). In the initial phase the kinetics of water uptake are similar to that of killed seed and it is therefore assumed that the uptake of water is non-biological and due mainly to adsorption, matrix hydration of solids, and capillary action. During this period membrane reorganization and proliferation occur and are followed towards the end of the period by an increased dependence on temperature as membrane-dependent processes such as osmosis become operative. The process of water uptake to start the initial phase of germination can be delayed by some desert species until there has been sufficient rainfall stored in the soil to ensure the successful completion of their life cycle. In certain leguminous species, for example *Medicago laciniata* water enters the seed through the hilar fissure which is coated with a water repellant. The degree of opening of the fissure is hygroscopically controlled. Initially it is only wide enough for the entry of water vapour, free water being excluded by the water-repellant coating. After a set time at a critical water vapour pressure the hilar fissure opens wide enough for the entry of free water and hence the initiation of the initial state of germination (Table 8.2). Fig. 8.17 shows the time course of imbibition of seeds that had been previously stored at different relative humidities (Koller, 1969). In *Medicago laciniata* there was a critical threshold in relative humidity which would allow germination to take place. In other genera such as *Lathyrus* there is no critical threshold level for the initiation of the capacity to imbibe. Instead, the greater the relative

Relative humidity during 2 months storage at 25°C. (%)	Imbibition*			
	L. hierosolymitanus		*L. aphaca*	
	Onset (days)	Rate (% per day)	Onset (days)	Rate (% per day)
13.0	4.2	11	7.9	14
44.5	1.9	13	5.3	11
49.0	1.3	13	4.5	15
61.0	0.6	20	2.4	14
74.0	0.4	25	1.9	13
80.5	0.3	60	0.8	20
83.5	0.2	94	0.2	93

Table 8.2 Effect of humidity during storage on kinetics of imbibition in *Lathyrus* seeds. (Reproduced from Koller, 1969.)

* Final percentages 100 in both spp.

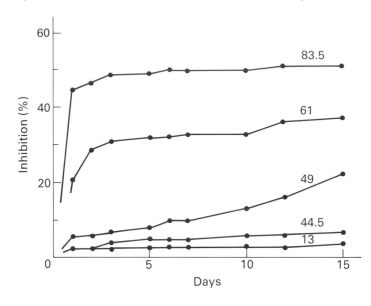

Fig. 8.17 Time course of imbibition as measured by the percentage of seeds swelling in *Medicago laciniata* after 2 months storage at various relative humidities. (Redrawn from Koller, 1969.)

humidity at which the seeds are stored, the quicker the onset of imbibition when given free water. In these species it is thought that the absence of a water-repellant layer around the hilar fissure may account for the absence of a threshold value for the relative humidity control and water uptake. In both ways the seeds have been able to sense the duration of wet weather and used this to control the onset of imbibition (Koller, 1969).

This initial phase of water uptake is then followed by a transition stage during which it is possible for the germination of the seed to be blocked. Depending on the environmental conditions, the balance between germination-promoting and germination-inhibiting reactions may lead either to a continuation or cessation of germination. Some inhibitory processes for germination are overcome only with high levels of hydration and when this is not achieved germination can be blocked and the seed can be dehydrated without loss in vitality (Koller, 1970). In some species seeds will germinate more rapidly if they are watered from above rather than from below. This flow of water past the seed is thought to remove water soluble germination inhibitors. This progressive removal of inhibitory substances acts as a form of 'rain guage' and helps the seed to ensure a propitious timing for its germination.

The third phase in water uptake is found only in viable seeds and begins with the protrusion of the radicle when the growth by cell extension has begun. In contrast to the transition period when moisture content, respiratory rate, and morphology of the seed remained relatively unchanged the third phase is one of ever-increasing activity. There is also an increase in embryo fresh weight which has been attributed to the vacuolation of existing cells. In some species this water uptake is responsible for the rupture of the testa while in others this does not develop enough force and the rupture of the seed coat is delayed until the rehydration of the scutellum mobilizes greater food resources (Wellington and Durham, 1961).

The uptake of water in the initial phases of germination is much influenced by the hardness and morphology of the seed coat. The testa is capable of rendering the seed gas-hard, water-hard or both. Water-hard seeds can be recognized as on soaking they will fail to swell. Gas-hard seeds give no visible sign of their condition and are therefore harder to recognize. The hardness of the testa is gradually eroded by weathering and germination is thus staggered. This spread of germination in a cohort of seeds however is not a random phenomenon. In legumes it is possible to number seeds in relation to their position in the pod. Koller (1969) has shown that in a number of leguminous species in which the seeds are observed separately in relation to their position in the pod (numbering the seeds from the base of the pod) then those in position one will germinate with a shorter period of weathering than those from positions nearer to the point of attachment to the parent plant (Table 8.3). This is a form of

Table 8.3 Effect of position in pod on susceptibility of water-impermeable seeds to weathering. (Reproduced from Koller, 1969.)

Position of seed in pod[†]		Imbibition after weathering (%)*		
		0 months	5 months	17 months
Hymenocarpus circinnatus	I	4	8	43
	II	8	69	95
Onobrychis cristagalli ssp. eigii	I	1	2	2
	II	0	2	2
	III	2	3	79
O. crista-galli var. *subinermis*	I	0	1	2
	II	0	1	6
	III	0	83	99
Medicago tuberculata	I	20	98	100
	II	0	10	95
	III	0	0	40
	IV	0	0	7
	V	0	0	2
	VI	0	0	0

* At Be'er Sheva, Israel, under 10 mm less, protected from rain.
[†] Starting from base.

somatic polymorphism which adds a degree of adaptive heterogeneity to any cohort of seeds in that they will not all germinate at the same time and face the same risks for survival.

The extent to which the seed coat exerts an effect is however dependent on soil texture, whether or not the seed is buried, and the level of evaporation. In coarse soils such as sand, the differences between seeds with coats which repel water or absorb it are seen to their greatest extent. Myxospermy (the possession of a seed coat that produces mucilage when wet) is found in the Plantaginaceae, Linaceae, Brassicaceae, Euphorbiaceae, Onagraceae, and Acanthaceae (Koller and Hadas, 1982). Seeds that produce a mucilage on wetting are least dependent on soil moisture content although they do run the risk of suffering from an oxygen deficit. It is interesting to note that some seeds of some myxospermous species combine this property with the ability to carry germination through to the third phase, i.e. the protrusion of the radicle in an entirely anaerobic environment. Thus to benefit from the advantages of mucilage the seed appears to have evolved an additional precaution against anoxic injury should the mucilage interfere with the uptake of oxygen.

In higher plants the ability to continue protein synthesis after periods of dehydration has been examined in relation to the dehydration of germinating seeds. When seeds are first imbibed and then immediately desiccated, growth will be resumed after a further period of re-imbibition. This ability is rapidly lost as the seedling develops, but initially it is of importance for the successful emergence of seeds that have to germinate in the fluctuating water regime of the upper regions of the soil or in desert conditions where short periods of rainfall are not sufficient for the completion of development and establishment of the seedling. The inability to recover from alternate wetting and drying coincides with the irreversible inactivation of the polyribosome fraction of the cell (Nir *et al.*, 1969). Similar results are observed on drying out the wheat embryo (Chen *et al.*, 1968) and in the damage caused by desiccation in the aquatic moss *Hygrohypnum luridum* (Bewley, 1974).

8.4 Drought avoidance and turgor maintenance

8.4.1 WATER UPTAKE

The ability of any plant to keep its tissues turgid depends on a balance between the external relationships in water uptake and water loss as well as on the internal osmoregulation of the plant cells. The most obvious starting point in this study is to examine whether drought-tolerant plants have any special means of obtaining water that are denied to drought-sensitive species. The phreatophytes and riparian plants with their roots reaching down to permanent water-tables are the most obvious examples of plants that have the ability to reach water sources that are unavailable to other plants. The creosote bush (*Larrea divaricata*) has been found with roots that can extend many metres to the phreatic zone. In most moderately drought-resistant species, water stress increases the root to shoot ratio. The organic nutrition of the root does not suffer greatly during drought and may even improve as shoot growth is reduced. Apart from the change in ratio there is also evidence that the absolute amount of root growth can be increased under increasing drought (Schultz, 1974). In a drying soil profile cotton roots not only grew faster than in a well-watered one but also achieved a substantially greater total length (Klepper *et al.*, 1973).

The manner of growth of roots of some drought-resistant plants, in addition to tapping a greater soil volume, also changes the porosity of the soil so that water infiltrates most rapidly along pores and cracks adjacent to the roots. The rate of infiltration of water round the base of an isolated mulga tree (Slatyer, 1962) was found to be substantially faster than on ground some distance away. As water-flow down stems can be a substantial contribution to total precipitation a permeable soil near the base of the plant is an advantage particularly during periods of heavy rain.

The most specialized roots of xerophytic species are aerial roots of epiphytic orchids (Fig. 8.18). These roots have a white appearance due to the development of the velamen which although dead, is prevented from collapsing by spiral and netted thickening. The cells are empty, and open to each other and the outside by pores which allow the escape of air when the roots take up water. Early studies by Goebel (1889−93) showed that water uptake is rapid and can be 40−80 per cent of root weight.

There has been considerable dispute as to whether drought-resistant plants are capable of absorbing water through their leaves. Many epiphytic species have water collection mechanisms where pools of water accumulate at the axils between leaf bases and the stem. Although these pools will reduce dehydration and may serve as funnels which channel water eventually to aerial roots it does not follow that water from such pools will enter the leaf. A critical acceptance of foliar absorption as a means of obtaining water requires the demonstration of four criteria, namely (1) a gradient of decreasing water potential from atmosphere to plant, (2) morphological specializations for water absorption, (3) significant absorption and redistribution of water,

Fig. 8.18 Aerial roots of *Dendrobium nobile* an epiphytic orchid of south-east Asia. The white appearance of the roots is due to the velamen, a tissue of 2−3 layers morphologically equivalent to the epidermis and, although dead, prevented from collapsing by spiral and netted thickening. The cells are empty and open to each other and the outside by pores which allow the escape of air when the roots take up water. Water uptake is rapid and can amount to 40−80 per cent of root weight.

and (4) a consequent reduction in plant water potentials (Rundel, 1982). Although many desert plants have had claims put forward for foliar absorption critical examination often suggests that ground water is the major source. The tree mesquite of the Atacama desert of northern Chile (*Prosopis tamarugo*) was described as a significant foliar absorber of water (Went, 1975) but later studies suggested that ground water is its major source of moisture (Mooney *et al.*, 1980).

However despite these controversies there is one group of plants where foliar absorption of liquid water clearly takes places and that is in the Bromeliads. Over 100 years ago Schimper (1898) described the specially adapted water-absorbing leaf hairs (trichomes) of the Bromeliaceae. Such trichomes exist in many members of the Bromeliaceae but are most spectacularly developed in the genus *Tillandsia* (Fig. 8.19). Fig. 8.20a shows the peltate form of these trichomes often described as shields or scales. They consist of a flat plate only one cell deep with a column of thick-walled cells at the centre (Benzing, 1976). The trichome forms a waterproof tube and

Fig. 8.19 Spanish moss *Tillandsia usneoides* (Bromeliaceae) an epiphyte which can absorb liquid water by specialized glands and also adapt from C₃ to CAM photosynthesis depending on the availability of water.

valve as all the cells are cutinized on their exterior surface. When water droplets land on the leaf the surface being wettable causes the water to spread out into a thin uniform film. The shield cells rapidly fill with water, the central cells of the tube become turgid and force this portion of the shield upwards while the outside edges of the shield bend downwards (8.20b). This creates a suction which draws the water in under the plate and down into the stalk cells. The shield then returns to its unflexed position on drying and is ready to repeat the cycle. In order to maximize the flexing and reflexing of the shield cells *Tillandsia* requires a wetting and drying cycle for efficient water absorption and survival. The genus *Tillandsia* has attracted much attention through its ability to inhabit the rainless Atacama desert of northern Chile and southern Peru. Despite the complete lack of precipitation and the only source of water being the fog that comes periodically from the cold Humboldt current, *Tillandsia straminea* has been shown to go through a clear water cycle with water absorption by night and transpiration loss by day and achieve a mean growth rate of 0.17 per cent per day. When sufficient fog is available this can give an annual rate of increase of 20 per cent (Alvim and Uzeda; in Walter, 1971). It is important to realize that the foliar uptake of water is dependent on absorption of liquid water which has condensed from fog and mist rather than water vapour which makes no significant contribution to water uptake. The only other plant group that has been clearly shown to absorb liquid water through its leaves is the Orchidaceae and the presence of absorbing trichomes has been shown in the tribe Pleurothallidinae (Pridgeon, 1981).

8.4.2 STOMATAL CONTROL UNDER WATER STRESS

In drought situations stomata may close either in response to internal or external stimuli. Closure as a response to an internal signal results when water stress within the leaf causes a loss of turgor which then affects the guard cells. However water stress within the plant can also cause stomatal closure without the delay that is necessary for loss of turgor. It is now well-known that abscisic acid formation is induced by drought and that this is then transported to the guard cells where it can bring about stomatal closure (Hiron and Wright, 1973). A mode of action of abscisic acid has been suggested by Raschke (1975) in which its transport to the guard cells alters

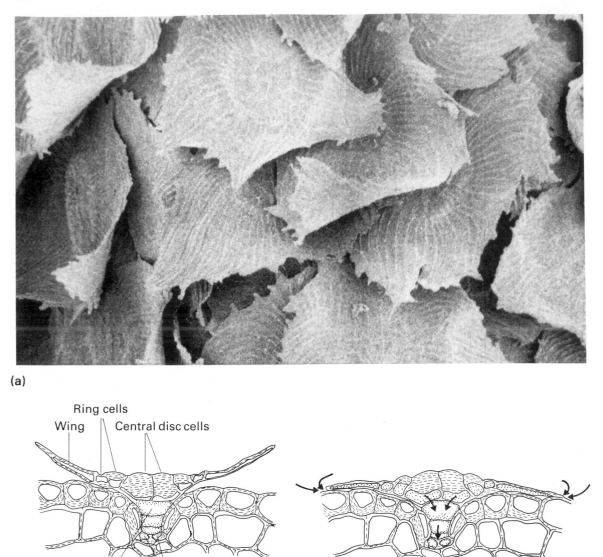

(a)

(b)

Fig. 8.20 (a) Surface view by scanning electron microscope of the water-absorbing trichomes of *Tillandsia usneoides* (Photo C. Studer). (b) Cross-section of a typical tillansoid in a dry condition (left) and in the process of water absorption (right). (Reproduced from Rundel, 1982 based on drawing by Benzing.)

the plasmalemma transport of H^+ and through the coupling of H^+ and K^+ transport results in a reduction in solute content in the guard cells and hence stomatal closure. This control of water loss has been described as 'feedback control loop' and is thought to become activated once a certain threshold has been passed. The threshold water potential that initiates this reaction varies both with species and the conditions under which they have been cultivated.

Any increase in abscisic acid either from endogenous sources for external application induces a closure of the stomata which persists for several hours after the water stress has been removed. Thus plants which suffer from drought will develop a stomatal opening and shutting rhythm which will maintain their stomata shut during the day. It is such a water-stress signal that triggers the change from C_3 metabolism to CAM in plants with the biochemical capacity for this type of metabolism. Thus in C_4 species such as *Portulacaria afra* (Ting and Rayder, 1982) where abscisic acid production reponse to drought causes the rhythm of stomatal movement to change so that the stomata close by day but not at night.

The reverse effect of abscisic acid on stomatal opening is brought about by cytokinin suggesting that plants may have a complete hormonal system which attempts to balance the degree of water loss. Reduction in plant turgor will be brought about by kinetin while an increase can be mediated by abscisic and indole acetic acid (Tal and Imber, 1971). An example of this two-way mechanism whereby the water content of the plants is affected by its environment through the action of plant harmones has been shown in tobacco (Mizrahi *et al.*, 1972). When the plants are grown in culture solution, reducing aeration of the root system causes wilting of the shoot. However if the plants are exposed to saline stress in the culture medium (($6g\,l^{-1}$) for 2 days before the aeration is turned off, then no wilting takes place as a higher abscisic acid level has been induced by the salinity stress and this has closed the stomata and prevented further wilting as a result of poor aeration.

Plants also possess the ability to sense changes in air humidity independently of the water status of their tissues (Fig. 8.21). This sensitivity allows a potential water shortage to be anticipated. The plant by sensing the dry atmosphere can close its stomata before there is a water deficit in its leaves. In contrast to the feedback mechanisms outlined above this mechnism has been named a 'feed-forward regulation' (Cowan, 1977). To avoid the effects of localized water deficits in the leaf the experiments have to be carried out with well-watered plants; then by changing air humidity and by measuring stomatal conductance and leaf water content simultaneously it is possible to show that stomata may open and shut in relation to changes in external humidity even although leaf water content remains constant. By using epidermal strips and observing them microscopically while passing saturated air over their inner surfaces, adjacent groups of stomata can be made to open and close in response to giving them alternately moist and dry air on the outer side of the epidermal strip (Lange *et al.*, 1971). In the experiments of

Lange and co-workers the stomatal movements of the fern *Polypodium vulgare* showed that closure was about twice as fast as opening. The feed-forward mechanism is predominantly sensitive to changes in air humidity but the magnitude and speed of the response can be modified by other environmental factors such as light, carbon dioxide, and temperature. Stomatal response to humidity change is usually greatest at saturating light intensities. In addition, the previous cultivation conditions can have an effect. In *Citrus* sp. the velocity of stomatal response declines after several days of exposure to progressively decreasing humidity. Plants showing Crassulacean acid metabolism (CAM) are particularly interesting as they tend to have a greater response to dry air at night (Lange and Medina, 1979).

The mechanism whereby plants can sense changes in atmospheric humidity is thought to be linked to the rate of guard cell transpiration (Lange *et al.*, 1971). The rate at which water is lost by cuticular transpiration is greater through the cell walls of the peristomal region than elsewhere in the leaf. This can be seen by following the movement of chemical tracers which tend to accumulate at the end-points of the transpiration stream. Substances such as lead-EDTA, monosilicic acid, and prussian blue accumulate in the guard cell spicules as well as along the their cell walls. Guard cell walls have uncutinized areas and these allow a preferential loss of water from the peristomal region something which can be observed by autoradiography with tritiated water (Maier-Maercker, 1979). The amount of turgor that a guard cell has to lose to effect stomatal closure is relatively small and as it has low wall pressure potentials this will partly counteract any tendency for water to flow towards the low matrix potentials in guard cell walls. However this closure of the stomata as a result of low atmospheric humidity will be sustained only if the solute content of the guard cells is subsequently reduced (Maier-Maercker, 1979). Thus certain changes in the guard cells can be seen after the closure of the stomata and these include a decrease in potassium content, a fall in pH and an increase in starch formation all of which will contribute to the reduction of turgor in the guard cells as compared with the subsidiary cells and make stomatal opening less likely. It is important to note that these events do not need to take place in order to initiate stomatal closure but if they follow closure they will ensure against premature opening. Thus the 'feed-forward mechanism' although it is initiated by a purely physical loss in water from the peristomal cells is followed

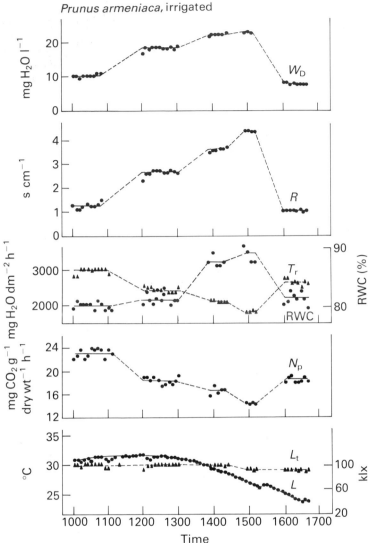

Fig. 8.21 Changes in water vapour concentration difference between the leaf and the surrounding atmosphere (W_D) together with changes in the total diffusion resistance (R), transpiration (T_r), net photosynthesis (N_P) and relative water content (RWC %) of the leaves in irrigated apricot trees (*Prunus armeniaca*). The leaf temperature (L_t) was kept constant and the light conditions (L) are natural. (Abscissa daytime). Note the close relationship between W_D and R which reflects the effect of atmospheric humidity on stomatal resistance. (Reproduced with permission from Schulze *et al.*, 1972.)

up by a series of metabolic events which will inevitably link the efficacy of this form of stomatal closure with temperature, light, and carbon dioxide concentration.

8.4.3 CUTICULAR ADAPTATIONS TO WATER STRESS

The cuticular resistance of terrestrial plants ranges widely from 2000–5000 s m^{-1} in mesophytes to 5000–40000 s m^{-1} in xerophytes (Cowan and Milthorpe, 1968). The properties which make up these differences in resistance can best be seen if the membranes which go together to make up this total

barrier are isolated and studies *in vitro* without the complications that arise from the presence of stomata and leaf hairs. Schönherr (1982) has made an extensive review of the properties of such isolated membranes. The permeability coefficient across cuticular membranes differs between species. The permeability of cuticular membranes depends on the water activity of the vapour phase on the outside of cuticle (a_{wv}). Fig. 8.22 shows that in onion the effect is least while in *Citrus aurantium* the permeability coefficient for transpiration (P_{tr}) was highly dependent on the water activity of the vapour phase. The bulk of the cuticular membrane is made up of a polymer matrix which supports the soluble cuticular lipids. These

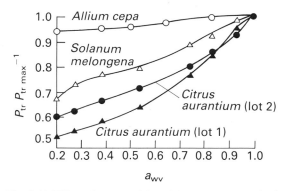

Fig. 8.22 Effect of water activity of the vapour phase (a_{wv}) on permeability coefficients of transpiration (P_{tr}) of isolated cuticular membranes. $P_{tr\ max}$ was determined at $a_{wv} = 1$. Note that in onion there is a minimal effect on transpiration of water deficit in the vapour phase by contrast with *Citrus aurantium*. (Reproduced with permission from Schönherr, 1982.)

lipids or cuticular waxes as they are commonly termed, are a complex mixture of fatty acids, fatty alcohols, esters, and paraffins and all have long hydrophobic chains. The amount of lipid by weight can vary from 2 to 30 per cent of the cuticular membrane. The water resistance of the membrane is highly dependent on the chain length and increases logarithmically doubling for every two − CH_2 groups (LaMer *et al.*, 1964). Although many xerophytic plants have thicker cuticles than mesophytes this in itself does not account for their higher resistance figures noted above. The thin cuticular membrane of onion bulb scales is just as effective as a barrier to water loss as the thick membrane of *Clivia* leaves (Schönherr, 1982). The major water barrier is the soluble lipid component of the cuticle and as this is a minor component, water permeability is not proportional to either thickness or weight of the cuticle. Thus, contrary to popular belief, a thick cuticle is not in itself a method of restricting water loss. Instead it may function as resistant matrix for supporting lipids and protecting both them and the plant against mechanical and microbiological damage (Schönherr, 1982).

8.4.4 OSMOREGULATION AND STRESS METABOLITES

Solute accumulation over and above that caused by varied water levels *per se* is generally referred to as osmoregulation. Provided plants are able to maintain some degree of hydration, osmoregulation can contribute much to the maintenance of turgor during periods of water stress. The increase in solutes which achieve the process of osmoregulation varies with both species and the plant organ in question (Morgan, 1984). In many species, sugars and amino acids are the principal solutes that contribute to osmoregulation. When tissues are expanding then osmoregulation is achieved mainly by the import of solutes. In expanded leaves however the increase can either come from a change in the levels of synthesis and export or else from the hydrolysis of insoluble reserves. One of the first symptoms of water stress in plants is a hydrolysis of starch and frequently this is also accompanied by the synthesis of the so-called 'stress metabolites' as discussed in relation to salt stress in Chapter 7. Their importance in drought resistance requires however that they also be considered in this chapter.

The two principal stress metabolites that accumulate in terrestrial plant cytoplasm as a result of stress are the betaines and the amino acid proline. The betaines are a broad group in which the most important members ecologicaly are the quaternary ammonium and tertiary sulphonium compounds. Betaines are usually found to accumulate under salt stress. Under drought proline appears to be the more important stress metabolite and its accumulation can be caused by relatively mild water stress (water potentials of −1.0 MPa)(Fig. 8.23). The extent to which proline accumulates depends on both the length of time and the severity of the stress. The source of the accumulated proline can be either synthesis from glutamate or else by the hydrolysis of protein. In the early stages of wilting it appears that the synthesis from glutamate is the principal source. When the tissues contain higher levels of proline there is an enhancement of proline oxidation and this can lead to a new steady state. However with severe stress this oxidation seems also to be inhibited and thus proline concentrations can reach very high levels. In *Triglochin maritima* the proline concentration in the leaves can reach 700 mM and in some species can reach 10 per cent of leaf dry weight (Stewart and Lee, 1974). By following the amount of tritium recovered in H_2O from 5−^3H−proline Stewart *et al.*, (1977) were able to follow the rate of proline oxidation in intact leaves and found that it was almost zero after leaves had been wilted for 3 hours. The initial effect of drought is to prevent the feedback inhibition of proline synthesis from glutamate. In addition the synthesis of protein from

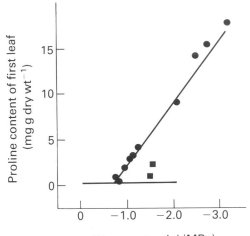

Fig. 8.23 Relationship between water potential and proline accumulation in the first leaf of 10-day-old barley seedlings subjected to water stress with polyethylene glycol solutions. The horizontal line is the proline level of unstressed plants. (Reproduced with permission from Aspinall and Paleg, 1981).

proline is also inhibited. The actual mechanisms which trigger these various methods of proline accumulation in relation to drought are not yet clear. There may be a hormonal action as in some species treatment with abscisic acid can induce proline accumulation.

Whether or not proline confers a greater degree of drought resistance on plants or whether it is just a consequence of the inhibitory effects of drought on nitrogen metabolism is still an open question. Obviously the large increase in proline will act as an osmoregulator. In addition proline in solution will affect the solubility of proteins and *in vitro* can protect bovine albumin from denaturation by ammonium sulphate. It is possible that proline acts by interacting with the hydrophilic surface residues on the proteins and increases the total hydrophilic area which will contribute to stability. Ecologically however, its role in drought-tolerant plants is not clear. Proline accumulation is a property of mesic plants just as much as it is of xerophytes. In comparing plants from the same desert habitat it is possible to find species such as *Carex pachystylis* which accumulate proline while in the same habitat *Artemisia herba-alba* does not (Pourrat and Hubac, 1974). Probably the best indication that proline has a role to play in drought resistance is its effect on

hardening plants against drought. Pre-exposure of plants to drought will increase their resistance and also accelerate the accumulation of proline. Nevertheless from a practical view of plant breeding it is the view of some investigators (Aspinall and Paleg, 1981) that it is premature to breed for proline accumulation as an aid to drought tolerance in crop plants.

8.4.5 DIURNAL VARIATION IN CO_2 UPTAKE

Maintaining open stomata for the entry of carbon dioxide for photosynthesis during the day is irreconcilable with restriction of water loss. It is therefore not surprising that in many desert plants carbon dioxide is absorbed at night. This displacement of carbon dioxide fixation from day to night was first studied in the Crassulaceae where it was shown to be accompanied by the accumulation of large quantities of malic acid. For this reason the process is usually referred to as 'Crassulacean acid metabolism' (CAM) even though it is now known to be a common feature in many homoiohydric species of diverse taxonomic groups. Species which are capable of CAM metabolism are found in the Bromeliaceae, Orchidaceae, Cactaceae, Aizoaceae, and Euphorbiaceae as well as the Crassulaceae. Detailed accounts of this pathway and of C_4 photosynthesis are available in all modern textbooks of plant physiology and will therefore not be repeated here.

The ability of plants with CAM photosynthesis (Fig. 8.24) to take up carbon dioxide efficiently at night is due largely to the properties of the enzyme phosphoenolpyruvate carboxylase. Phosphoenolpyruvate carboxylase has both a high affinity for carbon dioxide even at very low concentrations and an equilibrium constant which drives the reaction to completion even at low substrate concentrations. Using this pathway plants are able to trap large quantities of carbon dioxide into the malic acid pool. During the day when the stomata are closed a refixation of the carbon dioxide by ribulose bisphosphate carboxylase will however work efficiently in the stomata-closed leaf where the carbon dioxide concentration will be higher.

Ecologically it is of great interest that drought alone is sufficient to stimulate a change from daytime fixation of carbon dioxide to CAM dark fixation. When Spanish moss (*Tillandsia usneoides*)(Fig. 8.19), a common epiphyte in tropical and sub-tropical America, is subjected to water stress, the level of dark fixation of carbon dioxide increases four-fold

Fig. 8.24 The Joshua tree (*Yucca brevifolia*) one of the larger species to show CAM metabolism growing in the Mohave desert.

(Kluge *et al.*, 1973). Similarly, in *Opuntia basilaris* (Cactaceae) the pattern of carbon dioxide fixation between day and night alters before and after rainfall. With prolonged drought there is a progressive change to dark fixation (Szarek *et al.*, 1973). Likewise in the halophytic species *Mesembryanthemum crystallinum* and *Carpobrotus edulis* (*see* p.141) a ten-fold increase

in the level of dark fixation of carbon dioxide can be induced by treating roots with a 400 mM solution of sodium chloride (Winter, 1973).

Under the somewhat artificial conditions of floating leaf slices in various culture media, it is possible to show a correlation between the osmotic pressure of the external medium and the rate of malate synthesis in leaves. Malate synthesis in these conditions takes place only when the osmotic pressure differrence between the cells and the medium is low. Conversely, malate loss from the vacuole depends on a high osmotic pressure difference between the cells and the medium and is observed in media of low osmotic pressure (Lüttge and Ball, 1979).This observation has naturally led to the suggestion that the diurnal oscillations in malic acid levels in these plants are under osmoregulatory control. They also prompt the question of whether osmoregulatory control causes the change from day to night carbon dioxide fixation in plants such as *Tillandsia* and *Opuntia* when they are exposed to water stress.

At dusk, when the levels of malic acid will normally be low phosphoenolpyruvate (PEP) carboxylase will generate malic acid by dark fixation. This enzyme is 'feedback' inhibited by a malate concentration in the cytoplasm on only 1 per cent of the total amount synthesized in the dark period (Kluge and Osmond, 1972). During the night the levels of malic acid in the cell can rise to 100 mM so it must be concluded that the malic acid accumulation will be spatially separated from the site of enzyme action. The only possible site for such storage is the vacuole. According to Lüttge and Ball (1979) malic acid accumulation will be accompanied by an inflow of water into the vacuole and an increase in turgor pressure. In this manner Crassulacean acid metabolism may act not only as a carbon dioxide harvesting mechanism but also as a means of water conservation. The theory postulates that at a certain critical turgor pressure the properties of the tonoplast change from one of net malic influx to one of net efflux. It is also possible that changes in turgor pressure can cause alterations in the molecular structure of membranes (Steudle *et al.*, 1977) allowing a passive leak of malate from the vacuole. During the period of leakage PEP carboxylase will be inhibited and further synthesis of malate will not take place until malate itself is once again at a low level in the cytoplasm. This mechanism of controlling the changes in the level of malic acid synthesis and carbon dioxide fixation invokes a tension−relaxation system rather than changes in enzyme activity discussed by, for example, Queiroz (1974). The theory is simple and perfectly compatible with both the experimental evidence and the daily rhythms that are so characteristic of this aspect of carbon dioxide fixation.

During periods of intense or prolonged drought CAM plants exhibit a characteristic modification of the diurnal rhythm of day closing and night opening of stomata which is described as CAM-idling. In this condition the stomata remain shut both day and night. In spite of the prevention of dark uptake of carbon dioxide in this condition fluctuations in the levels of malic acid still take place but to a lesser extent. Apparently respiratory carbon dioxide is recycled through the CAM pathway so that some degree of photosynthesis is achieved with this internal pool of carbon dioxide during the day. Apart from maintaining metabolism through long periods of drought CAM-idling may serve the important function of preserving the photosystems from the photo-oxidation that would result if there was no regeneration of terminal electron acceptors. Another modification that arises is CAM-cycling. Here the stomata are open by day and closed at night yet organic acids still fluctuate as in CAM. The relevance of this behaviour is not yet clear but may be related to the tolerance of short-term drought which is experienced on an hourly basis by many epiphytes. CAM-cycling may serve as a method whereby the plants can quickly enter the CAM-idling mode of existence (Ting and Rayder, 1982).

8.5 The isotope discrimination ratio and photosynthesis in xerophytes

The observation by Bender (1968) that C_3 and C_4 plants differ in extent to which they discriminate against the heavier (^{13}C) isotope of carbon has provided a convenient method for studying the pathway of carbon dioxide fixation in field conditions. Until recently the list of species that has been studied in relation to their pathway of carbon dioxide fixation has remained a rather stereotyped selection which represents what is available in the glasshouse rather than species which are associated with varying habitat factors in nature. The isotope discrimination ratio is calculated as $\delta^{13}C$ value in parts per thousand with reference to a limestone standard:

$$\delta^{13}C(‰) = \frac{^{13}C/^{12}C \text{ sample}}{^{13}C/^{12}C \text{ standard}} - 1 \times 1000$$

When terrestrial C_3 plants fix carbon dioxide under conditions of an ample water supply so that stomatal resistance is low then $\delta^{13}C$ values of about -25 to -28 per thousand are found (Table 8.4). For C_4 plants the values are between -6 and -7 per thousand showing that these plants discriminate less against the heavier ^{13}C isotope. In CAM plants there is a much wider range of values from -9 to -25 per thousand. More recently it has been found that

Table 8.4 The $\delta^{13}C$ and δD values (metabolic hydrogen) of shoots of C_3, C_4 and CAM plants grown in different conditions. (Reproduced from Ziegler *et al.*, 1976.)

Species and pathway	Growth conditions	δ^3C value (‰)	δD value (‰)
Spinacia oleracea (C_3)	Controllled environment	$-22.1\,(7)^*$	$-134\,(7)$
	Cabinet, soil, Munich	$-21.5\,(3)$	$-129\,(3)$
Beta vulgaris (C_3)	Outdoors, soil, Munich	-27.4	-117
Triticum sp. (C_3)	Glasshouse, soil, Canberra	-26.8	$-\ 88$
Atriplex hastata (C_3)	Glasshouse, solution, Canberra	-26.9	-141
Mean C_3		-24.9	-122
Zea mays (C_4)	Glasshouse, soil, Munich	$-13.0\,(3)$	$-\ 84\,(2)$
		$-13.4\,(2)$	$-\ 70\,(2)$
	Glasshouse, soil, Canberra	-12.4	$-\ 69$
Sorghum bicolor (C_4)	Glasshouse, soil, Canberra	-11.8	$-\ 77$
Amaranthus edulis (C_4)	Outdoors, soil, Munich	-12.3	$-\ 87$
Atriplex spongiosa (C_4)	Glasshouse, solution, Canberra	-12.8	$-\ 98$
Mean C_4		-12.6	$-\ 81$
Kalanchoe daigremontiana (CAM)	Glasshouse, soil, Munich	-16.2	$-\ 37.4$
Bryophyllum tubiflorum (CAM)	Glasshouse, soil, Munich	-16.5	$-\ 22.0$

* Numbers in parenthesis refer to number of replicate experiments.

plants also discriminate with regard to hydrogen isotopes (Ziegler *et al.*, 1976).

Some of the complexities of assaying the organic content of plants for hydrogen isotopes are overcome if the analysis is performed on cellulose nitrate (Sternberg *et al.*, 1984). From a determination of the carbon, hydrogen, and oxygen isotopes in cellulose extract of plants that were grown side by side in the same glasshouse Sternberg was able to compare how C_3, C_4, CAM and CAM-cycling plants as well as plants that changed from C_3 to CAM metabolism behaved in relation to their isotope discrimination activities. From Fig. 8.25 it can be seen that in terms of $\delta^{13}C$ CAM and C_4 plants discriminate less than C_3 and CAM−cycling species. Only by combining the discrimination values for deuterium and ^{13}C however is it possible to distinguish four classes of plants. Not unexpectedly CAM-cycling plants are similar to C_3 plants which can change to Crassulacean acid metabolism. Sternberg and co-authors differ from Ziegler in their conclusion as they maintain that these differences manifest themselves as a result of the basic differences in the biochemistry of the plants. All their experimental material was grown under the same conditions. Ziegler suggested that the differences in the discrimination values arose due to the fact that the CAM plants maintain a greater degree of metabolic activity under stressed conditions than C_3 plants. Sternberg's analysis would support the view that these differences can arise without imposing a physiological stress. This is further supported by an examination of the $\delta^{18}O$ values. If there were differences due to differential exposure to evapotranspiration then increased water loss should result in an enrichment of ^{18}O in plant water and consequently in cellulose. This should give a correlation between the δD values for cellulose nitrate and $\delta^{18}O$ values for cellulose. Sternberg did not find such a correlation and therefore concluded that biochemical properties rather than different amounts of evapotranspiration are responsible for the varying amounts of isotope discrimination. Whatever the basis for the different values they do allow an elegant method, provided the researcher has access to a mass spectrometer for gauging the extent that plants in the field are exhibiting C_3, C_4, CAM or CAM-cycling activity. The method can even be used on plants that have been dead for centuries by taking samples from herbarium specimens. As an illustration of the type of ecological comparisons that can be

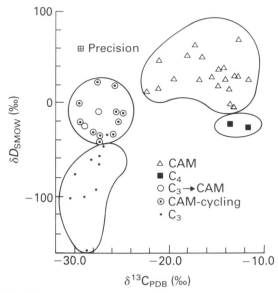

Fig. 8.25 δ D values of cellulose nitrate plotted against δ^{13}C values for cellulose nitrate in plants employing varying modes of photosynthetic activity (*see* text). (Reproduced with permission from Sternberg, 1984.)

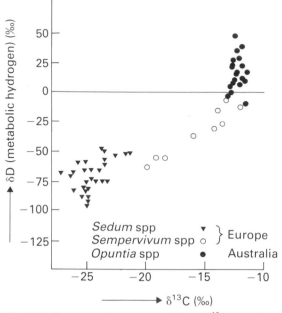

Fig. 8.26 Correlation between the δ D and δ^{13} C value in different CAM plants. The species of *Sedum* and *Sempervivum* were collected in the European Alps. (Reproduced with permission from Ziegler *et al.*, 1976.)

made with a mass spectrophotometer, Fig. 8.26 shows the extent to which *Sempervivum* species growing in the Alps differ in their isotope discrimination ratios from other species of the same genus in Australia. It can be seen that the Australian species exhibit a pattern that is more typical of CAM than those from the Alps.

8.6 Conclusions

The remarkable range of adaptations that plants exhibit for surviving drought leaves few habitats where plants are not able to survive on whatever water is available. The rain may be sparse or highly intermittent or can be limited to the condensation of sea fog, but if it comes at all then there are likely to be some species that have the means of using this water and growing. The fact that plants can exploit this resource even when it is extremely limiting shows that evolution has worked towards an optimization of carbon gain for the water available. The principle of optimization with regard to water loss can be summarized as balancing the risks of water use against the benefits of carbon fixation (Cowan, 1982). It is one of the marvels of adaptation and survival that plants have been able to find a positive solution to this problem in such a wide range of terrestrial habitats.

Fig. 9.1 *Veratrum lobelianum* (Liliaceae) a conspicuous and ungrazed herb of Alpine pasture which due to its alkaloid content is highly toxic to cattle. Extracts of rhizomes of the related *V. album* were used in past times as arrow-head poisons.

9

Surviving predation

9.1 Plant responses to predation

It is one of the marvels of nature that plants, while providing the original source of food for all animal and microbial life, are not themselves consumed to the point where they are are no longer able to support their predators. Much ecological research has been carried out on the mechanisms which determine the balance between success in maintaining stable population numbers, and failure due to over-exploitation of the habitat resources. Population studies reveal the host—predator relationship to be a complex self-regulating system where any species so imprudent as to extinguish its source of foods pays the inevitable price of suffering a collapse in the numbers of its own population. The ecological importance of effective population regulation mechanisms for long-term survival has been in the forefront of biological thinking ever since Darwin read Malthus's essay on the growth of population (Malthus, 1798). The threat of predation leaves its mark on the species under attack (Fig. 9.1) and this chapter is an examination, not of host—predator relationships (detailed accounts of this can be found elsewhere: Begon *et al.*, 1986; Crawley, 1983), but on the mechanisms that have evolved in plants in response to the effects of predation.

Predation by herbivores does not need to be viewed as a negative factor in plant survival. The plant—herbivore relationship has evolved because both have survived. Thus even with the minimal productivity of the Arctic, grazing can be ultimately to the advantage of individual plants (*see* Chapter 3, p.68) in that it will accelerate the return of nutrients to the soil. It can also alter the balance between species and some plants although grazed may benefit as they suffer less damage than their competitors. *Rumex obtusifolius* is the regular host of the dock beetle *Gastrophysa viridula* and is the preferred host plant providing a reservoir of beetles that can attack the potentially competing dock species *R. crispus* (Bentley and Whittaker, 1979). Grazing by the beetles reduces the root to shoot ratio in the latter species but increases it in the former. In some habitats the absence of *R. crispus* from sites where *R. obtusifolius* is heavily grazed may be due to the effects of beetle grazing because the least preferred species *R. crispus*

suffers the greatest damage from grazing. Thus, the more heavily-grazed species paradoxically benefits from the predation of the dock bettle as its competitor (*R. crispus*) is more sensitive to grazing. The hypothesis that grazers can maximize plant fitness has also to be considered in relation to the reproductive strategy of the plants, i.e. whether they are r-strategists or K-strategists as well as the stage in succession. Population models suggest that in the early stages of succession, r-strategists stand to benefit most from the rapid return of nutrients that can come from grazing (Stenseth, 1978).

Alternatively some authors prefer to classify plants in relation to the probability that they will be found by grazing animals. On this basis plants are described as unapparent (cryptic) or apparent (Feeny, 1975). Unapparent species are unpredictable in their occurrence and may exploit a number of habitats which are patchily distributed or else shift constantly as opportunities for colonization arise (as with r-strategists). By contrast, apparent plants are predictable in their occurrence in space and time and may be abundant in certain types of vegetation such as dominant grasses in pastures. Herbivorous organisms will have played a role in the evolution of such species and as a result such apparent plants may be expected to invest heavily in metabolic protection through substances that are weak in effect, quantitatively dose-dependent and relatively non-specific, for example polyphenols and fibre (Feeny, 1976). These substances will tend to reduce grazing by a wide variety of herbivores but the effectiveness will be limited so that the apparent species are exposed to relatively constant grazing of low intensity. On the other hand, cryptic plants are likely to invest relatively small amounts of resources in the synthesis of specific and potent secondary metabolites (for example alkaloids and toxic amino acids) which give a highly effective protection against particular herbivores most likely to be potential attackers (MacNaughton, 1983).

More recently Coley and co-workers (1985) have presented a more complete ecological approach to rationalizing the different methods used by plants to control herbivory. The Coley hypothesis links resource availability and growth rate of the plants with the degree of investment that plants devote to herbivore protection. When resources are limited plants with inherently slow rates are favoured (Montgomery effect, *see* p.166; Montgomery, 1912). Slow rates of growth will in turn necessitate a longer utilizable life for individual plant organs and hence favour large investment in herbivore defence. Slow turnover of plant parts in areas of low nutrient supply has already been discussed in Chapter 2 (*see* p.38). As a large proportion of the available nitrogen and phosphorus pool is lost when a plant part is shed there will be a selective advantage in areas of low nutrient supply for plants with long-lived organs. With increased leaf lifetime there will also be an advantage in defences with lower turnover rates as these minimize metabolic costs. By contrast, resource-rich environments favour plants with rapid growth rates (Grime, 1979) and such plants exhibit characteristic properties which facilitate the maintenance of high light-saturated photosynthetic rates, with the shedding of older shaded leaves and the rapid cycling and uptake of nutrients. Observations in a variety of plant communities show that herbivores prefer feeding on fast-growing species (Coley *et al.*, 1985). In boreal habitats it is the species which colonize recently disturbed areas, as along river banks where insect attack is greatest, rather than the slow-growing species of adjacent resource-limited sites (MacLean and Jensen, 1985). Similarly in temperate habitats fast-growing species from fertile

soils are preferred by snails (Grime, 1979). Table 9.1 illustrates the dichotomy between slow growers and fast growers in relation to their response to herbivore attack. A formal mathematical representation of this dichotomy is shown in Fig. 9.2. The Coley hypothesis assumes that in a world without herbivores maximum potential growth rates will be determined by resource availability. The model therefore supposes that when herbivores are present they remove a fixed amount rather than a fixed percentage of the plant mass, i.e. it is related to herbivore biomass and not plant biomass. Any plant that invests in defences reduces its losses to herbivores. The resultant growth rate is therefore the balance between growth reduction due to expenditure of resources on defence costs and better growth rate due to less herbivory. The family of curves (Fig. 9.2) illustrates that as inherent growth rate decreases (from upper to lower curves) the optimal level of defence increases and the actual level of herbivory decreases.

This model is original in that it makes two novel predictions concerning herbivory:

1 Herbivore damage will be less in slow-growing species.
2 Slow-growing species will benefit most from increased resource allocation to defence costs.

From the above model it is also possible to make

Variable	Fast-growing species	Slow-growing species
Growth characteristics		
Resource availability in preferred habitat	High	Low
Maximum plant growth rates	High	Low
Maximum photosynthetic rates	High	Low
Dark respiration rates	High	Low
Leaf protein content	High	Low
Responses to pulses in resources	Flexible	Inflexible
Leaf lifetimes	Short	Long
Successional status	Often early	Often late
Antiherbivore characteristics		
Rates of herbivory	High	Low
Amount of defence metabolites	Low	High
Type of defence (*sensu* Feeny)	Qualitative (alkaloids)	Quantitative (tannins)
Turnover rate of defence	High	Low
Flexibility of defence expression	More flexible	Less flexible

Table 9.1 Characteristics of inherently fast-growing and slow-growing plant species. (From Coley *et al.*, 1985)

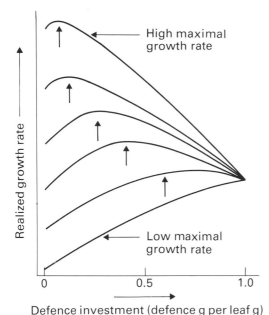

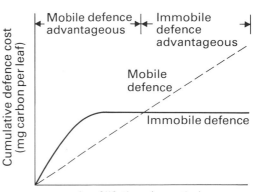

Fig. 9.3 The cumulative cost of defending a leaf with a large amount of an immobile defence with a low turnover rate as compared with a small amount of a mobile defence substance with a continuous and active turnover rate. (Redrawn from Coley *et al.*, 1985.)

Fig. 9.2 A mathematical model of the hypothetical relationship (*see* text) between defence investment and realized growth. (Redrawn from Coley *et al.*, 1985.)

certain predictions about the type of defence mechanisms most suited to slow- and fast-growing plants respectively. Defence compounds such as polyphenols and fibre (quantitative defences as defined by Feeny, 1976) are present in high concentrations and thus represent a high initial construction cost. Due to their metabolic inactivity their maintenance costs are low, but because of their immobility they will be retained in senescent leaves and will therefore be recycled only slowly. By contrast mobile defences such as alkaloids, phenolic and cyanogenic glycosides are not only more potent (qualitative defences as defined by Feeny) but also under constant metabolic turnover. This metabolic activity although increasing maintenance costs allows these secondary metabolites to be recovered from a leaf during senescence and utilized elsewhere. Fig. 9.3 illustrates a further dichotomy in behaviour in which the relative advantages of mobile or immobile defence substances differ in their degree of selective advantage depending on the expected life of the leaf. Further support for this hypothesis comes from studies on foliage palatability and longevity (Southwood *et al.*, 1986). These studies confirm the expectation of the Coley hypothesis that leaf palatability and herbivore damage are inversely related (Fig. 9.4).

Nitrogen-based compounds are probably the most expensive of the mobile defence agents and these too can be related to habitat preference and plant type, with the nitrogen-fixing legumes having a higher than average reliance on alkaloids. The following sections of this chapter examine these suggested patterns of behaviour in relation to the different defences used by specific examples of species from a range of plant communities and discusses their ecological advantages and disadvantages.

9.2 Host specificity

The greater the number of species sharing a habitat the more complex are the interactions and with increasing ecosystem complexity it is often assumed that this acts as a buffer promoting greater population stability. The ultimate in ecosystem complexity is reached in tropical rain forests, where not only the greatest number of plant species is to be found but also a predator community which is enormously rich in every kind of herbivorous organism from large, grazing mammals such as gorillas to leaf-cutting ants and termites.

The coexistence of plants and animals in this type of situation appears to be made possible by the astonishing degree of host specificity exhibited by the predator species, combined with territorial dispersion of the host plant. Thus when an animal feeding on a certain tree in a tropical rain forest seeks another tree of the same species it may have to explore a considerable area of forest before its search is rewarded. The restriction of immediate access to suitable food undoubtedly has an inhibitory effect in preventing explosions of animal populations in a world of seeming vegetable plenty. The control of populations by the dispersion of their food source has been elegantly demostrated in a recent study by Kareiva (1987) using different sizes of patches of golden rod (*Solidago canadensis*) and monitoring the effect of patch size on aphid populations. In the smaller patches there were more fre-

quent local explosions of local aphid populations. This unexpected result was explained in terms of the ladybirds that prey on the aphids as they appear to have more difficulty in locating their prey in the patchy environment. Similarly it can be expected that plants that have to be searched for will derive some protection from over-predation by their herbivores. Host specificity can therefore be construed as being of ecological benefit both to the herbivore, where it acts as a population constraint, and to the plant where it reduces the chances of extinction.

Host specificity is probably greatest in insects. Many species of aphid can be named unambiguously by referring to their principal plant host. Thus the willow aphid, the sycamore aphid, and the pea aphid are all insects which characteristically feed on these species. Others change their host with the season. The aphid *Macrosiphum euphorbiae* feeds on the leaves and buds of the rose (*Rosa odora*) in early spring but transfers its herbivory to potato in early summer (*see* MacNaughton, 1983). In a study of the feeding habits of seed-attacking beetles in a Costa Rican deciduous forest, Janzen (1980) showed that 75 per cent of the beetle species were each specific to a particular plant species. The specificity was also combined with a mutual exclusion behaviour among the beetle species for of the 100 plant species preyed on, 59 were attacked regularly by only one species of beetle. This degree of specialization however does not extend to crop plants. When new plants are introduced to an area through agriculture, it is the insects with polyphagous habits that are most likely to adapt to the new source of food. Thus in a study of the numbers of insects attacking crop plants in South Africa the majority were preyed on

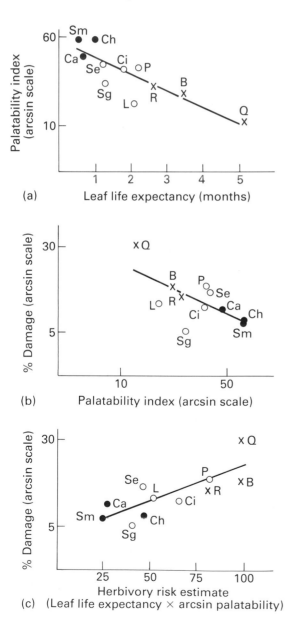

Fig. 9.4 Relationship of palatability to life expectancy and damage in plant species characteristic of secondary succession in southern England. (a) Leaf palatability to invertebrates (five species in bioassay) in relation to life expectancy. (b) Mean percentage of leaf area lost (damage) in relation to palatability. (c) Damage in relation to estimation of risk from herbivory (leaf life expectancy × arcsin palatability index). ●, early successional plants; O, mid-successional plants (old field); x, woodland plants. B, *Betula pendula*; Ca, *Capsella bursa pastoris*; Ch, *Chenopodium album*; Ci, *Cirsium arvense*; L, *Lotus corniculatus*; P, *Plantago lanceolata*; Q *Quercus robur*; R, *Rubus fruticosus* agg.; Se, *Senecio jacobea*; Sg, *Stellaria graminea*; Sm, *Stellaria media* (Redrawn from Southwood *et al.*, 1986.)

by a number of insects species that were polyphagous (Moran, 1983)(Fig. 9.5).

In herbivorous mammals the small number of plant species upon which many feed is very striking. The mountain gorillas of the Congo live in a plant world composed of thousands of species yet not more than 29 serve as food (Schaller, 1963). The koala bear is even more extreme in that it restricts its diet to a few species of a single genus, *Eucalyptus*. The ecological success of man is sometimes attributed in part to his omnivorous diet. However, with regard to his principal plant foods, he shows himself every bit as selective as the herbivores discussed above. Three cereals, rice, wheat, and maize, provide the main food supply for the human population in their respective areas. A fastidious palate may have survival advantages in ecosystems where host and predator live in equilibrium, but in modern man (currently enjoying a *temporary freedom* from natural constraints to growth) it threatens to destroy much of his natural environment. The limited number of plant species used by man has resulted in vast areas of the earth being given over to the monoculture of a very limited range of foodstuffs. These food plants, which are mostly annuals, occupy the habitat for only part of the year. They are less efficient as producers than the more complex natural ecosystems and there is therefore the constant danger of a reduction in productivity and a decline in soil fertility. Recent population growth in Africa and the over-exploitation of the land has made these areas all the more vulnerable during periods of drought, which not only bring famine but destroy what remains of the soil's potential for crop production. A selective diet now no longer serves as a survival mechanism. As a result of the invention of agriculture it operates in the opposite manner allowing astonishing population growth on monocultures of plants which will be ultimately unable to support the human population they have produced. It is not practical to suggest that we can change our eating habits after millenia of evolution, but it is instructive ecologically to examine the plant kingdom and see how this behaviour has evolved and the enormous variety of mechanisms which plants employ in their protection against the constant attacks of their predators.

Modern advances in forestry are also tending to increase the area covered by monocultures. Genetic selection of forest trees is expanding the use of cloned material from genetically improved stock. Although some precautions are being taken to use mixtures of cloned stocks the genetic heterogeneity

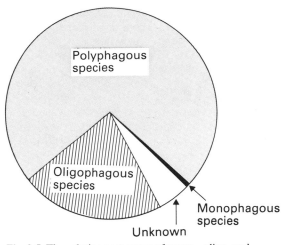

Fig. 9.5 The relative pest status of mono-, oligo- and polyphagous species of insects and mites on cultivated plants in South Africa. (Redrawn from Moran, 1983.)

of some recently-planted forests is undoubtedly being reduced. Thus both in agriculture and forestry human interference is likely to affect the plant herbivore relationship that has evolved in the past.

9.3 Secondary metabolites and plant palatability

However distinct the feeding habits of different herbivore species their qualitative nutritional demands in terms of amino acids and energy producing sugars are all remarkably similar. Some herbivores require these foodstuffs to be present in a concentrated form and feed on nutritionally-rich foods, such as seeds and fruits. Others have limited powers of digestion and live on sap and nectar, while species such as leaf-eating insects and ruminants can cope with larger quantities of roughage and modify its nutritional value by microbial fermentation. The limiting nutritional factor for many herbivores is the nitrogen content of the vegetation (Mattson, 1980). The nitrogen content of plant tissues can vary from 0.03 to 7.0 per cent dry weight with the highest concentrations being found in young growing tissues, and storage organs such as seeds. The lowest amounts for any living organism are in the sapwood and heartwood of trees where an average figure would be in the region of 0.2 per cent dry weight. Thus herbivores that are adapted to live on the lower nitrogen sources have low growth rates and long life cycles. The death watch beetle *Xestobium rufovillosum* has been known to live for 10 years on nearly sound dry wood. Species have therefore an *a priori* limitation on which tissues they can feed in that they have to have an adequate supply of nitrogen for their growth. It follows also that the plants richest in nitrogen will be the most likely to suffer the severest attacks, not just because they are a rich source of food, but because they can support species with a capacity for rapid growth and short life cycles.

Although the above limitations will affect the type of attack that a plant may sustain it is not the primary metabolites, sugars, proteins, etc. which affect palatability and selectivity in herbivores but the so-called 'secondary metabolites'. The term secondary metabolite is a blanket designation used to describe all substances that are not primary products of metabolism. They are the secondary elaborations of the primary metabolites and although they may not be necessary for the functioning of cell metabolism they are often essential for the continued survival of the whole organism. They are therefore by definition a very miscellaneous group of compounds. Some serve as growth regulators, while others have been variously regarded in the past as waste or storage products, or the unavoidable imperfections in the regulation of metabolism. In recent years however (for review, *see* Harborne, 1982) an even greater number of these compounds have been shown to have powerful effects as biological protectants and it is in this latter role that they are being examined in this chapter. Many secondary metabolites have marked toxic properties as in the alkaloids and cyanogenic compounds, while others react on the taste organs of predatory animals

causing sensations of bitterness and astringency. The particular effect of each compound depends not only on the amount ingested but on the other substances that are also present in the plant and which may modify the total effect on herbage palatability. To consider the role of secondary metabolites by their mode of protection as rationalized by Feeny and Colley (*see* p.205) would give only an overall ecological perspective of herbivore defence. The following account however is based on chemical affinities in order to give as complete an overview as possible to the range of mechanisms developed by plants to reduce predation. Although small alterations in chemical structure can change compounds from being strongly poisonous to completely harmless, this grouping on the basis of chemistry presents the most orderly arrangement possible in this diverse group of substances.

9.3.1 ALKALOIDS

The ecological consequence of alkaloid production by higher plants presents one of the most intricate examples of the extent to which co-evolutionary selection can operate in order to secure a protected ecological niche for particular species. Alkaloid production may protect a plant from grazing and the attack of certain insects, but for insects that have evolved an immunity to alkaloid poisoning, the alkaloid plant is a source of food and a breeding ground which carries a minimal risk of disturbance. The sequestration of plant alkaloids in the insect body can even transmit the metabolic protectant of the plant to the phytophagous insect and thus reduce its attractiveness as food to predating birds (Roths-

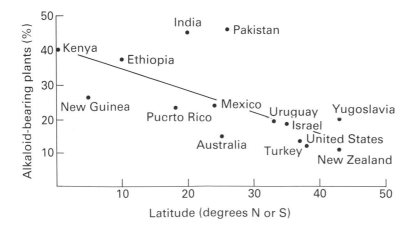

Fig. 9.6 Percentage of alkaloid-bearing plants in floras of 14 countries as a function of latitude. (Redrawn from Levin, 1976.)

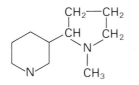

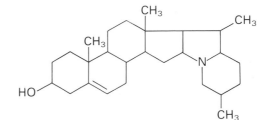

Nicotine from *Nicotiana tabacum*

Solanidine from potato

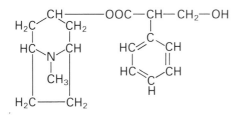

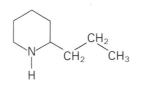

Atropine from *Atropa belladonna*
(Solonaceae)

Conine from *Conium maculatum*
(Umbelliferae)

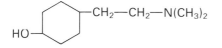

Hordenine produced in barley roots

Ephedrine from *Ephedra sinica*

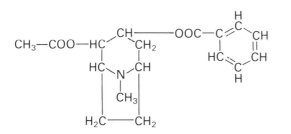

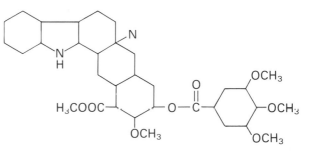

Cocaine from *Erythroxylum coca*

Reserpine from *Rauvolfia tetraphylla*

Fig. 9.7 Examples of different types of plant alkaloids and their plants of origin.

child, 1972). Alkaloids are commonest in the angiosperms and are found in approximately 20 per cent of species. They also occur in some gymnosperms such as yew (*Taxus baccata*). The incidence of alkaloid-bearing plants can be related both to plant habit and ecogeographical distribution. In the North American flora the occurrence of alkaloids in annual species is nearly twice that of perennials and among

tropical floras nearly twice that of temperate floras (Levin, 1976)(Fig. 9.6). Systematically the more primitive orders, Magnoliales and Ranales, have a higher percentage of alkaloid species than the remainder of the dicotyledons.

The basic unit of the alkaloid molecule is a heterocyclic ring containing nitrogen (Fig. 9.7). The nitrogen confers the base properties on the molecule and hence the name alkaloid. The term therefore describes a heterogenous collection of organic bases from which it is usual to exclude purine and pyrimidine. Chemically, the alkaloids can be divided on the nuclear structure of their bases. These various bases occur as the esters of amino alcohols and can be mono-esters, di-esters or cyclic di-esters. The mono-esters are usually the least toxic and cyclic di-esters the most toxic. Some alkaloids which have no double bond in their base structure as in the pyrrolizidine ring of platyphylline are not toxic.

Such is the potency of alkaloids as toxins that it has been suggested that the disappearance of the large dinosaurs at the end of the mesozoic period some 70 million years ago might be due to the evolution of higher plants containing alkaloids (Swain, 1977). Dinosaurs are now considered to have included warm-blooded species which should have been able to adapt to the cooler conditions at the end of the mesozoic period and therefore the climatic deterioration should not have caused their total extinction. Their abrupt exit from the world after 135 million years of undisputed hegemony has never been satisfactorily explained. The latter half of the mesozoic period did witness the most profound change that has ever taken place in the land flora of the earth, with the evolution of flowering plants and with them the extensive production of alkaloids. A large dinosaur must be presumed to have been a consumer of considerable quantities of vegetation. If a 5 ton dinosaur consumed 200 kg (450 lb) of herbage per day then it was probably not a selective grazer and with the evolution of many alkaloid-producing species could well have been exposed to plant poisoning (Swain, 1977).

Modern herbivores vary in their sensitivity to different alkaloids. The alkaloids in the genus *Senecio* are readily toxic to cattle. Man is easily poisoned by the deadly nightshade (*Atropa belladonna*)(Fig. 9.7) yet this same plant can be consumed in quantity by rabbits and pigs as they possess a metabolism that can detoxify these particular alkaloids (*see* p.213). Many insects have not only the ability to feed on alkaloid-containing plants, but as already mentioned can actually sequester these compounds in their own bodies to act as a deterrent to insect-feeding birds. Whether or not alkaloids prove poisonous or can be detoxified or even sequestered depends not only on the enzymatic armoury of the consuming species but also on the chemistry of the alkaloid itself (Fig. 9.8).

Alkaloid detoxification can be commonly affected by hydrolysis and N-oxidation and both these methods

are used by sheep and rats. In addition sheep are capable of demethylating alkaloids in their rumen which also renders them less toxic (Fig. 9.8). The distinction between alkaloid-sensitive and alkaloid-resistant species is heightened further by the toxicity enhancement that takes place on ingestion in some animals. Dehydrogenation of pyrrolizidine alkaloids results in the production of pyrrole derivatives which are much more reactive chemically than the original alkaloids (Mattocks, 1972). It is thought that it is the toxic properties of the dehydrogenated derivatives that are responsible for all the toxic symptoms associated with the ingestion of the original plant alkaloids.

The extraordinary degree of niche specialization in insects that can arise in relation to alkaloids is seen in the high degree of habitat dependence in certain fruit fly species that attack the rotting limbs of cactus plants. The senita cactus (*Lophocereus schottii*) is toxic to most species of *Drosophila* due to its alkaloid content yet one species *D. pachea* is able to breed in old stems of the plant (Kircher and Heed, 1970)(Fig. 9.9). This species of *Drosophila* is not only resistant to the specific alkaloids of this cactus (lophocereine and pilocereine) but has become dependent on a steroid produced by the cactus (schottenol) as a precursor of the moulting hormone (ecdysone). Another cactus of the same region, the suguaro, has a different alkaloid (carnegeine) and does not produce the steroid schottenol and is the preferred habitat of another species of fruit fly *D. nigrospiracula*. This example is a most striking instance of the extreme dependency of the fruit flies on specific hosts and the danger they would be in if a particular host species should disappear. The advantages of this niche specialization must outweigh the disadvantages, otherwise such a situation would not arise. Specific favouring of sites from which potential competitors are excluded is also found in aphid attacks on broom (*Sarothamnus scoparius*) which is actually stimulated by the alkaloid sparteine. The aphids adjust their feeding habits to follow low gradients of sparteine in the plant. Painting sparteine onto peas can even induce unusual aphid attacks (Smith, 1966).

As illustrated by these examples the toxicity of the varying alkaloids plays an important role in ecological relationships of herbivores. Alkaloids such as nicotine and ryanodine are particularly effective as insecticides. In tobacco possibly 10 per cent of the plant's carbon metabolism can be devoted to nicotine biosynthesis (Robinson, 1974). Nicotine, although it can be ingested safely by some insects, is apparently

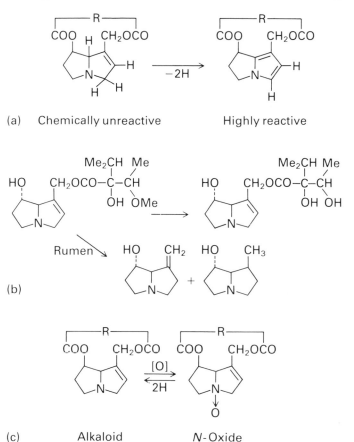

Fig. 9.8 Examples of alkaloid toxification and detoxification processes; (a) dehydrogenation of an unsaturated pyrrolizidine alkaloid to a dihydropyrrolizine ('pyrrole') derivative, (b) demethylation of heliotrine by sheep and conversion to non-toxic bases, (c) *N*-oxidation of a pyrrolozidine alkaloid. (Adapted from Mattocks, 1972.)

too toxic to be sequestered in the insect body. Insects which feed on tobacco immediately secrete the nicotine or else detoxify the alkaloid. To understand the different response of predators to plant alkaloids we need to examine the manner in which they exert their toxic properties. A very large number of alkaloids react on the central nervous system and this they appear to do directly without any modification of the ingested alkaloid molecule. Thus eserine (sometimes called physostigmine) the poisonous principal of the Calabar bean (*Physostigma venenosum*), the notorious West African ordeal bean, has its action in inhibiting acetyl choline esterase and thus interfering with the transmission of nerve stimuli. The milder effects on the nervous system of caffeine, nicotine, and theobromine are known to all. A number of alkaloids however produce their toxic effects by causing severe tissue damage particularly to the liver and lungs. Examples of alkaloids causing this type of poisoning can be found in the pyrrolizidine group which are so characteristic of

the genus *Senecio* (Fig. 9.8). These alkaloids in their native state in the plant are not reactive chemically but on ingestion can undergo reactions which can either detoxify them completely, or else transform them into compounds which are both chemically reactive and highly toxic (Mattocks, 1972). In this dichotomy between detoxification and toxicity enhancement, lies the explanation of why some insects can sequester the *Senecio* alkaloids while other animals find them highly toxic.

Insects do not however gain protection from predation merely by being able to sequester alkaloids. It is an interesting observation that insects which are poisonous to birds are also aposematic (warningly coloured). Protection appears to be conferred on the insect only when it carries some visual signal whereby its predators can recognize it as poisonous (Rothschild, 1972). The *Senecio*-feeding cinnibar moth is blue and birds learn to associate this colour with the chemical deterrent. It is therefore highly probable that in higher plants alkaloid possession on

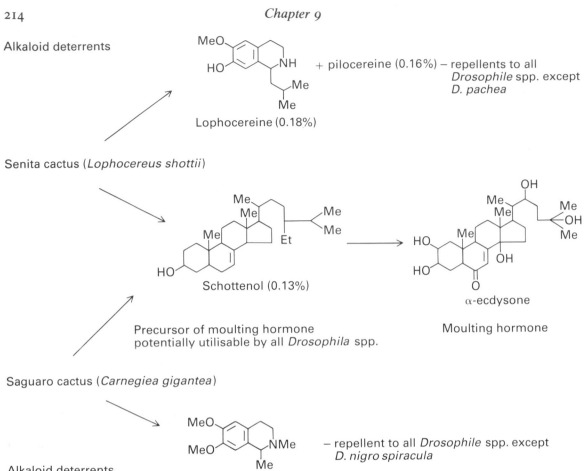

Alkaloid deterrents

Lophocereine (0.18%)

+ pilocereine (0.16%) – repellents to all
Drosophile spp. except
D. pachea

Senita cactus (*Lophocereus shottii*)

Schottenol (0.13%)

α-ecdysone

Precursor of moulting hormone
potentially utilisable by all *Drosophila* spp.

Moulting hormone

Saguaro cactus (*Carnegiea gigantea*)

Alkaloid deterrents

Carnegeine

– repellent to all *Drosophile* spp. except
D. nigro spiracula

Fig. 9.9 Combination of metabolic dependence of *Drosophila* spp. on the senita and suguaro cacti and niche specialization through the effects of alkaloid deterrents (Adapted from Kircher and Heed, 1970.)

its own is not a sufficient deterrent for a herbivore but is most probably associated with some disagreeable taste or odour which the animal learns to associate with unwholesome food. This phenomenon is also seen with other toxins. The accumulation of cardiac glycosides from milkweeds in monarch butterflies will deter bird predation due to the association of the colour of the insect with cardiac glycoside (*see* p.222). It is difficult for man to form any personal sensory judgement on the potency of taste and odour repellants to animals. The experimental addition of potentially repellant substances to herbage has little meaning as the effectiveness of these compounds depends on the particular cell condition under which they occur naturally. In this connection, cell pH and the enhancement effect of other metabolites have to be taken into account.

9.3.2 TOXIC AMINO ACIDS

Some of the most toxic compounds of vegetable origin are the soluble amino acids of unusual form that are produced mainly in seeds in certain plant groups, notably the Leguminosae. These amino acids are close analogues of the normal components of enzymes and structural proteins and it is therefore not surprising that when consumed by organisms which do not recognize them as such, they are incorporated in error into structures where they do not belong. Plants which produce these aberrant amino acids can recognize them as such and therefore do not incorporate them into their protein molecules. Aberrant amino acids are one of the subtlest of metabolic poisons; compounds to which the producing plant is completely immune but nevertheless are highly toxic to any species which consumes them.

A characteristic example of such a toxic amino acid is canavanine which can account for as much as 10 per cent of the dry weight of the seeds of the tropical vines *Dioclea* and *Canavalia*. Canavanine is a close analogue of arginine from which it differs by the replacement of a CH_2 group by oxygen (Fig. 9.10). Rats will starve to death rather than consume a diet that contains as little as 1 per cent of canavanine. This amino acid is not only toxic to mammals but inhibits growth in certain bacteria and is incorporated with inhibitory results into the protein of *Staphylococcus aureus*. Growth is also inhibited in fungi and algae and can even cause 50 per cent reduction in the growth of carrot tissue cultures at concentrations as low as 10 ppm (Bell, 1972).

The ability to synthesize these amino acids is largely confined to the Leguminosae. Even within the genus *Lathyrus*, different taxonomic subdivisions are characterized by distinct and rare amino acids (Bell, 1972). Fig. 9.10 shows the chemical similarities between some of these toxic amino acids and their normal counterparts. These similarities cause disruption of the normal enzymatic processes through false recognition of the different homologies. Thus a lower homologue of ornithine inhibits ornithine transcarbamylase of mammalian liver. Other groups have different toxic amino acids which act on the same principle of competitive inhibition of vital synthetic reactions involving the 'normal' amino acids.

The toxicity of these aberrant amino acids can be counteracted by increasing the intake of the relevant normal acid. As with most of these toxic situations there can be found examples of species that have adapted to the situation. Thus the larvae of the bruchid beetle *Caryedes brasiliensis* which in Costa Rica feeds on seeds of *Dioclea megacarpa* have evolved a resistance to canavanine. The resistance mechanism is two-fold; first there is no incorporation of canavanine into protein and secondly the insect has a high urease activity which detoxifies the canavanine (Fig. 9.11) and saves the nitrogen from the urea (Rosenthal *et al.*, 1977).

By far the most famous of these amino acid toxicities is the condition known as lathyrism which is caused by feeding on seeds of *Lathyrus sativus* and occasionally on the less common *L. cicera* and *L. clymenum*. Epidemics of lathyrism have been known in southern Europe since classical times. Typical symptoms include paralysis of the legs in man and the hind quarters of horses. If the diet is not corrected death is inevitable. Lathyrism was most commonly reported in southern Europe, Africa, and Asia. Hippocrates (*c*.460−*c*.375 BC) without understanding the root cause of the condition described a clear case of lathyrism where 'at Ainos, all men and women who continuously ate peas became impotent in the legs and the state perished'. This form of poisoning still occurs in the Indian sub-continent in years of famine. Famine doubly aggravates the condition as it increases the consumption of toxic seeds and reduces the intake of that portion of the diet which provides the antidote of normal amino acids.

The question has to be asked what is the ecological function of these amino acids and what factors bring about their selection. Even in such a pea-addicted society as the State of Ainos, man and his animals can only have eaten a small proprtion of the seed production of the toxic species of *Lathyrus* and the evolution of this property cannot be attributed to deterring human consumption. As with alkaloids, the major effects of non-protein amino acids are on insects. For example, the amino acid l-DOPA (3,4-dihydroxyphenylalanine) interferes with the synthesis of tyrosinase, an enzyme essential for the hardening and darkening of insect cuticle. *Mucuna* seeds can accumulate considerable concentrations of l-DOPA (6−9 per cent). When two

(a) $H_2NC(NH)NHOCH_2CH(NH_2)COOH$
Canavanine (toxic)

$H_2NC(NH)NHCH_2CH_2CH_2CH(NH_2)COOH$
Arginine (normal amino acid)

(b) $H_2NCH_2CH_2CH(NH_2)COOH$
α γ Diaminobutyric acid (toxic)

$H_2NCH_2CH_2CH_2CH(NH_2)COOH$
Ornithine (normal amino acid)

Fig. 9.10 Structural similarities between some toxic amino acids and their 'normal' metabolic counterparts.

$H_2CN(C=NH)NHOCH_2CH_2CH(NH_2)COOH$

$\longrightarrow H_2NOCH_2CH_2CH(NH_2)COOH +$

$H_2N(C=O)NH_2$

$\longrightarrow 2NH_3 + CO_2$

Fig. 9.11 Detoxification of the toxic amino acid canavanine and the scavenging of the nitrogen via urea to ammonia.

species of *Mucuna* grow side by side and one accumulates l–DOPA and the other does not, it has been noted that the seeds of the species that have the protectant are virtually free of attack from bruchid beetles while the species without chemical protection has its seeds heavily infested (Janzen, 1969).

From the above and other examples it is evident that the toxic amino acids in seeds preserve them from insect attack (Bell and Janzen, 1971) and probably also from microbial infection. In pulses, in common with other large seeded plants, there is a tendency for newly-imbibed seeds to leak soluble sugars and amino acids. If the leakage is severe, as seeds lie in water-saturated soil the efflux of soluble organic substances can promote microbial growth which may lead to failure to germinate due to rotting of the seeds. The possession of bacteriostatic and anti-fungal properties due to toxic amino acids could be of considerable survival value at this vulnerable stage of the life cycle.

9.3.3 TOXIC AMINES AND PEPTIDES

The number of higher plants containing toxic amines and peptides is not large compared with their occurrence in the fungi. The North American mistletoe (*Phoradendron flavescens*) contains toxic amines, phenylethylamine and tyrosine. The Mexican guajillo (*Acacia berladieri*) can similarly cause stock poisoning due to the presence of n-methyl-beta-phenylethylamine in its foliage. The toxic effect of these amines of simple structure is probably due to their confusion with other cell metabolites. However, the more complex amines and polypeptides which are common features of fungal metabolism, possess by virtue of their structure a variety of toxic effects which range from antibiotic activity and interference with hormonal regulation, to lysis of cell membranes. In the latter case the complexity of the polypeptide configuration confers on it a detergent-like quality, which produces effects similar to those encountered with the saponin toxins (*see* p.222). The peptides with the greatest toxicity are the fungal cyclic peptides of which amanitin and phallodin are two of the most notorious examples. Toxic peptides can however be found in higher plants. The European mistletoe (*Viscum album*) contains a cardiotoxic polypeptide viscotoxin. The fruit of the akee (*Blighia sapida*) contains toxic peptides, hypoglycin A and hypoglycin B which if eaten in improperly-cooked fruit cause vomiting, convulsions, coma, and death. The sensitivity of animal metabolism to peptides is evident from the large number of hormones which belong to this class of compound, for example insulin, glucagon, oxytoxin, and vasopressin. It is therefore a matter of surprise that there are not more cases of peptide toxicity caused by flowering plants. Histamine also plays a role in plant defence in that its injection into the skin along with acetyl choline by the stinging hairs of the nettle genus *Urtica* causes acute inflammation. Either compound on its own produces little reaction and it is the simultaneous exposure to both that produces the well-known burning sensation. (For review of mode of action of stinging hairs, etc. *see* Levin, 1973.)

9.3.4 PROTEINS AND PLANT PROTECTION

For nearly a century it has been known that the highly toxic nature of seeds of plants such as the castor oil plant (*Ricinus communis*) is due to their protein content (Stillmark, 1888). However it is only in the past 3 decades that the powerful role of proteins in protecting plants against microbial and viral infections has come to light. The two classic toxic plant proteins, ricin from *R. communis* and abrin from the precatory bean (*Abrus precatorius*) are now known to be particular cases of a phenomenon of much wider occurrence and ecological significance than just the capacity to poison animals. Ricin is extremely toxic to livestock and an amount equal to one ten-millionth of the animal's weight when injected is sufficient to cause death. Both abrin and ricin belong to the class of proteins to which the term lectin has been applied. These are proteins or glycoproteins of varying molecular weight, amino acid and carbohydrate composition which have a common property in being able to bind carbohydrate-containing molecules (Callow, 1975). When absorbed into animal tissues and localized at cell surfaces, plant lectins may produce toxic effects, cause agglutination of leucocytes or erythrocytes, or stimulate mitosis. They are also capable of binding with carbohydrates without producing any of these effects. Thus examination of the protein content of *Abrus precatorius* and *Ricinus communis* shows that the toxic and haemagglutinating properties belong to different lectins. There are toxic non-agglutinating lectins and non-toxic agglutinating lectins. Both classes of protein are lectins as they have the common property of being able to bind with carbohydrates. In their agglutinating properties lectins can be very specific and often react with only one particular blood group. This ability to pick out certain molecules has given them the name lectins (in Latin *legere* = to choose or select).

This property of selectivity in choice of binding site, seen in the artificial context of blood group selection, is important as it is the key to their natural function within the plant cell. The highly-developed degree of selectivity allows lectins to serve as recognition molecules which can play an important role in both herbivore—plant and host—parasite relationships. Cell surfaces possess a variety of carbohydrate-containing molecules, namely glycoproteins, glycolipids, and polysaccharides and to which the term glycocalyx has been applied. Lectins are of widespread occurrence in plants and are most readily demostrated in seeds, root surfaces, stems, leaves, and hairs (Callow, 1975). They play a wide role in recognition phenomena exhibited by plant tissues as in gamete recognition, breeding compatibility between pollen and stigma reactions, and resistance to microbial infection (Callow, 1975).

Fortunately not all toxic proteins are as potent as ricin and abrin. In cow pea and soybean there are proteins which are specific trypsin inhibitors. These are not toxic when fed to livestock although they will reduce the nutritional value of the foodstuff. Janzen *et al.* (1976) in a study of why bruchid beetles eat cow peas (*Vigna unguiculata*) and not black beans (*Phaseolus vulgaris*) showed that the cow peas are free of haemagglutin properties while the black peas are not. Both species contain trypsin inhibitors so these cannot play a role in deterring bruchid beetle attack although they may reduce the attractiveness of the seeds to other predators.

Toxic plant proteins cannot be left without mentioning bracken (*Pteridium aquilinum*). The fronds are highly poisonous while green and if cut when fresh and stacked remain toxic to cattle and horses. As well as being cyanogenic (*see* p.218) bracken contains another toxic agent, the enzyme thiaminase. This enzyme when ingested by livestock causes acute symptoms of vitamin B1 deficiency; the vitamin destroyed by this enzyme. If bracken is to be used for animal bedding it has to be cut late in the season when the fronds are dying when there will be no further danger of toxicity. Horses respond to treatment with vitamin B but treatment of cattle and sheep is usually unrewarding. This is thought to be due to the more complicated conditions arising from the microbial flora in the rumen of the latter animals. The ingestion of bracken does not cause an acute vitamin deficiency in ruminants as their intestimal flora does produce some vitamin B1. However, prolonged ingestion of bracken does cause digestive disorders, cessation of rumination, and eventually leads to damage to bone marrow and enteritis. The powerful effects of bracken consumption on the microbial flora of the rumen are highly suggestive of the possibility that in the bracken plant it will serve as a potent defence mechanism against pathogenic infection.

9.3.5 CYANOGENIC AND NITRO-GLYCOSIDES

Glycosides are conjugated molecules containing a sugar in combination with a non-sugar which is referred to as the aglycone. The sugar is frequently glucose but other sugars can take its place. Glycosidic compounds are widespread in plants and toxic glycosides are even more frequent than toxic alkaloids. The toxicity of any particular glycoside depends on the nature of the aglycone residue which is produced on hydrolysis of the original glycoside. For convenience these will be discussed under the four headings listed in Table 9.2.

Cyanogenic glycosides on hydrolysis liberate hydrogen cyanide. Cyanogenesis is a very widespread phenomenon in plants and over 1000 species have been reported as cyanogenic (Gibbs, 1974). It is curious to note the early confusion that concerned the ecological role of cyanogenic glycosides (Jones, 1972). Although their action in the protection of plants from snails was first suggested at the end of last century there was considerable scepticism of their antiphytophagous nature in view of the observation that certain cyanogenic plants such as *Prunus javica* actually appeared to attract leaf-eating parasites (Treub, 1896; *see* Jones, 1972). However, as already noted in relation to alkaloid production, the co-evolution of insects and plants has produced many complex situations. Over 60 families of flowering plants have species which contain cyanogenic glycosides. The protection afforded by the presence of these compounds is only partly effective in that it does not deter all phytophagous species. Many species of Mollusca are unable to feed on cyanogenic plants yet the common blue butterfly (*Polommatus icarus*) and the weevil (*Hypera plantaginis*) can feed on such plants with impunity. The are able to do so as they can detoxify the liberated hydrogen cyanide by means of the enzyme rhodanase (Parsons and Rothschild, 1964):

$$\text{rhodonase}$$
$$CN^- + S \rightarrow CNS^-$$

This reaction is very similar to the action of sodium thiosulphate the normal chemical antidote to cyanide poisoning:

Table 9.2 Secondary metabolites produced by plants which have a protective action in deterring grazing.

1 Alkaloids: found in approximately 20 per cent of higher plants and in many cases poisonous to animals which do not possess specially evolved detoxification mechanisms.

2 Amino acids: a number of soluble amino acids unique to plants are poisonous to animals as well as having bacterio-static properties. Some species also accumulate selenium in amino acids which makes them toxic to livestock.

3 Amines and peptides: certain rare peptides and amine groups occurring in plants are highly toxic to animals.

4 Proteins (lectins): certain plant protein fractions have powerful toxic effects on animal tissues as well as on fungal and viral infections. Particularly active are those proteins with the ability to combine with carbohydrates (the plant lectins).

5 Glycosides: the nature of the toxic reaction depends on the aglycone moiety of the molecule. These include:
    (a) cyanogenic and nitro-glycosides
    (b) coumarin glycosides
    (c) steroid and triterpenoid glycosides (cardiac glycosides and saponins)
    (d) irritant oils (glycosides which break down to liberate various irritant compounds)

6 Organic acids: a number of organic acids which do not play a part in the metabolic turnover of the plant cell act as protectants against predatory and pathogenic organisms.

7 Phenolic compounds: phenolic acids in the form of tannins are some of the most important compounds affecting herbage palatability and digestibility.

8 Volatile oils: various terpenoid compounds play a role in producing an odour or taste which can deter potential grazers.

9 Animal hormones: many plants produce compounds which are similar to animal hormones. These can interfere with the reproductive cycle of the herbivore and thus limit predation by restricting population increase or arresting development of the herbivore.

$$CN^- + Na_2S_2O_3 \rightarrow CNS^- + Na_2SO_3$$

In spite of the widespread occurrence in plants of cyanogenesis it does not appear to be as effective a defence mechanism as the possession of toxic soluble amino acids. The latter have an advantage over cyanogenic glycosides in that they are never toxic to the plant and are detoxified only rarely by predators. Cyanogenic glycosides on the other hand are readily detoxified by some insects and at the same time pose a constant threat of self-poisoning to the plant that possesses them. The most effective use of cyanogenic glycosides will be obtained when the ingested leaf contains both the glycoside and the glycosidase enzyme which when liberated by the attacking insect immediately releases hydrogen cyanide from the glycoside. The possibilities that exist are as follows:

Plant constitution	Phenotype
Cyanogenic glycoside and enzyme	Cyanogenic
Cyanogenic glycoside but no enzyme	Acyanogenic
Enzyme but no cyanogenic glycoside	Acyanogenic
No enzyme and no cyanogenic glycoside	Acyanogenic

This possession of both the glycoside and the hydrolysing enzyme however carries the risk that enzyme and substrate may be brought into contact with one another by conditions other than grazing. Bruising, freezing, and flooding are all environmental hazards that cause a breakdown in cellular compartmentation and result in the liberation of free cyanide if the plant contains both the glycoside and the hydrolysing enzyme. It is therefore not surprising that many species are polymorphic in relation to the possession of the glycoside and the hydrolysing glycosidase enzyme. In *Trifolium repens* and *Lotus corniculatus* both these properties are coded for by single dominant alleles which show no evidence of linkage. Cyanogenic activity in general decreases in areas where there is an increasing environmental risk of trampling, freezing, or flooding. Thus in ascending the mountains of New South Wales it has been shown that the genes for both glycoside and glycosidase in *Trifolium repens* are absent in areas where the July (winter) mean temperature is $-2°C$ (Jones, 1972). In areas where trampling is frequent along grass tracks the cyanogenic form of *Lotus corniculatus* becomes less frequent. In wetland sites *Lotus uliginosus* which is never cyanogenic. In oceanic climates there is greater all-year-round slug activity and not surprisingly in Britain, in areas where the January isotherm does not fall below 5°C, the frequency of cyanogenic forms of *L. corniculatus* is in the range 70–95 per cent compared with a central European figure of 25–50 per cent.

Bracken (*Pteridium aquilinum*) as well as having the toxic enzyme thiaminease (p.217) is normally cyanogenic. In a study of the annual course of cyanogenesis and phenolic content in a number of bracken populations Cooper-Driver and Swain (1976) noted that in most populations 96 per cent of the populations had both the glycoside (prunasin) and the corresponding hydrolase. In some populations however 95 per cent of the plants were acyanogenic, 85 per cent of the plants lacked the enzyme and substrate, and 15 per cent had the substrate but no enzyme. These acyanogenic populations were heavily grazed by deer and sheep. These results suggest that in the absence of the intense grazing pressure that normally exists in upland Britain from sheep and deer, there will be under natural conditions a selection in favour of acyanogenesis in bracken (Fig. 9.12). In general in the British Isles bracken is restricted in its habitat by frost and flooding, both of which are hazards for cyanogenic plants. Therefore although bracken has spread over many upland pastures it would be an even greater nuisance to hill farming if

it were not for the fact that the cyanogenic properties of the grazing-resistant populations limit its spread above the spring frost level and into the wetter pastures.

In tropical habitats cyanogenesis is very common and for many species it can be dispensed with only where some other form of defence is available. Many species of *Acacia* contain colonies of ants which live in hollow stems or thorns and obtain both food and shelter from their host. In a sense they are a form of secondary-metabolite defence in that they are sustained by the metabolism of the *Acacia* plant and attack any foreign object that invades the plant. Removal of the ant colonies causes a marked reduction in the life expectancy of the acacias. Individuals deprived of their ant colonies seldom live more than 3–12 months and never reach maturity (Janzen, 1975). These ant-inhabitated *Acacia* species are never

Fig. 9.12 Red deer in the Scottish Highlands. Bracken (*Pteridium aquilinum*) is mostly cyanogenic but when deer and sheep grazing is removed acyanogenic forms appear to be favoured (Cooper-Driver and Swain, 1976).

cyanogenic, whereas the non-ant species frequently have cyanogenesis as a means of chemical defence.

The evolution in plants of both the glycoside and the hydrolysing enzyme raises the question of the possible co-evolution of this defence mechanism along with detoxifying powers in snails (Dirzo and Harper, 1982). It is possible that possession of the glycoside on its own may have deterrent properties. The snail crop has suitable β glycosidase activity and this could possibly produce digestive disorders when feeding on glycoside containing plants. However Dirzo and Harper (1982) in a detailed study of the eating habits of snails on *Trifolium repens* demonstrated that possessing the glycoside on its own afforded no protection. The snail had to taste the glycoside to be deterred from grazing. On the basis of these observations they postulated the following pattern of evolution (Fig. 9.13). Originally snails would carry the glycosidase enzymes (related to food digestion). These animals would be common predators of initially acyanogenic plants. Mutation or recombination would lead to the production of cyanogenic glycosides which would give some protection against herbivory but this need not necessarily have been effective against slugs. In slugs such a food-acceptability barrier could have lead to the evolution of a detoxifying mechanism in the gut. This is suggested by the readiness with which snails eat plants that have the glycoside but not the enzyme.

The next step would be the evolution of plants with the glycosidase which would release cyanide as soon as the plant was grazed and by producing immediately an unpleasant taste provide an instantaneous and therefore more effective deterrent.

Similar in action to the cyanogenic glycosides are the nitroglycosides. Nitrite like cyanide is toxic to a wide range of potential herbivores. The genus *Astragalus* is noted for the accumulation of such glycosides which are toxic due to the release of nitrite on ingestion. The danger of this type of poisoning appears to be greatest in cattle.

9.3.6 COUMARIN GLYCOSIDES

Coumarins are formed from trans-cinnamic acid in a two-step process involving first an oxidation of the *o* position of the side chain and then the formation of a lactone ring. These two steps produce the parent molecule of a large number of compounds with powerful physiological and ecological activity (Fig. 9.14). Physiologically, coumarins are active as inhibitors of germination and cell elongation. Ecologically, they inhibit the growth of micro-organisms and may have a powerful effect in determining the grazing pattern of herbivores. To man coumarin smells sweet, but to sheep it is universally distasteful. Experimentation on this subject is difficult as addition of coumarins to herbage is unlikely to reproduce the

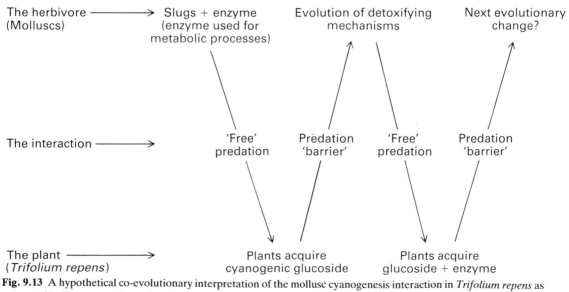

Fig. 9.13 A hypothetical co-evolutionary interpretation of the mollusc cyanogenesis interaction in *Trifolium repens* as suggested by Dirzo and Harper (1982).

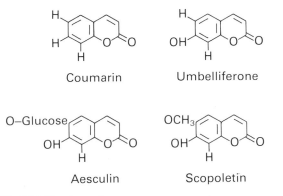

Fig. 9.14 Structure of some common coumarins.

effect of the naturally occurring compounds. It is possible, although as yet unproven, that selective grazing is influenced by coumarins. The sweet smell that man notices when walking on such characteristic coumarin-rich plant species as *Melilotus alba*, *Asperula odorata*, *Anthoxanthum odoratatum*, and *Hierochloe odorata* is due to the fact that crushing the leaves of these plants liberates a B glycoside hydrolase which liberates coumarinic acid. The liberated *o* coumarinic acid then spontaneously lactonizes to the sweet-smelling coumarin. There is some evidence that *Hierochloe odorata* differs from the other species in that it actually contains a free coumarin (Brown, 1981).

This unique distinctive property of *Hierochloe odorata* may be associated with its extensive use as an aromatic plant in churches during the mediaeval period. *Hierochloe odorata* is an extremely rare grass in Scotland and is found invariably on sites which had mediaeval monasteries. The name *Hierochloe* or 'Holy Grass' was given to this species as in some parts of Prussia it was dedicated to the Virgin Mary and strewn before the doors of churches on feast days, specifically for its sweet-smelling properties. It appears that its curious distribution in Scotland is due to its introduction to monastic houses during the mediaeval period.

Under certain conditions, such as when hay goes mouldy, coumarin can polymerize to produce dicoumarol which when consumed by animals reduces the level of prothrombin in the blood and can lead to death due to haemorrhage. Some coumarins are thought to be toxic in their native form. Thus aesculin from the genus *Aesculus* (horse chestnut or buckeye) and daphnin from the genus *Daphne* are suspected of being responsible for the toxic properties of these genera but other compounds may also be involved

(Kingsbury, 1964). The powerful effects of these compounds as physiological regulators and metabolic messengers are only now coming to light and further research may well reveal them to be of great importance as plant protectants.

9.3.7 STEROID AND TRITERPENOID GLYCOSIDES

A number of glycosides have a sterol group as their aglycone component. All such aglycones have the basic structure of sterane (Fig. 9.15). Among such steroid glycosides a number have long been noted for their stimulatory action on the heart and have therefore been termed cardiac glycosides. These cardiac glycosides can be divided into two groups; the cardenolides with 23C atoms and bufadienolides with 24C atoms. The latter, as their name implies, are found in the poisonous secretions of the warts of toads (*Bufo* spp.). They also occur in plants and can be found in the rhizome of the black hellebore (*Helleborus niger*). Over 400 cardiac glycosides are

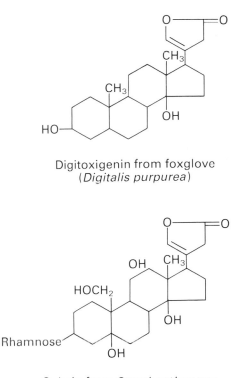

Fig. 9.15 The steroid cardiac glycosides digitoxigenin and oubain.

known in plants and most are found in the Scrophulariaceae, Liliaceae, and Apocynaceae (dogbane family). The most famous of all cardiac glycosides are those that form the toxic principal in the red foxglove (*Digitalis purpurea*). These belong to the cardenolide group with 23C atoms and of which digitoxigenin is but one example. The cardiac activity of the steroid glycosides is associated with the specific orientation of the unsaturated lactone ring (Fig. 9.15).

One of the most famous ecological case histories of the role of cardiac glycosides is in the relationship between the milkweeds (*Asclepias* spp.), the monarch butterfly, and the blue jay (Brower, 1969). The milkweed produces a number of cardiac glycosides which act as general deterrent to insect attack. The caterpillar of the monarch butterfly however has evolved not only to tolerate these glycosides but sequester them. By using the milkweeds as its preferred food the caterpillars have few competitors for this resource. The mature butterfly when it leaves the milkweed carries with it a sufficient concentration of cardiac glycoside to cause a blue jay, it if tries to feed on these insects, to spit them out. When presented with a second butterfly the jay will turn away in distaste as it has learnt to associate the bright blue colour of the butterfly with the bitter cardiac glycoside. In common with other toxins such as alkaloids and cyanogenic glycosides there is here a complex pattern of co-evolution as the predator needs to be immediately aware of the disagreeable nature of the food. In cyanogenesis this is done by the immediate hydrolysis of the glycoside. With alkaloids and cardiac glycosides this is achieved through the learning powers of the predator associating a bright colour with a distasteful food. This is a particularly powerful combination as it has been shown (Brower, 1969) that only 50 per cent of any butterfly population needs to carry the cardiac glycoside for the toxin to provide 100 per cent protection.

The triterpenoid glycosides are very similar to the steroid glycosides differing only in that their aglycone is related to the the triterpenoids rather than being based on a sterane. Their toxic function is also similar to that exhibited by the steroid glycosides that lack cardiac activity in that both act as surface agents. The steroid and triterpenoid glycosides are all large molecules which form colloidal solutions and produce a non-alkaline froth when shaken with water. In primitive societies such as the Indians of the Andean Highlands of Peru, the roots of plants containing such compounds are dug up and used for

their soap-like properties. A typical example is *Colletia spinosissima* which is still collected and used for washing in the remoter regions of Peru (*see* Fig. 8.13).

Both the triterpenoid and steroid compounds that exhibit this detergent-like quality are referred to as saponins. The aglycone moiety, which may or not be toxic on its own, is referred to as the sapogenin. The solubility of the complete glycoside and hence its biological activity is dependent on their sugar portion. Saponins are not readily absorbed into the blood stream and saponin poisoning in mammals usually depends on some injury occurring to the wall of the digestive tract which then permits their absorption. Once absorbed, the saponins react with the membrane-cholesterol and cause the disruption of cells and tissues. They are particularly active in causing the lysis of erythrocytes. The greatest toxicity of saponins is to fish where even in low concentrations they are capable of disrupting the cell membranes of the gill tissues. Fish poisoned in this way have not absorbed the saponin through their intestinal tract and are not toxic when eaten. Poisonous plants which contain saponins in the British flora include *Agrostemma githago* (the corn cockle), *Fagus sylvatica* (beech), and *Hedera helix* (ivy) which contains the saponin hederagenin (Fig. 9.16).

9.3.8 IRRITANT OILS

A number of plants contain glycosides which hydrolyse to yield as their aglycone components unstable and volatile oils which are often strong irritants. These irritants in the Ranunculaceae have long been used in folk medicine for raising hot blisters when applied as poultices. The genera *Anemone*, *Caltha*, *Trollius* (Fig. 9.17) and *Ranunculus* all contain in inocuous glycoside ranunculin which readily hydro-

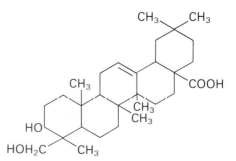

Fig. 9.16 The triterpene saponin hederagenin found in the leaves and berries of ivy (*Hedera helix*).

Fig. 9.17 Alpine meadow with the poisonous globe flower (*Trollius europaeus*). In common with other members of the Ranunculaceae this species contains the inocuous glycoside ranunculin which readily hydrolyses to form the vesicant (blister-inducing) protoanemonin (Fig. 9.18).

lyses to form a volatile and extremely irritant oil, protoanemonin (Fig. 9.18). This aglycone can cause poisoning of livestock but being unstable it is destroyed on drying and therefore hay containing buttercups is fortunately never poisonous on this account.

A similar situation occurs in the Cruciferae where mustard has long been used in the preparation of irritant poultices. Many cruciferous species contain the non-irritant sinigrin. In the presence of the enzyme myrosinase, this glycoside liberates allyl isothiocyanate (Fig. 9.19). This is a vesicant oil causing severe injury to livestock. However, its ecological importance does not stem from this property alone. The cruciferous oils, and presumably those of other families, play an important role in determining the attraction or repugnance of plants to aphids. The peach aphid (*Brevicoryne brassica*) is

stimulated to feed on plants containing this particular glycoside (van Emden, 1972). The feeding behaviour of aphids is however complex and can be modified by the sugar and amino acid content of the leaf. As plants mature the aphids can change from one host to another. As already noted (p.208) the aphid *Macrosiphum euphorbiae* feeds on rose leaves

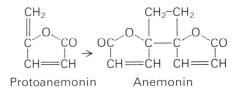

Fig. 9.18 Structure of protoanemonin and its dimeric product anemonin.

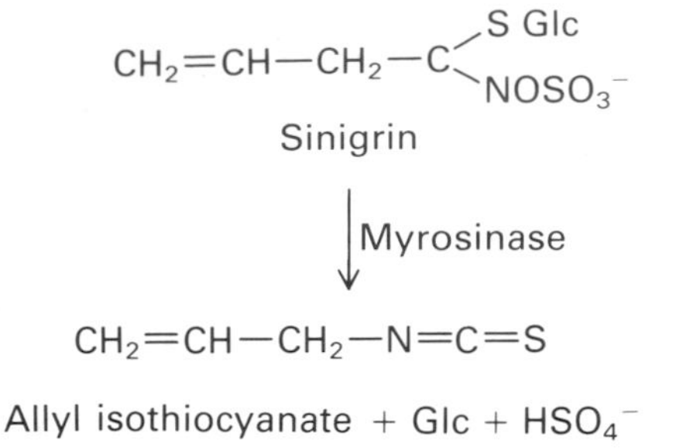

$$CH_2{=}CH{-}CH_2{-}C\!\!\begin{array}{l}S\ Glc\\[4pt]NOSO_3^-\end{array}$$

Sinigrin

Myrosinase ↓

$$CH_2{=}CH{-}CH_2{-}N{=}C{=}S$$

Allyl isothiocyanate + Glc + HSO_4^-

Fig. 9.19 Structure of the glycoside sinigrin which when acted on by myrosinase liberates the vesicant oil allyl isothiocyanate which can cause severe injury to livestock.

and buds in the spring and turns to potato later in the season.

9.3.9 ORGANIC ACIDS

A noticeable feature of the metabolism of plants is the high concentrations of certain soluble organic acids. In some species the accumulation may be so great as to increase the acidity of the expressed cell sap to pH 2–3. Apart from making tissues generally acid and the role of certain acids in the tricarboxylic cycle there is also a special protective role that certain acids play by virtue of their very high concentrations in specific tissues. Examples of such accumulations include tartaric acid in grapes, oxalic acid in sorrel and rhubarb leaves, isocitric acid in blackberries (*Rubus fruticosus*), and malic acid in apples. Malonic acid is a feature in many of the Leguminosae and quinic acid is prominent in species of *Vaccinium*. The high concentrations of these acids in fruits undoubtedly protect them from being eaten before the fruit is ripe and the seeds are ready for dispersal. Oxalic acid in leaves of sheep's sorrel (*Rumex acetosella*) can cause poisoning in sheep when eaten in quantity and rhubarb leaves (*Rheum rhaponticum*) can induce nausea, violent emesis, diarrhoea, and impaired clotting of the blood when eaten by man even in small quantities. The toxic properties of oxalate-containing plants appear to be highest when grown on calcium-deficient soils as this causes its accumulation as the free acid. On absorption by the animal the oxalic acid is precipitated as calcium oxalate which reduces the calcium content of the blood and gives rise to the symptoms noted above. The disagreeable nature of oxalate-containing plants to the palate of herbivores can be imagined when it is remembered the use made of *Dieffenbachia seguine*

(the dumb cane) in the punishment and torture of slaves. The intense irritation caused to the mucous membranes of the mouth by the oxalate crystals of this plant cause such severe discomfort that speech is impossible. Other plants which produce intense irritation on chewing and where oxalates may be the cause (this is disputed in some cases) include species of *Alocasia*, *Philodendron*, and *Symplocarpus foetidus* (the skunk cabbage). Careful study of the herbage preferences of sheep suggests that organic acids may play a role in affecting the palatability of many plants. High concentrations of malic acid when present in *Dactylis glomerata* discourage consumption, whereas the opposite effect is found with malonic acid. The effectiveness of any particular acid in deterring grazing depends not so much on the acid itself but on the species in which it occurs. Thus, there are conflicting reports on the role of shikimic and citric acids in determining the attractiveness of vegetation to sheep (Arnold and Hill, 1972).

No uncertainty however surrounds the toxic effects of monofluoroacetic acid which occurs in the South African plant *Dichapetalum cymosum*. This acid is highly toxic to livestock as it is a powerful inhibitor of the tricarboxylic acid cycle and when fluoroacetic acid enters the cycle instead of acetic acid inhibits the action of aconitase and causes the cycle to stop.

9.3.10 PHENOLIC ACIDS AND TANNINS

One of the most striking roles of phenol-based molecules in plant physiology is their synthesis within resistant plants in response to fungal infection. The resistance of certain plant species to pathogenic fungi has long been known to be associated with a hypersensitivity in which the necrosis of the infected cell successfully limits the invasion of the pathogen. The limitation of the pathogen has subsequently been traced to the production of specific metabolites, termed phytoalexins (warding-off compounds). These substances are produced on the death of the host cell and are highly toxic to the fungal invader. Careful experimentation in which droplets containing ineffective fungal spores were introduced into the seed cavity of leguminous pods has led to the isolation, crystallization, and subsequent identification of a number of phenolic-based compounds exhibiting phytoalexic behaviour. Orchinol (Fig. 9.20) is synthesized in tubers of the orchid *Orchis militaris* in response to fungal infection. Pisatin and phaseolin are slightly more complex molecules which are produced in resistant tissues of pea and bean respectively when

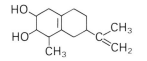

(a) Rishitin

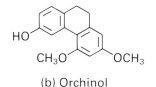

(b) Orchinol

$CH_3CH_2CH=CHC\equiv CCOC=CHCH=CCH=CHCOOR$

(c) Wyerone, R = CH_3; (d) Wyerone acid, R = H

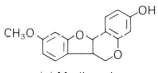

(e) Medicarpin

$CH_2OHCHOHCH=CHC\equiv CC\equiv CC\equiv CCH=CHCH_3$

(f) Safynol

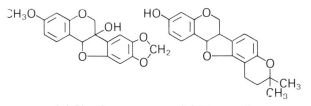

(g) Pisatin   (h) Phaseollin

Fig. 9.20 Phenolic compounds with phytoalexic behaviour. (From Deverall, 1977.)

phthora infestans infection in potato. Saynol, a poly-acetylene with phytoalexin properties has been isolated from safflower (*Carthamus tinctorius*) and a similar compound, wyerone, has been found associated with resistance of broad bean (*Vicia faba*) to infection by chocolate spot disease (*Botrytis cinerea*) (Deverall, 1977).

Plant extracts which cure leather are termed tannins. This property is due to their content of polymerized phenol derivatives which are able to precipitate protein. The ability to precipitate protein is governed

Hydrolysable tannins

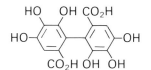

Hexahydroxydiphenic acid
(linked to glucose)

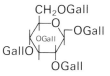

Pentagalloylglucose
(Gall = galloyl residue)

Condensed tannins

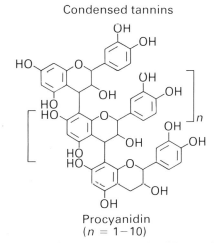

Procyanidin
($n = 1-10$)

Fig. 9.21 Tannins of oak showing the difference between hydrolysable and condensed tannins. (Reproduced with permission from Harborne, 1982).

exposed to infection with *Sclerotinia fruticola*. Ipo-meamarone is a substituted isocoumarin which is produced in sweet potato and carrot under similar conditions.

Other substances of differing chemical form are now joining the list of phytoalexins. One of the most important from an agricultural standpoint is rishitin, a terpenoid which can prevent the spread of *Phyto-*

by the molecular size of the phenolic polymer. The molecule must have sufficient phenolic groups to form links with either single protein chains or between two or more chains. If the molecule is too large it will not be able to orientate itself with the protein and linkage will not be achieved. For this reason most tannins have molecular weights between 500 and 3000. The tannin complement of plants can be broadly divided into two components, the hydrolysable and the condensed tannins. The former are esters of sugars and phenolic acids or their derivatives, and are easily hydrolysed chemically by acids and bases and biologically by enzymes. The two commonest forms are the gallotannins, which on hydrolysis yield gallic acid, and the ellagitannins which on hydrolysis give gallic and ellagic acids. The condensed tannins are however, the commonest in plants and contain catechins (flavan 3, diols) and leucoanthocyanins (flavan 3, 4 diols). Tannins are found in all plant tissues but the highest concentrations, particularly of condensed tannins (Fig. 9.21), are found in woody plants. Tannins afford protection to plants from herbivores in that their presence in high concentrations reduces palatability and therefore lessens the attacks of herbivores when other forms of herbage are available. Sheep grazing on the leguminous crop serica (*Lespedeza cuneata*), a crop much grown in North America, reduce their consumption of this plant by 70 per cent if the tannin content rises from 4 to 12 per cent (Arnold and Hill, 1972). Tannins act as soon as the plant material is ingested. Man is aware of the astringent properties of tannins when they precipitate the saliva proteins. The property of enzyme precipitation continues to act as herbage passes through the digestive system. Plant material rich in tannins is in consequence poorly digested by ruminants as well as other animals.

9.3.11 VOLATILE OILS

One further group of terpenoid compounds which can influence the extent of animal predation on plants is the volatile oils. In this class are found mono and sesquiterpenes and these include both open chain and cyclic compounds. Open chain monoterpenes include citral, geraniol, and linalol (Fig. 9.22), while cyclic examples are menthol, thymol, and cineole. Bicyclic examples of monoterpenes include camphor and pinene. The compounds are synthesized by special glands under the cuticle or in some cases in glandular hairs on the leaf surface. In peppermint (*Mentha piperata*) the oil droplets or

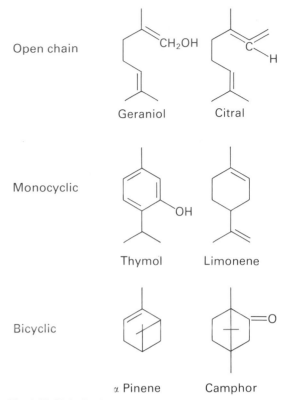

Fig. 9.22 Volatile oils commonly found in plants.

vacuoles can be observed in the cytoplasm which are then discharged into a space under the cuticle. These accumulations are large enough to be easily viewed under the light microscope. Whether the oils are secreted under the cuticle or into glandular hairs they are eventually liberated from the leaf and exert a number of varying functions, which include the inhibition of fungal and bacterial growth and signals to insects. Camphor has long been known for its ability to discourage the egg-laying of certain moths. The volatile oils in peppermint are also claimed to be the reason why geese are reluctant to feed on this plant. Although non-toxic the volatile oils probably play a very great role in determining the feeding habits of selective grazers whether it be large herbivores such as sheep, geese, or insect feeders. Volatile oils also play a role in the phenomenon of alleleopathy where their distillation onto the soil surface surrounding a plant in warm dry climates can inhibit the germination of competing species. The sagebrush (*Salvia leucophylla*) and other aromatic chaparral species such as *Artemisia californica* can

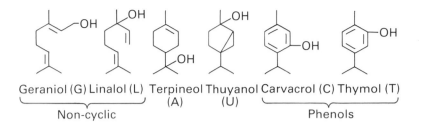

Fig. 9.23 Monoterpenes that have been related to seedling choice by grazing slugs (Gouyon *et al.*, 1983).

Geraniol (G) Linalol (L)   Terpineol (A)   Thuyanol (U)   Carvacrol (C) Thymol (T)

Non-cyclic   Phenols

have characteristic bare zones from 1 to 2 m in width around the base of the bushes. Volatile terpene toxins (Muller, 1966) such as camphor, pinene, and thujone appear to exert their alleleopathic effect by acting as germination inhibitors under the dry conditions of the Californian chaparral.

The interaction between grazing pressure and terpene content is however more complex than just a simple taste aversion to a particular compound. In wild thyme (*Thymus vulgaris*) the terpene composition varies from plant to plant and is goverened by six genetic loci. The structure of the six possible terpenes is shown in Fig. 9.23. In any one individual thyme plant only one of these terpenes will be found. When the palatability to slugs (*Agrioclimax reticulatus*) of four of these different terpene types of thyme was measured (Gouyon *et al.*, 1983) it was found that the order of preference was:

terpineol = carvacrol > thymol > thuyanol

The slugs appeared to distinguish the plants depending on the terpene they contained. For man to smell the terpenes it is necessary to crush the plants. However it was found in these experiments that the slugs were able to recognize some terpenes they did not like without tasting. Thus the plants which contained thuyanol (which has a peculiar smell to man) were less frequently killed because they were attacked least. By contrast thymol-containing plants could not be recognized without tasting by the slugs. The situation was further complicated by the observation that the slugs changed their grazing preferences after experiencing some of the chemotypes. Although further work is needed in order to determine exactly which genotypes will be favoured under particular circumstances, it is clear that there is here a curious polymorphism which divides and diversifies the attacks of the predators and in doing so apparently increases the survival chances of certain individuals. As with other polymorphic characters, the particular individuals which will be favoured must vary according to changing environmental circumstances. For the polymorphism to persist these environmental changes must take place relatively frequently.

9.4 Hormonal defences

One of the most astonishing discoveries of the last two decades on the ecological relationships between plants and animals has been the role played by plant-synthesized animal hormones in their defence against predation. Plants are capable of mimicing the male and female hormones of mammals, and the juvenile and moulting hormones of insects. An awareness of the hormonal activity in plant foodstuffs arose in the west during the Second World War when Dutch women, forced by famine to eat tulip bulbs, suffered upsets in their menstrual cycles. Other peoples had been aware previously of the potency of certain plants to affect human reproduction, as Thai women have been reported to use extracts of the leguminous tree *Pueraria mirifica* to cause abortions. The active substance miroestrol, isolated by Bounds and Pope (1960) from the roots of this tree, is very similar to the female hormone oestrone but with even greater activity (Fig. 9.24). Subterranean clover (*Trifolium subterraneum*) contains the isoflavonoids, genistein and formononetin (Fig. 9.25), which can reduce lambing percentages in sheep to less than 30 per cent (Bradbury and White, 1954). Isoflavone accumulation is a characteristic of the Leguminosae but other substances may play a similar role in other groups. However it is in the Leguminosae that we have the best-known examples of oestrogenic activity in plants. Similarly the isoflavones in pasture plants appear to affect the population control of quails (Leopold *et al.*, 1976). In years of

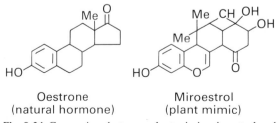

Oestrone (natural hormone)   Miroestrol (plant mimic)

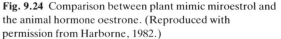

Fig. 9.24 Comparison between plant mimic miroestrol and the animal hormone oestrone. (Reproduced with permission from Harborne, 1982.)

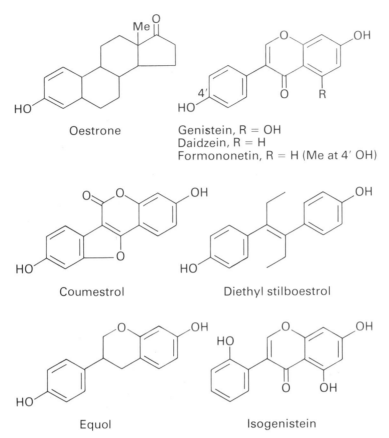

Oestrone

Genistein, R = OH
Daidzein, R = H
Formononetin, R = H (Me at 4' OH)

Coumestrol

Diethyl stilboestrol

Equol

Isogenistein

Fig. 9.25 Plant isoflavanoids which act as oestrogen mimics. (Based on Harborne, 1982.)

good rainfall and abundant food supply the legume species grow well and are low in isoflavone content. However in drought years, the isoflavone content rises and reduces the egg-laying capacity of the birds and consequently improves the survival chances of smaller broods. Whether or not a direct hormonal relationship between plants and animals has evolved as a result of grazing pressure is still a subject of debate. Possibly birds which respond to the increased isoflavone content of their food supply by reducing their fecundity are responding to an environmental signal which increases their chance of reproducing in years of low rainfall. Changes in isoflavone content may be just a consequence of water shortage in the plant to which the birds respond and thus anticipate the food shortage that will arise as a consequence of the drought. In this case there is no benefit to the plant in this relationship. Therefore the oestrogenic effects of isoflavones are only incidental and their production may serve for disease resistance as the isoflavone skeleton also forms the basic structure of the phytoalexins.

Although a direct hormonal relationship between

plants and mammals is still a matter of some controversy there is a much greater certainty on the role of insect hormones by plants as natural insecticides. Man has been aware of the existence of natural insecticides in plants for a considerable time as is evident from the extensive use that has been made of pyrethrum, an extract of chrysanthemum flowers (*Chrysanthemum cinerariifolium*) (Fig. 9.26) which has a high 'knock down effect' on a wide range of insects. The last two decades however have brought to light a hormonal control of insect populations that is brought about by plants producing substances similar to the juvenile and moulting hormones (ecdysones) (Fig. 9.27). The effect of plants on insect juvenile hormones came to light when it was found that the European bug *Pyrrhocoris apterus* failed to mature in Harvard laboratories when fed on American paper and persisted in moulting into forms intermediate between nymphal and adult stages. The same insect suffered no harm when grown on European paper (Slama and Williams, 1965). Subsequently it was shown that the source of the American paper, the balsam fir (*Abies balsamea*), contained a compound

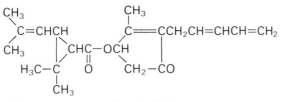

(a)   Pyrethrin I

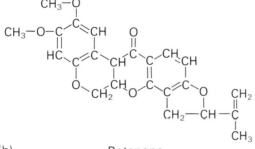

(b)   Rotenone

Fig. 9.26 Natural insecticides: (a) the monoterpene carboxylic acid ester pyrethrin I found in the flowers of pyrethrum (*Chrysanthemum cinerariifolium*), (b) rotenone the active constituent of derris, a powder obtained by grinding up the roots of certain tropical plants, mainly *Derris eliptica*. Both these substances have the advantage of being insecticidal but harmless to human beings.

with a very high juvenile hormone activity (Bowers *et al.*, 1966), while the European fir contained a much less active compound.

The juvenile hormone is necessary as well as the moulting hormone in the early stages of insect development while the moulting hormone has to be present at certain critical concentrations as the insect progresses through the various larval stages that lead to pupation. When either of these hormones is not present in the correct amount then the successful development of the imago stage of the insect is prevented. It was soon realized that this form of insect control would provide a means of producing insecticides that would be non-toxic to mammals and plants as they were specifically aimed against insect metamorphosis. The greatest concentration of these substances have been found in the gymnosperms and ferns. The common yew (*Taxus baccata*) contains as much β ecdysone in 25 g of dried leaf or root as half a ton of silkworms (*see* Harbone, 1982) as well as the highly toxic alkaloid taxine.

In addition to be the substances which act in a manner similar to the juvenile and moulting hormones (*see* Fig. 9.28) plants also have evolved anti-hormonal strategies to protect themselves from insects. A series of simple substituted chromenes (Fig. 9.29) have been isolated which act as anti-juvenile hormones (Bowers, 1984). These substances which have been named precocenes cause a cellular disruption of the corpora allata (the glands that secrete juvenile hormone). The precocenes undergo an oxidative activation into highly reactive

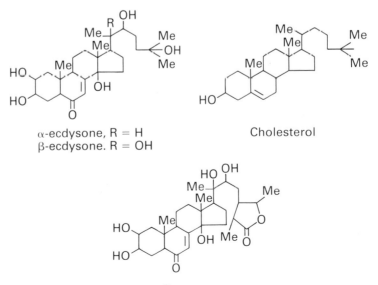

α-ecdysone, R = H
β-ecdysone. R = OH

Cholesterol

Cycasterone

Fig. 9.27 Structure of insect moulting hormones compared with cholesterol. (Reproduced with permission from Harborne, 1982.)

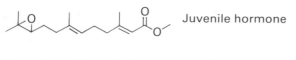

Juvenile hormone

Plant-derived juvenile hormone mimics

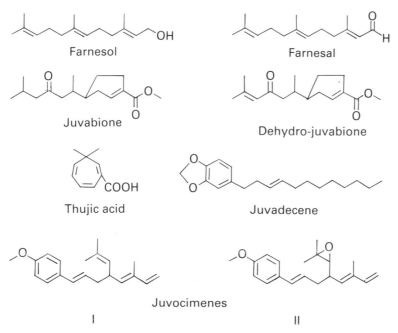

Farnesol

Farnesal

Juvabione

Dehydro-juvabione

Thujic acid

Juvadecene

Juvocimenes

I   II

Pro-allatotoxins

I   Precocenes   II

HMG CoA Reductase inhibitors

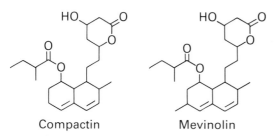

Compactin   Mevinolin

Fig. 9.28 Insect juvenile hormone and structural mimics found in plants. (Redrawn from Bowers, 1984.)

Fig. 9.29 Plant-produced anti-juvenile hormones which cause cellular disruption of targeted organs (the corpora allata) in the insect body. (Based on Bowers, 1984.)

alkylating reagents which destroy the corpus allatum of the insect (Bowers, 1982). As with our cases of hormonal interference in animals it is not certain whether the process of natural selection has given rise to this as a specific means of defence or whether it is a fortuitous consequence of evolution in response to other factors. Plants have evolved an enormous range of substances to protect themselves from predation and herbivorous insects have evolved enzyme systems which can degrade many of these substances. This in turn has led to plants developing substrates which on oxidation are transformed into alkylating agents capable of destroying the insect herbivore's protecting system which is usually in the form of oxidizing enzymes. The final result in this co-evolutionary sequence is the destruction of specific tissues such as the corpus allatum thus inhibiting development or alternatively damaging gut tissue which will reduce feeding. The oxidative activation of the precocenes is a similar strategy to that of toxic soluble amino acids in that the substance only becomes toxic on incorporation into the animal body.

9.5 Ecological consequences of predator protection

In the introduction to this chapter it was pointed out that some authors have argued that different ecological situations call forth distinct defence strategies.

Both growth rate (Coley *et al.*, 1985) and distribution (Feeny, 1976) have been related to the production of different chemical defence systems. On the one hand there are those plants which escape predation by the production of nitrogen-based compounds that are highly toxic and specific. These species can be found in the Gymnospermae, the class Magnoliopsida (Dicotyledonae) and some non-graminoid members of the class Liliopsida (Monocotyledonae). On the other hand there are those species that rely more on feeding deterrents rather than highly active toxins and these include families such as the Poaceae (MacNaughton, 1983). These species which are sometimes referred to as apparent plants (Feeny, 1976) may be selected through the impact of herbivores so that they can survive sustained grazing provided the pressure is not too great. In such cases non-specific deterrents which cause the herbivore not to graze too heavily on any one species may have a collective action in favouring survival. It is possible to show in model systems that in such cases the palatability of polymorphic species can vary depending on the intensity of grazing pressure as one morph is favoured against another.

The greatest degree of chemical defence against predation should be expected in those tissues in which the plant has placed the greatest investment of its resources, namely seeds and immature fruits. The toxins themselves can be grouped into two classes; (1) the digestibility reducing allelochemicals (DRA) and (2) the toxic alleleochemicals (TA). The latter which are largely nitrogen based, occur with greatest frequency in nitrogen-rich tissues. These are the highly effective, but resource-expensive deterrents that have to be used to protect resources of considerable value. The DRA substances are in a sense 'the poor plant's toxin'. In areas where nitrogen is severely limiting nitrogen-rich toxins would be an additional drain on a resource that is already in short supply. By contrast the plants of oligotrophic habitats such as bogs, heaths, gravels, and tropical leached sandy soils tend to use phenols, tannins, and terpenoids as non-specific grazing deterrents to ensure as long a life as possible for their leaves.

Some of these non-specific substances also have the advantage that their presence is easily reduced when it becomes ecologically advantageous for the plant tissue to be eaten, as when fruits ripen and are ready to have their seeds dispersed. When fruits ripen deterrents such as tannins and organic acids become converted to attractive foodstuffs for animals. The immature banana fruit (*Musa sapientum*) contains 17 per cent tannins by dry weight and is in con-

sequence relatively free from attack by chewing herbivores even although it is a rich source of starch (Janzen, 1983). In the wild there is a potential advantage for this change from non-edible to edible to be spread over a period of time so that the dispersing animals do not suffer a glut and leave the seeds to rot within the over-ripe fruit. Thus the numbers of animals available for the dispersal stage will have a bearing on the rate of maturation of the fruit. In cases where many animals disperse the fruit then there should be a rapid ripening over a short period. This is the case in many species of wild figs (Janzen, 1979) where one wild fig tree may mature millions of fruits over a 2–4 day period. The opposite situation can be found in tropical trees which mature a few fruits each day over several months and are visited daily by animals that know where the tree is to be found (McDiarmid *et al.*, 1977).

Seeds can be divided ecologically into whether they are chemically defended or not. Among the edible seeds are the fruits of the Gramineae and the tree species that show mast-flowering years. In this type of behaviour, survival depends largely on satiating the appetite of the potential herbivore. As an evolutionary strategy to gain dispersal at the cost of loss of seed it appears to be successful. The opposite strategy of less copious seed production but protecting the asset with considerable chemical defences has been illustrated in the plants of more productive sites where highly toxic alkaloids or soluble amino acids accumulate in the seed.

9.6 Conclusion

Whether or not a portion of a plant be it root, leaf, fruit, or seed is likely to be eaten by a predator is clearly not a matter of chance. Plants that allow themselves to be eaten have evolved so that even this seemingly negative event will maximize some aspect of their survival either by spreading their seeds, recyclng their nutrients, or prolonging their vegetative life more efficiently than that of their competitors. Like all aspects of environmental stress the advantages of any particular strategy are only relative and depend on the need to out-perform potential competitors. None of the defence mechanisms described in this chapter directly improves survival in terms of fitness to the physical environment. It only pays a tobacco plant to invest so much of its nitrogen resources in nicotine if by doing so it will be eaten less than its competitor. The defence strategies of plants in their struggle to retain their niche or dominate their habitat are only too similar

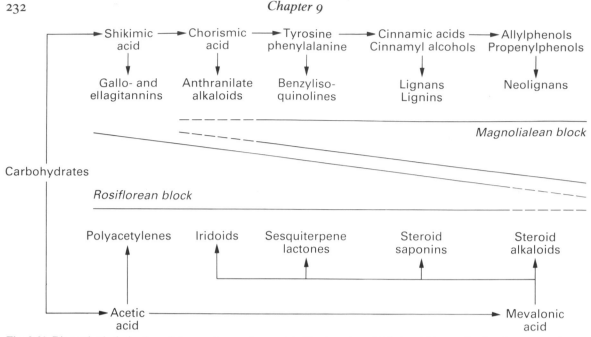

Fig. 9.30 Biosynthetic derivation of classes of secondary metabolites based on shikimic acid (upper line) and acetate and mevalonate (lower line). The products originating from the full expansion of the shikimate pathway characterize the more primitive members of the magnolialean block, whereas a stepwise contraction of this pathway is noted in the more derived members of both the magnolialean and rosiflorean block. In the most highly evolved members of the rosiflorean block the acetate pathway is elaborated to the exclusion of the shikimate-derived metabolites. (Redrawn from Kubitzki and Gottlieb, 1984.)

to the constant up-grading of defensive weapons for the defence of freedom or political domination among the nations of the world. Species that invest heavily in their offspring, whether they are plants or animals, the so-called K-strategists, cannot afford to risk the minimal defence strategy of the highly reproductive grasses and mast trees.

The perpetual need for biochemical defence against predation can be traced back through the entire evolutionary history of the angiosperms. It has been suggested from chemotaxonomic studies that any particular type of repellant will lose its effectivenes with time as insects and other predators evolve resistance mechanisms. The ecological advantage then will pass to other groups with new repellants and thus permit their evolutionary expansion. Thus the nitrogen-expensive isoquinoloid alkaloids of the Magnoliidae gave way to the tannins of the Hamamelidae, Rosidae, and Dilleniidae and these in their turn lost their supremacy to the iridoid compounds of the Asteridae (Cronquist, 1977). It has even been suggested that the largest evolutionary gap within the angiosperms is not the anatomically defined division between dicotyledons and monocotyledons

made by John Ray (1628–1705), but in the biochemical repellants that are used by the magnoliid, caryophyllid and monocotyledonous families (the magnolilean block) as opposed to the rest of the dicotyledons (the rosiflorean block)(Kubitzki and Gottlieb, 1984). Fig. 9.30 illustrates how these two blocks may be classified in relation to their reliance on the shikimate pathway for the synthesis of chemical repellants. The magnolialean block exploits principally compounds derived from the shikimate pathway while the rosiflorean block replaces this progressively with compounds derived from the acetate pathway. This view of evolutionary history, although speculative, does serve to emphasize the importance of biochemical defence in the angiosperms. The floral biology of the higher plants co-evolved with the insects and it is therefore to be expected that there was also a need for the development of a defence against insect predation. The expansion of the shikimate pathway at the beginning of the evolutionary history of the angiosperms (Kubitzki and Gottlieb, 1984) may have been just as important a feature as insect pollination in the evolutionary success of the angiosperms.

(a)

(b)

(c)

(d)

10
Surviving pollution

10.1 Introduction

Industrial societies release large quantities of substances into the environment that otherwise would not be there. Fortunately it is only rarely that these constant outpourings of industrial waste and energy result in the total extinction of plant life. Usually it is certain groups of species that are damaged, as in the lichen deserts that used to characterize our cities before steps were taken to reduce atmospheric sulphur dioxide content. Trees are particularly sensitive to the gaseous discharge from metal smelters and forests that are downwind from their smoke stacks can be completely killed. In the last 10 years the damage to forests in central Europe has again highlighted the sensitivity of trees to pollution. Forest damage from industrial pollution is not just a modern phenomenon; in the latter half of the nineteenth century German foresters were already recording damage to conifers which they attributed to coal smoke (Schröder and Reuss, 1883; *see also* Kandler, 1983 and Forest decline p.253). Fortunately, plant life is very resiliant and although some species may suffer, there are others that are either not affected or else can adapt and even evolve specialized resistant ecotypes in direct response to the particular type of pollution being imposed on the area. Consequently it is uncommon for pollution to cause a total vegetation kill. More usual are situations where the addition of pollutants produces no immediately obvious effects on plants or animals. In these cases there may only be a reduction in growth or small changes in species composition. The frequent lack of noticeable outward signs of injury with much pollution damage makes it essential to have careful ecological monitoring to ascertain if there is a long-term risk to life or a danger of upsetting the balance of sensitive ecosystems.

The extent of environmental pollution can be examined conveniently under the headings of atmosphere, water, and soil. The atmosphere differs from water and soil in that despite its large volume and constant circulation it appears to be particularly sensitive to pollution as seen in the steady and inexorable rise of its carbon dioxide content this century (Fig. 10.2). The atmosphere does not have dispersed within it an active micro-flora and fauna as found in water and soil and therefore lacks any power of biological homeostasis. Purification of the atmosphere is entirely physical and largely dependent on washing with rain and transferring the pollutants to the land and oceans. From the current rise in

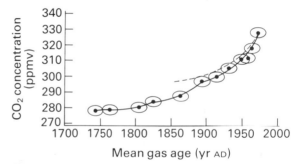

Fig. 10.2 Measured carbon dioxide concentration plotted against estimated gas age from polar ice-cores. The horizontal axis of the ellipses indicates the close-off time interval of 22 years. The dotted line represents the model-calculated back-extrapolation of the atmospheric carbon dioxide concentration assuming an input exclusively from fossil fuel. The difference between this and the observed values represents biomass burning and shows that processes such as deforestation were already contributing to carbon dioxide increase in the early nineteenth century. (Redrawn from Neftel *et al.*, 1985.)

Fig. 10.1 (*opposite*) Eighty years of change in a mountain forest in southern Germany; (a and b) detail and general view of the Gindelalm near Hausham (altitude 1242 m) from a postcard stamped 1900 showing an ageing and diseased forest canopy; (c and d) the same view on the 23 October, 1985. The forest was felled in 1905 and the already-established under-storey together with new seed established the present 80 to 100-year-old stand with a healthy closed canopy and good needle retention. Note that despite the increased use of the region by motor traffic the more recent picture shows an improved ecological condition. The forest is healthier and the mountain tracks show less erosion. (Photograph by courtesy of Professor O. Kandler, Munich.)

atmospheric carbon dioxide concentration, the re-establishment of the atmosphere-ocean carbon dioxide equilibrium is proving to be a slower process than had hitherto been anticipated. In both water and soil, biodegradable pollutants can be acted upon by micro-organisms and provided the ecosystem is not overloaded, a certain degree of pollution can be rectified due to the property of homeostasis. Thus in former times it was possible to discharge sewage into the upper levels of Lake Constance (and many other lakes) and at the same time extract drinking water from a depth of 40 m. Now however since the homeostatic (i.e. self-regulating) properties of the lake have been overtaxed, the entire ecosystem has become polluted and it is no longer a source of drinking water unless treated. Nevertheless, polluted lakes are not sterile and if the pollution is removed it is often surprising how quickly they can recover. Lake Washington in Seattle made a remarkable recovery within 6 years of the Seattle sewage being diverted elsewhere (Edmondson, 1970).

Some forms of atmospheric pollution will disappear rapidly even without biological action if not constantly replenished. The ozone produced in photochemical smog will disappear overnight if not regenerated by the next day's motoring. Acid rain could be reduced if the sulphur dioxide and nitric oxide emissions from power stations and internal combustion engines were reduced. Unfortunately however, not all forms of pollution will disappear so readily, even when steps are eventually taken to control the original source. The lead that has been added to the terrestrial ecosystem in recent years (*see* Fig. 10.28) is likely to remain there for centuries. Similarly, the oceans have so far been able to absorb only half the carbon dioxide that has been added to our atmosphere this century. Consequently, even if we were able to stop augmenting the carbon dioxide in the atmosphere (which is extremely unlikely) the present level would probably not fall for some considerable time.

Although the problems of pollution of the atmosphere and those of water and soil are linked, as in the action of acid rain and forest decline (*see* p.254) their mode of entry to the plant differs and it is therefore convenient to examine them separately. The detoxification mechanisms of leaves and roots are quite distinct as are the evolutionary responses of plants to soil and atmospheric pollutants. The fact that vegetation can adapt to pollution either by changing species composition or by evolving resistant ecotypes, does not obviate the necessity for careful monitoring of pollutant levels in plants. These pollutants may not be as lethal to plants as to animals.

The latter with their position at the top of the food chain can accumulate pollutant concentrations which may bring about premature death or sterility. With vegetation it is the insidious nature of pollution in producing sub-toxic effects reducing growth and changing species compostion that needs careful monitoring. Plants are the initial accumulators and a knowledge of their response to pollutants is important not just for maintaining productivity but for the general salubrity of the environment.

In reports on atmospheric pollution concentration units are either gravimetric ($\mu g\ m^{-3}$) or volumetric (ppm = parts in 10^6 by volume or ppb = parts in 10^9). For the purposes of comparison volumetric units are used for preference in this chapter. Figures for conversion of the different pollutants from gravimetric to volumetric and vice-versa can be found in Table 10.1.

10.2 Gaseous pollutants

The choking smogs that were until recently common in many western European and North American cities have now been greatly reduced as a result of clean-air legislation. As with most political acts however the legislation has been sufficient only to reduce the amounts of obnoxious substances to a level where they cease to be offensive to the electorate. The lower odour detection level of sulphur dioxide to the human nose lies between 200 and 1100 ppb. Below these levels the air may be pure to human senses but it is still highly toxic to lichens and coniferous trees. Coniferous trees in Sweden have their growth reduced when the sulphur dioxide annual level is only 20 ppb (Nash, 1973) (*see* Fig. 10.4). Fortunately the ecological awareness of the electorate is increasing and steps are being gradually taken by most European countries to reduce the sulphur dioxide emissions of power stations and nitric oxides from vehicle exhaust. The impetus for this improvement is due to public reaction to environmental deterioration such as the disappearance of fish from acidified lakes in Scandinavia and Scotland and the damage to forest trees that can be seen in many areas in Germany and neighbouring central European countries. The forest decline (*Waldsterben*) that is currently causing concern in Germany and other continental areas in Europe and North America appears to be linked to atmospheric pollution, either acting directly on the foliage, or else in combination with the effects of acid rain on soils (*see* p.253).

The increased level of atmospheric carbon dioxide is also a serious aspect of gaseous pollution. Although

Table 10.1 Conversion factors between gravimetric and volumetric concentrations of pollutant gases at 20°C, 101.3 KPa pressure. (Reproduced with permission from Lendzian and Unsworth, 1983.)

Gas	Molecular wt g mol^{-1}	To convert from ppb to μg m^{-3}, multiply ppb by	To convert from μg m^{-3} to ppb multiply μg m^{-3} by
Fluorine, F_2	38.0	1.58	0.633
Hydrogen fluoride, HF	20.0	0.83	1.202
Hydrogen sulphide, H_2S	34.1	1.42	0.705
Nitric oxide, NO	30.0	1.25	0.801
Nitrogen dioxide, NO_2	46.0	1.91	0.523
Ozone, O_3	48.0	2.00	0.501
Peroxyacetyl nitrate (PAN), $CH_3COO_2NO_2$	105.0	4.37	0.229
Sulphur dioxide, SO_2	64.1	2.67	0.375

not toxic like sulphur dioxide, carbon dioxide accumulation is probably in the long term even more dangerous than sulphur dioxide ecologically, due to its global warming effect and the fact that it will be very difficult to reverse the trend. It is also ecologically worrying that carbon dioxide is not the only gas which is increasing its concentration in the atmosphere and contributing to the so-called 'greenhouse effect'. Methane, nitrous oxide, and the halocarbons (CCl_4, $CCIF_3$, CF_4, $CHCIF_2$, $CBrF_3$, $CBrCIF_2$, etc.) are showing very rapid rises and are likely to contribute significantly to the global-warming trend in the next few decades. By AD 2050 concentrations of atmospheric nitrous oxide of 350–450 ppb are likely (Dickinson and Cicerone, 1986). Based on these projections it has been suggested that the total contribution of trace gases (other than carbon dioxide) is more likely than not to exceed the contribution from carbon dioxide to global warming.

Such large changes in atmospheric composition and their effects on the blocking of the escape of thermal radiation are widely accepted as likely to cause marked climatic changes in the coming century. Global climate models predict not only changes in temperature but also of rainfall and the frequency of extreme conditions such as frost and drought. It is already apparent that the establishment of the global-warming pattern will not take place uniformly. Although the oceans of the world already show an average increase in temperature (Jones *et al.*, 1986a) (Fig. 10.3) there is a reversal of this trend at latitudes greater than 50°N (Lamb, 1982). The last 25 years have shown a marked cooling of seas in sub-Arctic regions and a greater frequency of late springs in

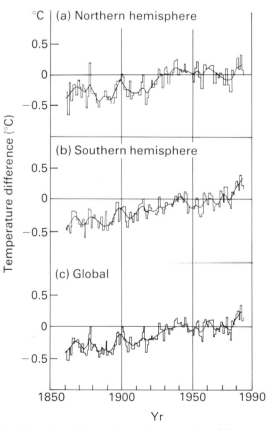

Fig. 10.3 Global (c) and hemisphere (northern [a] and southern [b]) annual mean temperature variation since 1861 based on sea-surface temperature data. (Reproduced with permission from Jones *et al.*, 1986a.)

northern Europe. It is therefore to be anticipated that the next few decades will herald an era of greater climatic variability for both terrestrial and aquatic ecosystems.

Clearly increases in carbon dioxide will prove advantageous for plant photosynthetic activity. However the importance of climatic changes consequent on an increase in the atmospheric concentration of carbon dioxide have not yet been systematically studied. These changes are likely to have ecological consequences greater that the direct effects of carbon dioxide on plant photosynthesis. It can be expected that relatively small alterations in temperature will have marked effects on plant distribution. A fall of 1°C for instance would be sufficient to move the cultivation of the vine south by 470 km (Lamb, 1967) and depress the timber-line in central Europe by 100–200 m (Firbas and Losert, 1949) (*see* Chapter 2, p.33). Recent dendrochronological studies combined with radiometric data (Scuderi, 1987) have confirmed the sensitivity of trees to minor temperature changes (*see* Chapter 4, p.89).

10.2.1 SULPHUR DIOXIDE

Early attempts to estimate the concentrations of gaseous sulphur dioxide that were harmful to plants nearly all underestimated its toxicity. When long-term experiments are carried out it is seen that repeated exposure can reduce growth even although no visible signs of damage are seen in the foliage. Numerous species of grasses as well as coniferous and broad-leaved trees show reduction in growth when exposed to about 150–500 ppb SO_2 concentration (Lendzian and Unsworth, 1983). Earlier work had estimated toxic levels of sulphur dioxide by measuring the concentration necessary to produce lesions in either leaves or needles. A mere dose effect however is insufficient as sulphur dioxide toxicity is a function of length of exposure as well as concentration. The mean annual concentrations of sulphur dioxide over Europe in the early 1980s are shown in Fig. 10.4. This data serves to indicate the areas most likely to experience high sulphur dioxide concentrations and shows that the areas for sulphur dioxide and ozone contamination are not identical (*see* Fig 10.12). During periods of peak pollution the sulphur dioxide concentrations that occur can be many times the annual mean values. Because of the long-term exposure factor in sulphur dioxide toxicity it is among the evergreen species that the most

sensitive plants are found. In particular *Pinus sylvestris* and *Picea abies*, the two most important species of northern European coniferous forests, are very intolerant of sulphur dioxide. Many European cities became not only the so-called 'lichen deserts' but were also devoid of the sensitive *Picea abies* by the end of the nineteenth century (*see* Kandler, 1987), In North America the white pine (*Pinus strobus*) is particularly sensitive being eliminated entirely within 15 km of smelters (Hutchinson, 1984). These species normally retain their needles for up to 5 years and therefore if needle life is reduced, or their photosynthetic efficiency is impaired as they age, then tree growth will inevitably be affected. In grasses and cereals tillering can be reduced as well as the partitioning of dry matter between shoots and roots. Some of the most tolerant species which only at concentrations of about 2000 ppb show signs of damage include members of the Atlantic deciduous forest such as the sessile oak (*Quercus petrea*), honeysuckle (*Lonicera periclymenum*), and box (*Buxus sempervirens*). Surprisingly some conifers are also relatively tolerant, namely *Juniperus sabina*, *Thuja orientalis* and *Chamaecyparis pisifera* (Ranft and Dässler, 1970). Lichens and in particular the species of moist and shady habitats are very sensitive to sulphur dioxide poisoning. The connection between moisture and sulphur dioxide sensitivity would appear to be an obvious consequence of the high solubility of sulphur dioxide in water. One volume of water can contain 80 volumes of sulphur dioxide ($228 \, g \, l^{-1}$ at 0°C). Therefore films of moisture on the thalli of lichens or leaf surfaces will concentrate the sulphur dioxide levels acting on the plant.

Water probably also plays a significant role in determining sulphur dioxide sensitivity through its effect on stomatal opening. A study begun in 1916 for the American Smelting and Refining Company (Lotfield, 1921) showed that plants could tolerate up to five times more sulphur dioxide at night than they could by day. The greatest sensitivity to sulphur dioxide toxicity is found in the morning when stomatal opening is greatest in most plants. Some authors have questioned the role of stomata in determining plant sensitivity to sulphur dioxide poisoning. Drought stress however, can reduce injury and this would suggest that plants with closed stomata are more resistant. The connection with plant water relations however will probably depend on the intensity of pollution. When sulphur dioxide levels are low the epidermal cells adjacent to the stomata lose turgor (Black and Black, 1979) and the stomata open. At

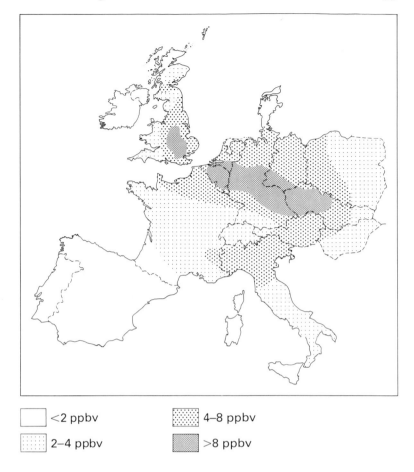

Fig. 10.4 Annual mean concentrations of sulphur dioxide in ppbv (parts in IO⁹) for Europe for the early 1980s; note that these mean concentrations are considerably less than the maximum doses that plants will be exposed to and serve only to indicate the areas most likely to suffer from sulphur dioxide pollution. (Data compiled and supplied by courtesy of J.N. Cape and D. Fowler, Institute of Terrestrial Ecology, Edinburgh.)

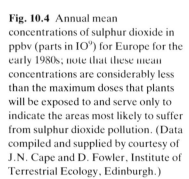

higher concentrations stomatal closure is usually caused by disorganization of guard cell structure (Wellburn *et al.*, 1972).

Once absorbed by the leaf the toxic effects of sulphur dioxide depend on the pH value of the cytoplasm and the rate at which it is removed by the metabolic activities of the plant. When dissolved in water a number of ionic states are possible. At the pH values which prevail in plant tissues there can be large changes in the concentrations of HSO_3^- and SO_3^{2-}. The lower the pH, the greater the oxidizing power of the sulphur species. Sulphur dioxide is also readily oxidized to sulphate by illuminated chloroplasts (Ziegler and Libera, 1975). As it is the membrane structures of the leaves that are attacked by sulphur dioxide, high levels will destroy the plant's detoxification mechanism. When leaves are gassed with SO_2 there can be detected an emission of ethane and ethylene indicating a peroxidative damage to fatty acids (Kimmerer and Kozlowski, 1982). The initial

effect appears to be on the CO_2 fixing enzymes, ribulose bisphosphate carboxylase and PEP carboxylase. A destruction of chlorophyll will eventually take place, but the time course of the reaction shows that this is not the initial site of injury.

In a study of sulphur dioxide injury in the Cucurbitaceae (Sekiya *et al.*, 1982) it was noted that young leaves were resistant to acute exposure while mature leaves were sensitive. This difference in sensitivity was not due to absorption differences between young and old leaves. When illuminated after a period of exposure to SO_2 young leaves emitted hydrogen sulphide much more rapidly than mature leaves. Using $^{35}SO_2$ it was shown that young leaves converted approximately 10 per cent of absorbed sulphur dioxide to H_2S. With mature leaves this figure was less than 2 per cent, suggesting a biochemical basis for the resistance of the young leaves to sulphur dioxide.

Apart from the differences between young and

old leaves, there are also differences between populations in their tolerance of SO_2 poisoning. In the annual Carolina cranesbill (*Geranium carolinianum*) populations growing close to a power station in Georgia, USA were significantly more tolerant than those from rural controls (Taylor and Murdy, 1975) (Fig. 10.5). To test for sulphur dioxide resistance populations were raised from seed collections whose parental response to SO_2 was known. The heritability of SO_2 tolerance was found to be about 50 per cent. In the control population very few tolerant individuals were found (Fig. 10.6). Similar effects have also been found in the dock *Rumex obtusifolia* and in a number of grasses, including Italian rye grass and cocksfoot. In Britain there is a high level of SO_2 tolerance in grasses around our industrial cities. These plants, although they survive in these polluted conditions, nevertheless suffer from chronic SO_2 poisoning as when they are grown in air that has been purified by passing through activated charcoal,

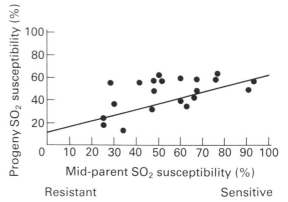

Fig. 10.6 Hereditability of sulphur dioxide tolerance in *Geranium carolinianum* as shown by a regression of sulphur dioxide susceptibility of progeny on the mean susceptibility of their parents. (Data of Taylor, 1979, as quoted by Hutchinson, 1984.)

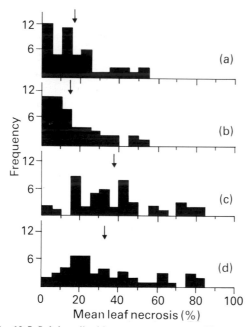

Fig. 10.5 Sulphur dioxide toxicity as measured by mean leaf necrosis of 1972 populations of *Geranium carolinianum* from sites at different distances from power plants. Note that populations nearest to the power plant have the lowest level of damage. The distance from the power plant is: (a) 0.4 km; (b) 0.64 km; (c) 12.2 km; (d) >40 km. The solid black arrow represents mean of respective population. (Redrawn from Taylor and Murdy, 1975).

they can increase their growth by almost 50 per cent (Bell and Clough, 1973). In some areas with a long history of coal-smoke pollution it has been found difficult to establish the new improved varieties of grasses such as the commercial S_{23} strain of *Lolium perenne*. Instead greater success is achieved using the indigenous strains which possess a resistance to SO_2 that has probably developed on a time scale dating back to the dawn of the industrial revolution.

10.2.2 CARBON DIOXIDE

In the pre-industrial era the concentration of carbon dioxide has been variously estimated between 274 ppm (Stuiver, 1978) and 290 ppm (Keeling, 1978). Recent studies of air trapped in bubbles in Antarctic ice suggest that in the sixteenth century the carbon dioxide content of the atmosphere may have been as low as 260 ppm (Fig. 10.7; *see also* Fig. 10.2). By 1986 the concentration had risen to 345 ppm and it is expected that it will rise to between 400 and 600 ppm by the year 2050 (Dickinson and Cicerone, 1986). This increase in carbon dioxide stems only partly from the burning of fossil fuels. Analysis of the relative proportions of the carbon isotopes suggests that a large contribution also comes from the extensive deforrestation that has taken place this century. Apart from the effects this may have on global climate this is likely to have a direct effect on plant growth.

In studies which use elevated carbon dioxide con-

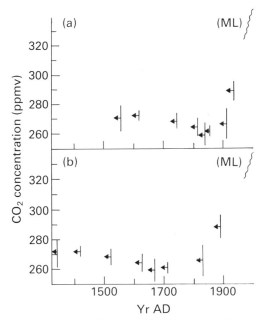

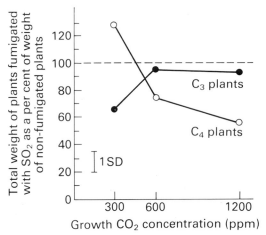

Fig. 10.8 Average total growth of C_3 and C_4 plants fumigated with 0.25 ppm sulphur dioxide expressed as a per cent of the growth of the non-fumigated plants at 300, 600 and 1200 ppm CO_2. (Redrawn from Carlson and Bazzaz, 1982.)

Fig. 10.7 Atmospheric carbon dioxide concentrations during past centuries as deduced from ice-core analysis. Two different approaches were used to date the ice-cores; (a) dating ice by modelling ice-flow and (b) dating based on the assumption of a constant snow accumulation. Note the low level of approximately 260 ppm in the seventeenth century; ML = atmospheric data from Mauna Loa since 1958. (Reproduced with permission from Raynaud and Barnola, 1985.)

centrations at levels that might be obtained by the middle of next century (600 ppm), C_3 species showed an ability to increase their growth rate with additional carbon dioxide while C_4 plants gave very little response (Carlson and Bazzaz, 1982). Sulphur dioxide resistance was also increased at high levels of carbon dioxide in the C_3 plants (Fig. 10.8). Conversely the superior resistance of C_4 plants to sulphur dioxide at normal carbon dioxide concentrations was lost at elevated CO_2 concentrations. These varying responses may be due to differential reactions of the stomata of C_3 and C_4 plants to carbon dioxide concentrations (*see* p.198). In C_3 crop plants increased levels of CO_2 can cause significant rises in fruit and seed yield (Kimball, 1983). Most significantly, it would appear that the competitive advantage of C_4 plants is reduced when the level of carbon dioxide is increased even to just 600 ppm (Bazzaz and Carlson, 1984); an atmospheric concentration that might well be reached sometime next century. In natural communities differential reproduction as a result of

changing CO_2 levels may also bring about significant ecological and evolutionary changes (Strain and Bazzaz, 1983). Increased CO_2 levels also accelerate life cycles and lead to earlier flowering (St. Omer and Horvath, 1983) with untold consequences for the competition between species and the process of insect pollination. The increase in leaf size that is brought about by high CO_2 levels will increase biomass for the species concerned in open habitats. However, in closed-canopy systems increased self-shading and the consequent further attenuation of radiation would have adverse effects for sub-canopy species.

One of the most general findings concerning the growth of plants at increased levels of CO_2 is their capacity for greater drought tolerance. Wheat plants grown at 1000 ppm CO_2 as compared with controls grown at 350 ppm, were completely able to compensate for a drought stress (Sionit *et al.*, 1980). Both C_3 and C_4 plants have been reported as increasing their net photosynthesis when the CO_2 levels were increased from 300 to 1000 ppm. However, certain adverse cytological effects have been found. Starch grains can reach excessive sizes and cause the disruption of chloroplasts (Gates *et al.*, 1983).

10.2.3 PHOTOCHEMICAL SMOG

The term smog was first used at the beginning of this century to describe the combination of smoke and fog which until the introduction of the Clean Air Act was a serious hazard to all animal and plant life in and around our major cities. The abatement of coal-smoke pollution has however coincided with an increase in oil consumption by power stations and motor vehicles. The nitric oxides from oil-fired power stations and the exhaust from vehicles, although less visible than the smoke of coal fires, respond to sunlight to give rise to a mixture of gases sometimes referred to as photochemical smog which is highly dangerous to plants. In oceanic climates some of the worst effects of this form of pollution are avoided as winds are frequent and high cloud cover reduces the light intensity. However, in areas with frequent temperature inversions (Fig. 10.9) as well as bright sunlight, the photo-oxidation of the nitric oxides in vehicle exhaust and power station smoke greatly increases their toxicity. Any form of combustion causes the formation of some nitric oxides from the nitrogen and oxygen of the atmosphere. The hotter the flame, the greater the production of nitric oxides. The oxygen in the atmosphere then reacts with the nitric oxide to produce nitrogen dioxide. In the dark the reaction ends here with the production of nitrogen dioxide. However in light, nitrogen dioxide is split into nitric oxide and atomic oxygen which then combines with molecular oxygen to form ozone. Some photochemical reactions which take place in polluted atmospheres are listed in Table 10.2.

Normally ozone generated from nitrogen dioxide would have transitory existence as it would back react with the nitric oxide. However as internal combustion engines also produce hydrocarbons, and aldehydes which photo-oxidize to generate free radicals, such as the hydroxyperoxyl (HO_2) and the alkylperoxy (RO_2) these compete and the balance of the back reaction is disturbed. Thus ozone production instead of being limited to the equilibrium concentration will continue to accumulate as the reaction proceeds to the right as the nitric oxide is removed by hydrocarbon trapping. Because the free radical production is dependent on light, the trend is reversed at night and ozone disappears. Ozone pollution therefore has to be renewed daily. In unpolluted areas normal levels of ozone in the atmosphere average about 20 ppb. This gas is so toxic to plant life that it needs only to slightly more than double its normal concentration to 50 ppb to cause tissue damage. Although the level of ozone generation is highly dependent on the hours of sunshine there can be sufficient sunny weather, even in the oceanic climate of the British Isles, for ozone injury to be detected, using tabacco as an indicator plant (Ashmore *et al.*, 1978) (Fig. 10.10). Fig. 10.11 shows the annual course of daily maximum ozone levels (hourly mean values) recorded in 1978 at Devilla forest, Scotland. The figure also indicates the contributions to these levels from the major ozone sources. It can be seen that even in the relatively unpolluted air of Scotland the 50 ppb threshold, which can be deleterious over prolonged periods, as well as the 100 ppb level which can cause damage in shortlived episodes, both occur. Overall, the probability of ozone injury however is best estimated, not from mean concentrations, but from the length of time that plants are exposed to elevated levels. Fig. 10.12 shows the regions most at risk in Europe when the period of exposure is taken into account. When compared with the areas at risk from sulphur dioxide pollution (*see* Fig. 10.4), the ozone danger zone is further south, reflecting the need for sunlight for ozone production.

In contrast to sulphur dioxide injury, ozone poisoning causes the leaf stomata to close. Stomatal closure before ozone exposure can increase resistance and the application of abscisic acid can be used as a preventive treatment (Jones and Mansefield, 1970). When stomatal closure is brought about by ozone this does not act as a self-protecting mechanism and

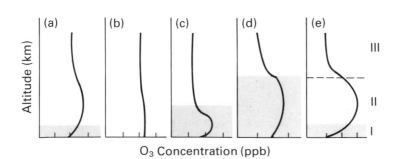

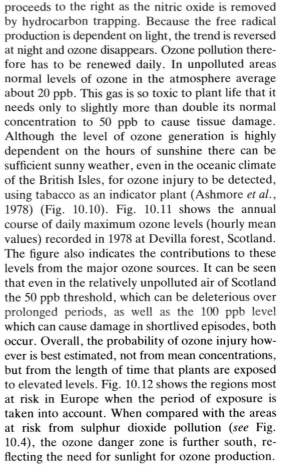

Fig. 10.9 Possible behaviour of the vertical ozone profile in polluted areas over a 24-hour period: (a) at night; (b) after sunrise when vertical exchange starts; (c) before and (d) after mixing is complete; (e) during photochemical episodes at night with an aged smog layer between the subsidence inversion and the radiation inversion layer. I, radiation inversion layer; II, aged smog layer; III, free troposphere. (After Van Duuren *et al.*, *see* Fuhrer, 1985.)

Table 10.2 Some photochemical reactions which take place in polluted air.

$$NO_2 \xrightarrow{h\nu} NO + O^{\cdot}$$ Generation of superoxide radical

$$O^{\cdot} + O_2 \rightarrow O_3$$ Formation of ozone

$$NO + O_3 \rightarrow NO_2 + O_2$$ Back reaction of ozone, blocked by hydrocarbons and PAN generation

$$O_2 \xrightarrow{h\nu} 2O^{\cdot}$$

$$O_3 \xrightarrow{h\nu} O^{\cdot} + O_2$$

$$O^{\cdot} + H_2O \rightarrow 2OH^{\cdot}$$

$$CH_3CHO + OH^{\cdot} \rightarrow CH_3CO + H_2O$$

$$CH_3CO + O_2 + NO_2 \rightleftharpoons CH_3COO_2NO_2 \text{ (PAN)}$$
Peroxyacetyl nitrate

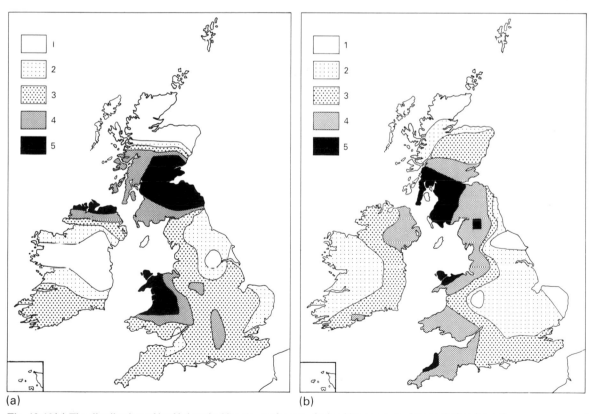

(a)　　　　　　　　　　　　　　　　(b)

Fig. 10.10(a) The distribution of leaf injury in *Nicotiana tabacum*, derived by computer interpolation from the mean values of weekly leaf injury index at 53 sites for the period 13 June − 5 September 1977. Five classes are distinguished (denoted by increasing degrees of shading): 1, 0.01−0.54%; 2, 0.54−1.06%; 3, 1.06−1.59%; 4, 1.59−2.11%; 5, 2.11−2.54%. (b) The distribution of mean weekly sunshine hours for the period 13 June − 5 September 1977. The five classes represent: 1, 28.9−33.2 hours; 2, 33.2−37.5 hours; 3, 37.5−41.9 hours; 4, 41.9−46.2 hours; 5, 46.2−50.5 hours. (Reproduced with permission from Ashmore *et al*., 1978.)

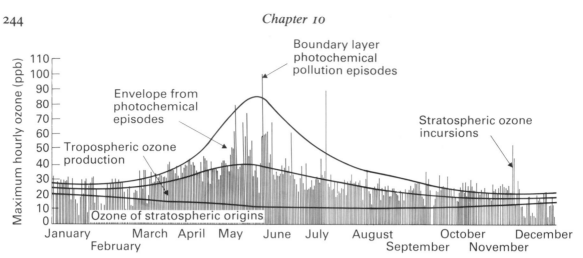

Fig. 10.11 Daily maximum ozone concentrations (hourly mean values) recorded in 1978 in Devilla Forest, Scotland, together with tentative contributions from major ozone sources. (Redrawn from *Ozone in the United Kingdom*, Department of the Environment.)

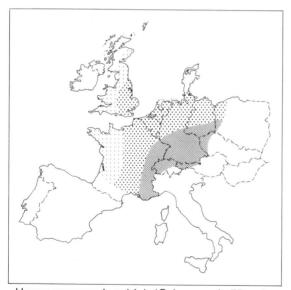

Hours per year in which (O₃) exceeds 75 ppbv

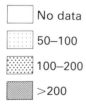

Fig. 10.12 Hours per year in which ozone levels in Europe exceed 75 ppbv in the early 1980s. (Data compiled and supplied by courtesy of J.N. Cape and D. Fowler.)

tissue deterioration continues as long as the leaves are exposed to the pollutant. An early effect of ozone damage is to reduce the water potential of the leaf. Within 75 minutes of exposure to an ozone concentration of 60 ppb the primary leaves of *Phaseolus vulgaris* suffer a drastic water loss (Evans and Ting, 1973). The abaxial surface of the leaf is capable of recovering resistance, provided the fumigation treatment is not too prolonged. The tissues which specifically suffer are the palisade and the upper epidermal layers. Ozone being a highly reactive substance is capable of combining with very many organic molecules. Most observations, however, indicate that it is the unsaturated fatty acid residues of the membrane lipids that are the primary site of attack (Heath, 1975), although Mudd (1982) argues that ozone attacks the protein of cell membranes and then passes through the membranes to attack the cytoplasm. In either case it is this attack on membranes which is presumed to cause the leakage of the palisade cells, the rapid loss of water, and efflux of potassium, and hence the stomatal closure that typifies the response of most plants to ozone fumigation. Electron microscope studies have also detected a granulation of the stromas (Heath, 1975) and this has been attributed to a precipitation of fraction I protein. This protein which contains the ribulose bisphosphate carboxydismutase activity of the chloroplast is particularly rich in sulphydryl groups, containing over 90 such groups per mole. The coagulation could therefore be due to the ozone-mediated sulphydryl oxidation.

As with sulphur dioxide injury, ozone poisoning is accompanied by a large number of metabolic changes which are probably secondary reactions and although they can aggravate damage it is seldom certain if they are a prime cause of injury. Membrane leakage which gives rise to loss in water potential can give rise to accumulations of free amino acids together with increases in soluble protein levels. The increased level of amino acids as a result of sulphur dioxide exposure (Godzik and Linskens, 1974) can be brought about by any situation which restricts growth. Damaged plants, with their higher soluble nitrogen content, tend to attract greater densities of aphid attack than healthy plants. The same phenomenon can be observed when young forest plantations have gaps in the intially planted trees filled with replacements. The newly-planted trees are growing more slowly than the established plants and therefore have higher amino acid and amine contents. Consequently the forest aphid population converges on the more nutritionally-rewarding, newly-planted trees. Ozone injury to trees also increases insect attack for possibly similar reasons. Ponderosa pine has been noted in California as being more prone to bark beetle attack after ozone exposure (Miller and Elderman, 1977).

Recently, stress-ethylene production has been suggested as a contributing factor to ozone injury (Mehlhorn and Wellburn, 1987). When plants experience environmental stress they respond by producing large amounts of ethylene, which is often referred to as stress-ethylene (Abeles, 1973). When young pea plants are fumigated daily with ozone no visible injury appears after 3 weeks. By contrast severe leaf injury develops when plants that were grown in clean air are suddenly exposed to a similar concentration of ozone. The latter plants also produce more ethylene and pretreatment with an inhibitor of ethylene biosynthesis aminoethoxyvinlyglycine (AVG) reduces subsequent ozone injury. Ozone and ethylene are known to react together to produce free radicals which could bring about the peroxidation of membrane lipids. This in turn can also cause the liberation of ethane and more ethylene. In reviewing this work Unsworth (1987) points out the attractiveness of a hypothesis that links ozone injury with the action of environmental stress on plants, but also indicates weaknesses in that the experimental treatments may have altered the rate of entry of the ozone into the leaf.

Free radicals however produced, are a highly probable source of cell damage from ozone exposure. It is therefore not unexpected that protection against ozone damage can be achieved to some extent by the use of certain anti-oxidants which are used as fungicides. The most effective have been found to be the dithiocarbamates and derivatives of mecaptobenzothiozole. Superoxide dismutase has been found also to have a protecting effect against ozone injury (Lee and Bennet, 1982). In experiments with ozone-sensitive beans (*Phaseolus vulgaris*) the synthetic urea complex N-(2-[2-oxo-1-imidazolidiny]ethyl) -N-phenylurea (EDU) was found to be an effective protectant against chronic or acute foliar injury due to ozone, either when sprayed on leaves or applied through the soil. This EDU-enhanced tolerance of ozone was associated with increased levels of superoxide dismutase (SOD) and catalase activities in the leaves. Greater SOD activity in young leaves as compared with older leaves was also correlated with lower sensitivity to ozone in the younger leaves. It may be therefore that tolerant plant tissues have greater enzyme scavenging capabilities for protection against toxic oxy-radicals. Superoxide dismutase catalyses the reaction:

$$O_2^{\cdot -} + O_2^{\cdot -} \rightarrow H_2O_2 + O_2$$

This enzyme is essential to protect photosynthesizing leaves against the toxic effects of oxygen or oxy-radical-reaction products (*see also* p.125). As the phytotoxic effects of ozone may be due to formation of active oxygen intermediates such as superoxide anion, hydroxyl and perhydroxyl radicles or H_2O_2, superoxide dismutase could act in conjuction with catalase as an enzymatic detoxification system (Lee and Bennett, 1982).

As noted above ozone accumulates in the presence of hydrocarbons due to trapping nitric oxide and thus preventing the back reaction to nitric dioxide and oxygen. This same process of hydrocarbon complexing produces the accumulation of an additional pollutant, which although it occurs only at one-tenth of the concentration of ozone is capable of causing severe irritation to the eyes as well as being highly toxic to plants. The nature of this compound for a time eluded identification but has now been found to be a mixture of peroxyacetyl nitrate (PAN) together with a series of homologues, viz. peroxyprionyl nitrate (PPN) and peroxybenzoyl nitrate (PBN) all of which have similar properties. The degree of toxicity increases with the size of the molecule and therefore PBN > PPN > PAN in damage caused to plants. Peroxyacetyl nitrate (PAN) can be produced in the laboratory using trans-2 butene as the hydrocarbon (Kerr *et al.*, 1972).

hv
$$CH_3CHCH_3 + NO + NO_2 \rightarrow CH_3COO_2\,NO_2 + HCHO + CH_3ONO_2 + CO_2 + CO$$

Typical symptoms of the damage due to PAN and its associated homologues include a glazing or bronzing of the lower surface of the leaves . Experimentation on the tolerance of plants to PAN has not been extensive, probably due to the explosive nature of this compound when kept in concentrated form and under pressure.

The fact that PAN pollution causes eye irritation immediately suggests that it will have some biochemical affinities in its mode of action with other synthetic lacrimators. Research on this subject has been carried out in relation to riot control, where the most effective tear gases are found also to be excellent sulphydryl reagents and are inhibitory against enzymes which depend on SH groups for their activity. The entry of PAN into the plant appears to be through the stomata as with the other air-borne pollutants already discussed. Once inside the leaf the PAN molecules attack the mesophyll cells immediately bordering the mesophyll spaces. Recovery is possible, but if fumigation persists the leaf tissues collapse. As might be expected of a pollutant which acts against sulphydryl groups, the chloroplast and its membranes are also effected as with ozone poisoning and in some cases it is possible to confuse the symptoms of these two pollutants. Injury by PAN has been studied extensively only on crop plants. However, even in this limited range of plant material there is considerable variation in the sensitivity of the tissues to the pollutant. Concentrations of as little as 30 ppb arising under natural conditions for a few hours can produce toxic effects next day in sensitive species. Sensitive crop plants include spinach, oats, celery and tobacco, while greater resistance is found in cabbage, cauliflower, rhubarb, carrot, onion, and cucumber. Resistance in these experiments was defined as being able to withstand 100 ppb for 2−4 hours as compared with the upper limit of 30 ppb in the sensitive species.

The dangers from photochemical smog to plants are therefore very great and in areas such as the Los Angeles basin a once flourishing citrus industry has now in many places been destroyed. Fortunately none of these compounds persist in the atmosphere and if greater efforts were made to reduce the emission of nitric oxides from motor vehicles and power stations, their danger to plant and animal life could be much reduced.

10.2.4 FLUORIDES

The toxicity of soluble fluorides has been well known since last century. Their production has increased in the latter half of this century through expansion of aluminium and steel smelting as well as the manufacture of bricks and phosphate fertilizer. Most factories that produce flourides either as volatile gases (HF and SiF_4) or as the less troublesome dust particles, take steps by means of scrubbers, filters, and electrostatic precipitators to reduce the fluoride content of waste emissions. It is sometimes possible to place fluoride-emitting industries in sites where the level of toxic-waste emission does not cause any serious damage to the natural vegetation. The aluminium smelting plant near Fort William in Scotland caused little visible damage to the surrounding vegetation. This was partly due to the steps taken to reduce fluoride emission but was also greatly aided by the strong winds that are a constant feature of the west coast of Scotland. Also the vegetation in the vicinity of the smelter was mostly grass which is relatively resistant to fluoride poisoning. Nevertheless, the fluorine content of some of the herbage near the smelter was as high as 1 mg F g^{-1} dry weight of plant. Had this smelter been sited in a continental area with less frequent winds and a native vegetation of forest, a totally different situation would have developed. In a 50-square-mile region around Spokane in Washington, fluorides were responsible for extensive needle damage and death in *Pinus ponderosa*. In the vicinity of the smelter, pine was eliminated as the dominant species and the vegetation was reduced to herbaceous species. Likewise, in other areas other tree species including *Fagus sylvatica*, *Pinus sylvestris*, *Carpinus betulus*, *Pseudotsuga menziesii* and various species of *Quercus* have all been reported as being damaged and suppressed, while annual and herbaceous plants become the major element of the vegetation (Treshow, 1970).

Fluorides act as general enzyme inhibitors. Certain enzymes are particularly sensitive to fluoride inhibition and fluoride has been used as a specific inhibitor of enolase where it blocks the transformation of 2-phosphoglyceric acid to phosophenolpyruvate in glycolysis. Phosphoglucomutase and succinic dehydrogenase can also be inhibited by fluoride. Fluoride prevents the action of peroxidase, catalase, and cytochrome oxidase due to its ability to combine with the ferric iron in the heme group of these enzymes (Hewitt and Nicholas, 1963). A further toxic action can be expected from the removal of available calcium and then magnesium (CaF_2; MgF_2) thus depriving chlorophyll molecules of their essential metal ions. Fluoride is a cumulative poison and it is therefore understandable that the greatest effects will be found in those species with long-lived foliage. Fluoride poisoning acts mainly on leaves where the typical symptom is tip burn and scorching of the margins. The leaves are attacked as most fluoride pollution enters the plant through the stomata and once within the leaf only a minor portion is translocated elsewhere. Fluoride can be taken up via the roots in which case it is distributed throughout the plant, but it is only rarely that concentrations in soil are high enough to adversely affect plant growth. If a factory emitted 1 ton of elemental fluorine daily and half of this settled within a 10-mile radius, then the average accumulation in the surface 150 mm (6 inches) of soil would be only 1 ppm per annum (Treshow, 1970). At this rate of pollution it would take several decades to produce a concentration of 25–30 ppm which experimentation shows to be necessary to reduce plant growth through root uptake.

Ecologically fluoride pollution is not a serious problem. Sources of contamination are usually localized. As the poison is cumulative, careful monitoring of the contamination can give ample warning of when damage can be expected. No damage is likely until the concentration accumulated in the leaves exceeds 50 ppm dry weight and this is most likely to be noticed in plants such as evergreen conifers where needles persist for 3–6 years. The greatest danger from fluoride to livestock is found in those plants which grow on fluoride-rich soils and have evolved the capacity to accumulate fluoride in large quantities. This flouride accumulation makes them highly toxic and its role as a possible deterrent to grazing was discussed in the previous chapter (*see* p.224). When sources of contamination are removed as when aluminium smelters are closed the level of fluoride accumulating in moss and lichen species falls rapidly (Kenworthy, personal communication).

10.3 Water and soil-borne pollutants

Pollution of oceans, lakes, rivers, and the soils of dry land differs from the pollution of the atmosphere in that the contact between the toxin and the environment is much more complex. In atmospheric pollution plants suffer, either by direct exposure to lethal concentrations of poisons, or else by the accumulation of sub-lethal doses. In soil and water systems there is a more complex interaction in which micro-organisms, higher plants and their decomposing remains all play a part in modifying the effects of the pollutants. In some cases the toxins can be chemically altered or accumulated and thus rendered more dangerous, but by far the commonest situation is for the toxin to be degraded to an inocuous form. Soil and water systems thus possess great powers of homeostasis which are not found in the atmosphere. This power of homeostasis, the maintenance of the constancy of the internal environment of a system due to its metabolic activity, can however be overtaxed and it is in these situations that we see the tragic consequences of destroying a biological system which if not abused would have helped to maintain a salubrious environment.

Ecologically it is in the cold regions of the world that the greatest care has to be taken in not overtaxing the homeostatic powers of the ecosystem. In these areas metabolic activity is at a minimum and quantities of pollutants, which could be decomposed and volatilized in temperate and tropical climates, accumulate and poison the micro- and macro-flora. In polar habitats this can produce biological changes which are virtually irreversible. There is therefore considerable need for ecological awareness of the dangers of pollution when opening up the oil fields in the Arctic or establishing large industrial townships in areas with prolonged winters, as has recently taken place in the Soviet Union in an effort to extract the natural resources of the Siberian Taiga.

10.3.1 ACID RAIN

In 1872 Robert Angus Smith, the first British Air Pollution Inspector, first used the term 'acid rain' when he observed that the rain falling around Manchester contained sulphuric acid. The term 'acid rain' is now used more generally to describe the total deposition of atmospheric pollutants on vegetation and soils. This deposition can take three possible forms:

1 'Wet deposition' as rain, hail or snow.
2 'Dry deposition' in which the pollutants as gases or particles are captured or absorbed onto plant or soil surfaces (Fig. 10.13).
3 'Occult deposition' where the pollution is incorporated into cloud droplets which are then deposited on the vegetation when it is enveloped in cloud (Fig 10.14).

Fig. 10.13 Dry deposited acidity in Great Britain.
(Redrawn from Fowler *et al.*, 1986.)

This latter type of precipitation can act as a concentrating mechanism in areas where orographic cloud sits for a considerable time over upland areas. Polluted air rising into the cloud can leave its pollutants dissolved in water droplets which then deposit a concentrated solution on surfaces with which they come in contact. Similarly, rainfall in the form of drizzle has a greater total drop surface area than an equivalent amount of heavy rain. The wash-out efficiency of this light precipitation is therefore greater and pH values between pH 2.5 and 3.0 can be recorded (Vermeulen, 1979).

Rain-water is generally considered 'acid' if the pH value falls below 5.6 (the equilibrium of carbon dioxide with distilled water). Studies of ion balances in rain water preserved from glacier ice in the Cascade Mountains record minimum pH values of about 5.6 indicating a pH of rain-water saturated with carbonic acid. The arbitrary use of this value does not allow for natural sources of sulphur and nitrogen. Gaseous sulphur compounds of terrestrial and marine origin include dimethyl sulphide, carbonyl sulphide and hydrogen sulphide all of which can be oxidized to sulphur dioxide. Similarly the release of nitric oxide by natural processes of denitrification can contribute nitric acid to the precipitation. Thus the pH of industrially unpolluted rain water may vary between pH 4.6 and 5.6 with a mean value about pH 5.0 (Charlson and Rodhe, 1982). Such values are found in remote areas as in the Scottish Hebrides, northern Norway, and the Indian Ocean (*see* Fowler *et al.*, 1986). Rain water with pH values of 5.6 are however rare.

There is little clear evidence as to whether the

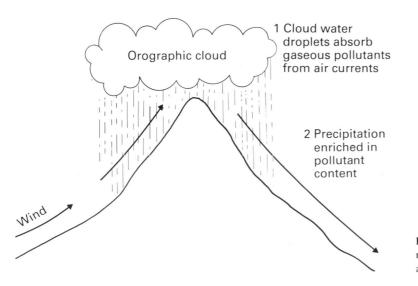

Fig. 10.14 Diagram showing the manner in which 'occult deposition' of acid rain falls over mountain regions.

acidity of rainfall has altered significantly over the past few decades, despite the increased consumption of fossil fuels. At a time when the European emissions of sulphur increased by about 35 per cent (between 1963 and 1975) wet deposition actually decreased at many European monitoring sites and Swedish monitoring has shown that by 1977 wet deposition of sulphate has remained constant or has decreased over the previous decade (Rodhe and Granat, 1984). The British emission of sulphur dioxide into the atmosphere fell by 40 per cent from 6.09 million tons in 1970 to 3.67 million tons in 1983. At the same time however there has been an increase in nitric oxide production from traffic, both as a result of the increased numbers of vehicles and hotter more efficient internal combustion engines. Thus although the sulphuric acid component of the rain-water may be falling the nitric acid content is rising and rainfall pH values in industrial areas remain below 4.5 (Fig. 10.15). Apart from the acidity of the rainfall the

total acidity in terms of kg of hydrogen ions needs to be considered. Fig 10.16 compares rainfall acidity and wet deposited acidity for the British Isles. Although the actual rain falling is more acid in the industrial eastern regions the total amount of acidity deposited is greater in the west due to the greater rainfall.

Irrespective of whether or not the pH value of rainfall has changed over the past decades there is no disputing the increased acidity of many freshwater lakes in Scandinavia and Scotland (Figs. 10.17—

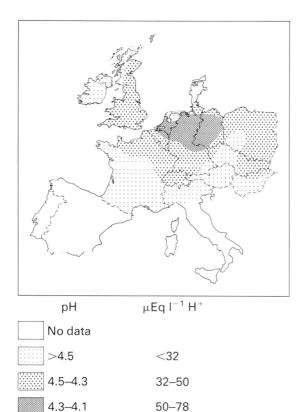

Fig. 10.15 Volume weighted annual mean acidity in precipitation from 1972 to 1982. (Data by courtesy of J.N. Cape and D. Fowler, Institute of Terrestrial Ecology, Edinburgh.)

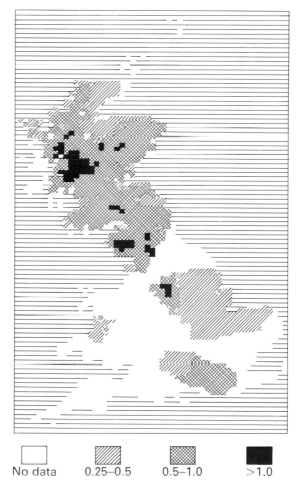

Fig. 10.16 Rainfall acidity in Great Britain; total wet deposited acidity in areas of high rainfall accumulate the greatest yearly acidity totals although the hydrogen ion concentration of the rainfall may be less than areas with lower rainfall. (Map redrawn with permission from Fowler *et al.*, 1986.)

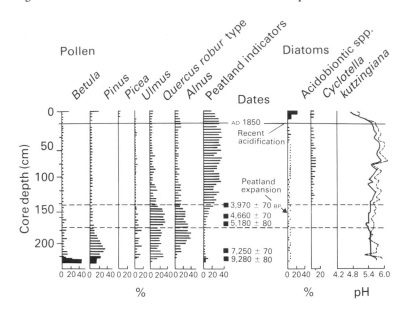

Fig. 10.17 Stability of pH values during peat accumulation as seen in the composite pollen and diatom diagram from core from the Loch of Glenhead (Galloway, Scotland) using data based on ^{210}Pb and ^{14}C. The reconstruction of pH values is based on an index derived from the diatom flora. Peatland indicators are: *Calluna vulgaris, Sphagnum,* Cyperaceae, *Succisa* and *Potentilla.* (Reproduced with permission from Jones *et al.,* 1986b.)

10.19). The progressive transfer of sulphate and acidity into surface waters has been taking place for several decades. This progressive acidification now appears to have reached a level in many lakes at which the threshold for fish survival has been passed. Over a longer period natural acidification has been removing calcium from soils in upland regions which

has resulted in this element also being depleted in lakes and rivers. Fig 10.20 shows the relationship between calcium concentration and acidity in Norwegian lakes with and without fish populations. Most of the fishless lakes have calcium concentrations of less than 50 µEg l^{-1}. In an experimental study of the effect of controlled acidification in which the pH

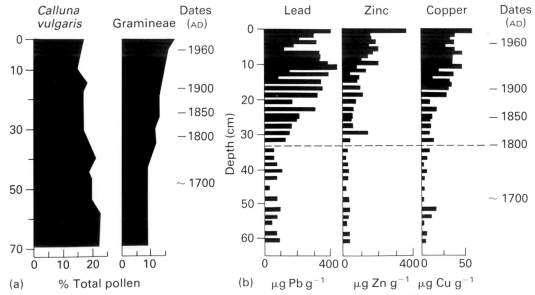

Fig. 10.18 Progressive reduction in heather (a) and increase in pollutants (b) as seen in the sediment core at Loch Enoch, Galloway, Scotland. (Reproduced with permission from Battarbee *et al.,* 1985.)

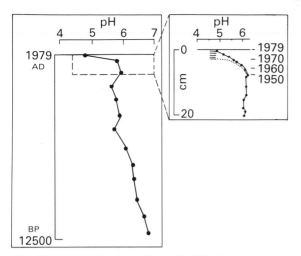

Fig. 10.19 Past pH values of lake Gårdsjön (Sweden) inferred from the diatom composition of sediment cores. (Reproduced with permission from Renberg and Wallin, 1985.)

of a small lake was reduced from pH 6.8 to pH 5.0 over 8 years it was found that no species of fish reproduced at pH values below 5.4 and that the lake would therefore become fishless within a decade due to hydrogen ion changes alone (Schindler *et al.*, 1985). High acidity on its own interferes with reproduction through inhibiting the process for softening the hard egg case and thus preventing hatching. Fish are also killed by a secondary consequence of acid run-off from soil, namely soluble aluminium. Aluminium becomes readily soluble at pH 5 and below.

Soluble aluminium poisons fish by upsetting the operation of the gills, injuring their epithelial cells with the result that gills become clogged with mucus and the oxygen content of the blood falls. Aluminium toxicity is greatest when the calcium content of the water is low. In a study of acid run-off and snow melt in the Adirondack Mountains (Schofield and Trojnar, 1979) it was found that lakes unable to support brook trout all had soluble aluminium contents over 0.29 mg l^{-1}, whereas lakes which still supported fish had aluminium contents of not more than 0.1 mg l^{-1}. Thus soluble aluminium ions will kill fish directly and low pH on its own will interfere with reproduction.

In upland areas acidification and calcium depletion through the leaching of the soil is a continuous long-term process. Some upland catchment areas in the English Lake District show a progressive change indicating a certain degree of acidification dating back over 5000 years (Pennington, 1984). This acidification can also be attributed to anthropogenic causes. The arrival of man, the consequent removal of deciduous forest and the spread of *Calluna vulgaris* and other heath vegetation coupled with increased *Sphagnum* activity would all contribute to soil acidification and mineral impoverishment of upland areas. Soil acidification alone cannot account for the very low pH values now found in fishless lakes. Natural organic acids can provide a significant fraction of the anion in soil solution, but as they are typically precipitated as organo-aluminium complexes in the deeper soil horizons they are unlikely to be responsible for the low pH presently being recored in many

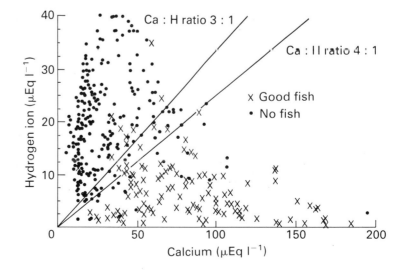

Fig. 10.20 Acidity and calcium concentration in lakes in southern Norway. The lakes are classified according to fish status. (Reproduced with permission from Jones *et al.*, 1986b.)

lakes and rivers. However it is possible that key soil processes together with an input of strong acids can account for the observed trends in soil and water pH values in North America and Europe. Thus although acidic soils contain exchangeable hydrogen and aluminium ions in amounts equivalent to thousands of years of acid deposition, it is only with the addition of strong acids from industrial pollution that they are likely to add these ions to freshwaters in amounts that are toxic to fish (Reuss *et al.*, 1987). This effect of industrial acid input on top of natural soil acidification has been shown in lakes both in Scotland and Sweden. In some Scottish Lochs in Galloway where blanket-mire developed over the catchment area in the mid post-glacial period a pollen and diatom study of a sediment showed that the lake was acid (pH 5.5−6.0) before peat formation began and that no further acidification took place while the peat was developing (Jones *et al.*, 1986b). Acidification levels to less than pH 5 occurred only after 1850 AD during the period of increasing acid emission from fossil fuel combustion (Fig. 10.17). These results suggest that soil acidification alone cannot be considered as a sufficient cause for the very low pH values found in many lakes today. It also shows that industrial acidification has been acting on soils and waters for over a century. A similar study at lake Gårdsjön in south-west Sweden (Renberg and Wallin, 1985) showed that the pH of the lake decreased from 7 to 6 over 12 500 years due to natural causes and that the decrease to the present level of pH 4.5 has only taken place in the past few decades (Fig. 10.19)

The extensive planting of conifers in many upland catchment areas has produced a new set of conditions in relation to soil and water acidity causing concern to some environmentalists. The large surface area of the needle foliage in coniferous forests provides one of the most efficient methods of trapping and transferring to the soil the concentrated acidity of occult precipitation (*see* Chapter 4 on dew condensation, p.138). The dense, closed canopy of plantations of spruce and pine also produce large accumulations of needles which generate a new source of acidity. Coniferous needles thus provide two sources of acid input, the first from their trapping of fog and mist and the second from the acidity generated in their own litter. The latter would be relatively harmless on its own, however when coupled with the additional sources from normal and occult precipitation polluted with strong acids they can augment the total acidity being added to drainage water. Reafforestation also provides an additional hazard as ploughing for drainage and tree planting is bringing back into the upper layers of the soil the leached aluminium deposits of the past 10 000 years. Thus the tree plantations which have replaced grazing in many upland areas in both Britain and Scandinavia might be expected to increase the input from natural and industrial acidity and depending on soil and cultivation conditions accelerate aluminium input into the catchment area. Hence changes in land use coupled with the industrial acidification of the rainfall are potential hazards in relation to the pH status of lake and river water. The effects are likely to be particularly noticeable in those areas with low buffering capacity and where oligotrophic conditions are unable to contribute any neutralizing ions. The removal of calcium has been progressing steadily for millenia as upland calcium reserves are not being replenished, either by natural weathering or agricultural practice. Consequently, many Norwegian lakes now have insufficient calcium for fish survival (< 1 ppm), irrespective of the presence or absence of toxic levels of aluminium (Fig. 10.20). Although these arguments suggest that reafforestation in areas exposed to acid rain could aggravate the acidification of lakes there is as yet no experimental evidence that this takes place. In a careful study of six lakes, three on unafforested sites and three on afforested sites in Scotland, there was no clear trend to suggest that tree planting had contributed to the acidity of the lake water as judged by the diatom flora in the lake sediment (Flower *et al.*, 1987) (Fig. 10.21).

Irrespective therefore of whether tree planting is taking place or not, the absence of calcium together with high aluminium content produces an environment that is unsuited for fish survival. Thus over large areas of northern Europe and in particular in the regions of hard metamorphic rocks as in the Highlands of Scotland and in Norway and Sweden, fish are dying as a consequence of the fall in pH values of lake and river water and the increased amounts of soluble aluminium that accumulate in rivers and lakes as a result of acid soil leaching. Changes in land use with emigration from marginal areas have resulted in a run-down of ecosystem productivity in many upland areas. The much discussed reduction in fish stocks is but one symptom of this decline. Others noted in Scotland in recent years include the decline in grouse populations and lambing percentages. It is on these impoverished habitats that acid rain now falls and has to be considered as yet one more negative factor in their general ecological deterioration.

There has been considerable controversy as to

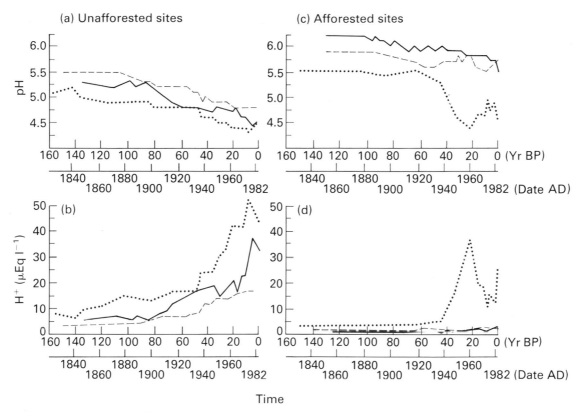

Fig. 10.21 The recent history of lake-water acidity changes in three Scottish Lochs in unafforested sites compared with three afforested sites. (Reproduced with permission from Flower *et al.*, 1987.)

whether acid rain causes any damage to vegetation. The degree of potential damage is not just a function of the acidity of the rain but is also influenced by the type of precipitation (*see* p.247) The oxidation of SO_2 to SO_4 although causing low pH values in rainwater only produces a very dilute solution of sulphuric acid. Direct damage to foliage from such acid precipitation has been shown only in simulated laboratory studies. Scanning electron micrographs of leaf surfaces that have been exposed to simulated acid rain develop lesions initially near trichomes and stomata on the adaxial surfaces (Evans, 1979) (Fig. 10.22). Once a lesion develops it can cause a small depression for the collection of subsequent rains and this causes the lesion to spread to adjacent cells. In general newly expanded leaves are most sensitive to acid rain. The most sensitive species to this direct foliar injury are herbaceous plants such as bracken (*Pteridium aquilinum*), soybean, and sunflower. Deciduous broad-leaved forests also contain many sensitive species. However oak leaves are relatively resistant due to the hyperplastic and hypertrophic reactions

of the spongy mesophyll when the overlaying palisade cells collapse (Evans, 1979).

10.3.2 FOREST DECLINE

The disappearance of fish from acidified, calcium-deficient lakes is matched chronologically by the damage to needles and growing shoots that has attracted public attention recently in many European forests. This damage is usually referred to as 'forest decline' or '*Waldsterben*'. Although public awareness of the danger to trees of atmospheric pollutants has increased in the last decade, this does not mean that forest decline is an entirely new phenomenon, nor that it is always due to atmospheric pollution. Fig. 10.23 shows a forest in southern Germany in 1905 with silver fir (*Abies alba*) with the 'stork's nest crown' form that is a typical sign of unhealthy trees clearly visible, as well as two unhealthy spruce trees. The stork's nest crown is a common deformation in silver fir (*Abies alba*) with loss of older needles and the formation of adventitious shoots and branches.

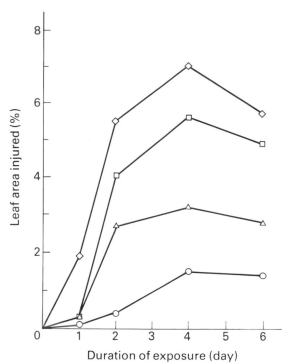

Fig. 10.22 Relationship between per cent leaf area injured and number of daily exposures of 20 minutes each to simulated acid rain on the middle leaflets of soybean (*Glycine max*). ◊ = pH 2.7; □ = pH 2.9; △ = pH 3.1; ○ = pH 3.4. (Redrawn with permission from Evans, 1979.)

It can also be found, but with smaller numbers of adventitious shoots, in spruce and beech especially after drought. Fig. 10.1 shows the varying state of health of a mountain forest in the Bavarian Alps. The upper photographs taken in 1900 show a forest made up of trees mostly over 100 years old in which, by modern criteria, practically every tree would be considered unhealthy. The forest was felled in 1905/06 and the already established undergrowth then developed to produce the present vigorous forest with an intact canopy. In another 10 years the present forest will have the same age as the unhealthy one photographed in 1900. This example shows that global surveys of forest decline will be liable to include examples of forest decline that are due solely to the interaction of climate, forest age, and disease as well as atmospheric pollution.

European silver fir is particularly sensitive to water stress and with trees over 50 years old the 'stork's nest' symptom can be found irrespective of the presence of atmospheric pollutants. German landscape artists from the pre-industrial era at the beginning of the nineteenth century (for example Karl Frommel, 1832) have depicted trees with just this morphology at their summits. However, silver fir suffered a serious decline in Germany towards the end of the nineteenth century. This early phenomenon of forest decline, referred to in Germany as '*Tannensterben*' (fir decline) reached epidemic proportions towards the end of the nineteenth century (Neger, 1908) and occurred in both industrial and non-industrial regions and was probably a coincidence of climatic stress, forest management and fungal disease. These may have been years of climatic stress as far as silver fir was concerned, but there seems no doubt in the opinion of those working with trees at the time that the atmospheric pollution from cities adversely affected the growth of the more common spruce (*Picea abies*). In Munich it was found impossible to grow this tree in the Botanic Garden and the lack of success was attributed to coal-smoke (Goebel, 1899, Guide to the Royal Botanic Garden, Munich).

Since the early 1970s there has been an increase in 'forest decline' in Germany and neighbouring areas in central Europe. At first damage was confined to silver fir, which was already a minority species and known to be prone to periods of decline. However, since the late 1970s damage has also been noticed on spruce (*Picea abies*), pine (*Pinus sylvestris*) and beech (*Fagus sylvatica*). This type of injury is distinct from the 'stock's nest' syndrome which typified the earlier forms of forest decline and has therefore been described as the 'new type of forest decline' (*neuartigen Waldschaden*) (Blank, 1985). The symptoms include needle or leaf shedding, canopy loss, and in many cases an acute chlorosis or yellowing of second year needles. The chlorotic injury is seen most clearly on needles on the upper side of the branches, in trees at the edge of forests, or in clearings (Fig. 10.24). Initially chlorosis damage was restricted to trees at higher altitudes (above 500 m with younger trees (less than 60 years old) being more severely affected than older ones. Although trees in the upper regions of valley sides are one of the commonest sites of damage, trees of all ages in many different types of habitat can be affected. The initial distribution of chlorotic injury to trees on high mountain valley sides with the top-side needles showing the first symptoms indicates the role of light, at least in making damage apparent. The mountain valley habitats are the regions where atmospheric pollution can be trapped and where high

Fig. 10.23 Photograph taken in 1905 in the Fichtelgebirge (Germany) showing evidence of forest decline in both European silver fir (*Abies alba*) and spruce (*Picea abies*); X = injured spruce trees. (Photograph: from Neger (1908) kindly supplied by Professor O. Kandler – Munich.)

Fig. 10.24 Branches of spruce in the Bavarian Forest showing the typical needle chlorosis of upper second year needles. Note the presence of lichen on the trunk of an adjoining tree indicating that the forest is not subject to high sulphur dioxide concentrations.

light intensity will facilitate the production of ozone. The visible symptoms at first sight have all the appearance of a photochemical bleaching of light-exposed needles. Ozone has been the pollutant most commonly connected with this type of injury. Other pollutants may also give rise to this same chlorophyll bleaching by sunlight. In particular the anthropogenic volatile halogenated hydrocarbons are a possible cause of this phenomenon (Frank and Frank, 1985). Similarly, exposure to elevated concentrations of natural hydrocarbons such as β-pinene which could be released from needles after pollutant damage can reduce chlorophyll and carotenoid content in spruce needles and cause a severe reduction in photosynthesis (Wagner *et al.*, 1987) (Fig. 10.25). The bleaching however does not follow the normal pattern of pigment photo-oxidation where it would be expected that first carotene, then chlorophyll and xanthophyll would be progressively reduced. Instead there is a general overall reduction in pigment content which suggests some blockage of their synthesis rather than a true photochemical bleaching (Kandler *et al.*, 1987).

A typical consequence of damage leading to pigment loss is the enhancement of anti-oxidant activity (Fig. 10.26). However, if atmospheric pollution alone was the prime cause of injury then it would be expected that in any given area subject to toxic levels of ozone or other toxic gases that there would be a uniform incidence of damage between trees of a similar age, class, or exposure. Long-term observations of damaged needles however reveal a number of curious factors which question whether atmospheric pollution alone can be the principal cause of forest decline:

1   In any stand only some trees are affected and one tree can show serious damage while its neighbour remains healthy.
2   In cases of chlorotic injury alone the symptoms are frequently reversible with needles that are chlorotic in one season recovering the following year.
3   Soil treatment with essential nutrients (N, Mg, K, P, S, and Fe) can frequently, but not always, restore a healthy colour to the needles. The precise nature of the mineral deficiency varies with soil type. In the calcareous soils of the Alps potassium and manganese deficiencies are found in chlorotic

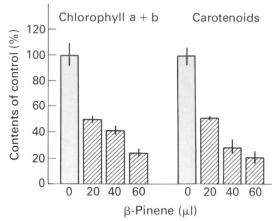

Fig. 10.25 Effect of β-pinene on chlorophyll and carotenoid content of spruce seedlings. (Redrawn with permission from Wagner *et al.*, 1987.)

needles while on acid soils as in the Blackforest and Bavarian Forest magnesium deficiency is more typical.

These observations suggest that although the damage may be made apparent by photochemical bleaching with the aid of ozone, the sensitivity of the trees to this pigment loss may depend on the nutritional status of the soil. This could account for the variation between trees as even neighbouring trees might have differing mycorrhizal developments. Seasonal recovery and response to mineral nutrients is highly suggestive of a soil-mediated factor in the 'new type' forest decline symptoms. Purely atmospheric damage would be expected to be more uniform in action. It is important to note however that the

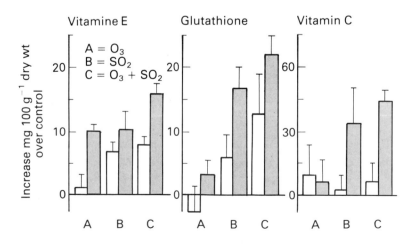

Fig. 10.26 Enhancement of anti-oxidant concentrations in two-year-old spruce needles: □ Needle year 1985; ■ Needle year 1984. (Redrawn from Kunert and Hofer, 1987.)

more general symptoms of needle shedding and canopy loss have not so far been correlated with any soil factors (Kandler, personal communication).

Increased nitrogen input into forest soils from anthropogenic sources has also been suggested as a contributing factor to soil acidification (van Breemen *et al.*, 1982). In the Netherlands it has been observed that high concentrations of ammonium can be found in the rain-water falling through the forest canopy. The ammonium (volatalized from manure) is in the form of ammonium sulphate with the anion coming from aerial sulphur dioxide pollutants. On reaching the soil this is rapidly oxidized to nitric acid and sulphuric acid producing very low pH values (2.8–3.5). Increased nitrogen together with increased acidity is likely to induce greater mineral deficiencies as the nitrogen will enhance growth and thus increase the demand for nutrients which are likely to be even further depleted due to the increased acidity. The Dutch study may be an extreme case as the forests were in areas with intensive animal husbandry. However there is a general increase in atmospheric nitrogen input to ecosystems both from agriculture due to increased use of nitrogen and from fossil fuel burning. There are therefore grounds in areas such as Holland for suspecting a doubly damaging effect from nitrogen pollution as it both increases the demand for mineral nutrients and reduces their supply. In some cases a stimulation of tree growth has been noted before the onset of 'forest decline' symptoms. This stimulation of growth coupled with leaching by acid rain of essential nutrients could provide a possible explanation of the damage. Needles which are nutrient deficient are likely to be more prone to photo-oxidation damage and a correlation has been demonstrated in one investigation at least between high nitrogen and low magnesium content in spruce needles (Schulze *et al.*, 1987) (Fig. 10.27).

Clearly, finding a single cause for 'forest decline' is proving elusive. The many facets to the damage reflect variations in geography, soil chemistry, climatic stress, pollutant exposure, and possibly most important of all forestry practice. Different experiences in different forests have resulted in a polarization of views as to whether the new type of forest decline is a direct effect of atmospheric pollution on the aerial parts of the trees or whether it acts via the soil, making the aerial parts prone to an environmental stress which can include both drought and ozone injury. There is little evidence of 'forest decline' in the British Isles where trees are grown on short 40-year rotations, are fertilized from the air, and only rarely exposed to hot dry summers. German

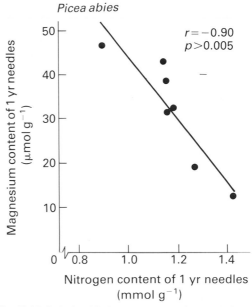

Fig. 10.27 Relationship between magnesium and nitrogen content of spruce needles in the Fichtelgebirge (South Germany) 1985. (Adapted from Schulze *et al.*, 1987.)

forestry by contrast, has a long tradition of natural management which has always endeavoured to maintain mature forests of a greater age than would be found in commercial plantations in the British Isles. Part of this tradition is the avoidance of the use of fertilizers. This again contrasts with the British scene where aerial dressings of fertilizer are common. The unfertilized mountainside soils of Germany are therefore likely to be strongly influenced by the effects of acid rain, both in leaching and the addition of nitrogen. They are also geographically the sites (*see* Fig. 10.12) where high ozone levels are likely to be most frequent. The widespread incidence of forest decline in many diverse habitats and species makes it impossible to attribute any one cause to the current epidemic. The lack of a precise scientific diagnosis of this disease clearly causes the public much disquiet. Confidence in environmental protection is undermined, as despite much research there is no simple ecological explanation that fits all cases. The explanation of the enigma of forest decline probably lies in the fact that like the acidification of northern lakes, there are a number of adverse factors which have unfortunately coincided in recent years. The increase in ozone levels, the depletion of minerals, the addition of nitrogen, and the adverse climatic factor of some extremely hot summers in the late

1970s can all be linked with 'forest decline'. Changes in management practices might reduce the incidence of injury. The German forestry tradition is one that is to be admired ecologically with the large areas of mature and varied forest providing a wide diversity of ecological habitats. To adopt the British practice of short rotations and aerial fertilizer application although it might reduce the incidence of damaged trees would be to admit defeat and further increase the eutrophication of rivers and lakes.

10.3.3 HYDROCARBONS

The catastrophic effect of oil spillage on marine life such as seaweed, molluscs, and crustacea as well as bird populations, although ecologically unfortunate and a nuisance to those using the seashore either for business or pleasure is not the worst form of biological disaster. In temperate regions provided the pollution is not repeated conditions can be expected to return to normal within a year. Many hydrocarbons undergo complete auto-oxidation without any biological mediation provided they are activated by light (particularly u.v. below 400 nm), heat, or metal ions. The ability of fungi to attack hydrocarbons has long been known and about 20 per cent of all microbial species examined exhibit some ability to digest these compounds. The greatest activity is found in soil bacteria and fungi which can metabolize both branched alkanes as well as alkenes, aromatics, and acyclic compounds. Due to this activity crude oil spillage on soils in temperate and tropical climates will disappear in a matter of months. In cleaning the oil spillage from railway tracks it is now being found much more economical to rely on microbial degradation than to use physical cleaning. When the oil spillage is heavy and in danger of swamping the microbial flora, treatment with sulphur-coated aurea granules can provide a suitable stimulus for microbial growth. The sulphur coating ensures the slow release of the nitrogen which is needed to stimulate microbial growth (*see* Biotechnology in Glasgow, Newsletter, 1987 − University of Strathclyde).

The most serious hydrocarbon pollutants of soil and water systems are the synthetic derivatives which have been produced in recent years by the detergent and insecticide industries. Recalcitrant hydrocarbon molecules include the so-called hard detergents, so named because they are resistant to enzyme attack. The detergent alkyl benzene suphonate (ABS) consists of a benzene ring with an alkane side chain terminating in a hydrophyllic sulphonic group. The alkane chain contains all possible branched arrangements of carbon atoms with C_{12} being the commonest. The branching of the side chain inhibits the destruction of the detergent by microbial β-oxidation and from 1945 onwards many of our rivers showed the effect of this indestructible detergent in the great quantities of foam they carried. The foam was not only unsightly but obliterated plant life, spread the eggs of parasitic worms and caused fatal accidents to boating parties. Now with the substitution of un-branched chains, the detergent problem is not nearly so serious as the detergents are readily degraded by microbial action.

Hydrocarbons are also made resistant to degradation by the formation of cross linkages and polymerization. The pollutants left by these compounds, the plastics, polyethylene, polyvinyl chloride, and teflon, although an aesthetic nuisance are not chemically toxic to plant life. Another group of hydrocarbons which due to artificial substitution are resistant to microbial degradation are the halogenated pesticides. The ecological consequences of this accumulation were brought to the attention of the world at large by Rachel Carson's *Silent Spring*.

10.3.4 METAL POLLUTANTS

Metals which as a result of industrial activity occur in increased concentrations and cause phytotoxic effects include zinc, copper, lead, and mercury. Contamination of soils with high concentrations of zinc and copper, although they present serious impediments to plant growth as in the recolonization of industrial wastelands, are usually localized in their occurrence to the spoil heaps of old mines. Lead and mercury on the other hand deserve special mention for their ubiquitous and insidious increase as environmental pollutants.

Lead

Lead has long been used by man in many forms. In the past lead carbonate ($PbCO_3$) or white lead was used in paint and is still employed as a wood primer. Red lead, triplumbic tetroxide (Pb_3O_4) and calcium plumbate ($CaPbO_4$) are used in the prevention of rust. Lead is also used in pottery glazes and as a stabilizer in plastics. Recently its most extensive mode of entry as an environmental pollutant has been in its use in tetraethyl and tetramethyl lead as anti-knock agents to improve the octane-rating of petrol. It is therefore not surprising that the lead content of the environment and principally our soil

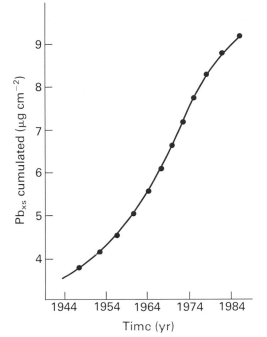

Fig. 10.28 Average inventory of the input of anthropogenic lead over the north Atlantic including lead accumulated before 1940. Data derived from atmospheric flux rates of pollutant lead from gasoline exhaust. (Data reproduced with permission from Veron *et al.*, 1987.)

is showing a steady rise. Antarctic soils, which may be taken as uncontaminated, normally have an average lead content of 10 ppm whereas near roads and lead smelters soils frequently have concentrations of 1000 ppm in their surface horizons.

It is possible to follow the chronology of lead pollution by analysing the ice horizons in the Greenland ice-cap and in ocean sediments (Fig. 10.28). During the first millenium BC it was no higher than 0.2 g Kg^{-1}. Similar studies of the lead content of herbarium specimens have shown that for mosses collected in Sweden, the lead content was only about 20 ppm in the period 1860–1875. This then doubled during the following 25 years (1875–1900) and remained constant until 1950 when a dramatic rise brought the content to its present level of 80–90 ppm. The rise in the nineteenth century has been attributed to the increase in coal consumption and that of the last 30 years to the combustion of lead-treated petrol (Higgins and Burns, 1975).

Higher plants are affected by lead only in high

concentrations and even in soils rich in this element, as in areas of mine waste, ecotypes are found which can tolerant up to 1 per cent of lead. The most susceptible part of the habitat to lead-poisoning is the soil micro-flora. Micro-organisms are relatively sensitive to lead-poisoning and organic decay and litter incorporation will be retarded when lead content increases. This problem is most commonly met with in the biological treatment of sewage, where lead levels varying between 2000 and 8000 ppm can be encountered. This not only produces difficulties in the treatment of sewage but means that the treated sludge will contaminate agricultural soils if used as a fertilizer. Detoxification of lead-containing sewage can be achieved by the addition of sulphate under anaerobic conditions. The sulphate under anaerobiosis is reduced to sulphide and this causes the lead and other heavy metals to be precipitated as the insoluble sulphide. Although this technique does allow the biological treatment of sewage to continue it does not counteract the spread of lead as a contaminant in the environment.

Lead at present does not present a problem to higher plant life, but the ability of plants to accumulate this element provides yet one more source of potential lead-poisoning to animals and man. It is a potentially dangerous pollutant as once it reaches toxic levels in our soil and water it will be difficult to reduce it speedily to a safe level. Although adult human beings can tolerate levels of up to 0.8 g l^{-1} in their blood, children are much more sensitive and as little as 25 µg dl^{-1} in fetus blood has been suggested as being the cause of subsequent learning difficulties in young children.

When lead toxicity is examined in higher plants its effects are seen most strikingly on cell division in root tips. Accumulation takes place in the nuclei as well as along cell walls. At 0.125 ppm (6×10^{-7}M) lead can cause a decrease in the rate of mitosis in root tips of *Zea mays* and *Allium cepa*. Lead can also form complexes with sulphydryl groups and from histochemical studies it is this action which is thought to interfere with the mitotic activity of the root tips.

In mammals a characteristic symptom of lead poisoning is the inhibition of porphoryin biosynthesis. A comparison of δ-amino-laevulinic acid dehydratase in erythrocytes and chloroplasts (Hamp and Ziegler, 1974) showed that this enzyme is equally sensitive to lead inhibition in plants and animals. This is not the only point where haem synthesis is inhibited by lead. Fig. 10.29 indicates a number of places in haem

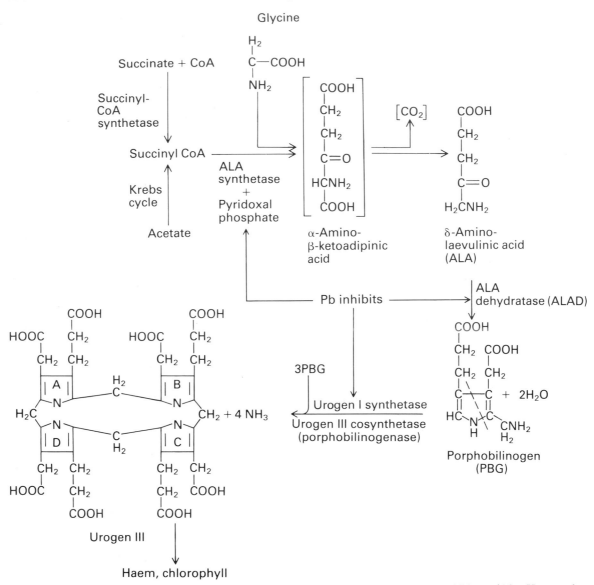

Fig. 10.29 Outline of chlorophyll biosynthesis showing steps where lead has been reported as inhibitory. (After Hamp and Ziegler, 1974.)

synthesis where lead has been found to be inhibitory and will prevent chlorophyll synthesis in plants just as it prevents haem synthesis in animals.

Mercury

Mercury is another element whose widespread use in industrial processes means that it is constantly being brought from its natural ore-deposits and distributed in increasing amounts on the surface of

the earth. The largest industrial use at present is in the chlor-alkalı process for chlorine production where mercury is released in the waste water. Organic and inorganic forms are also used as fungicides in agriculture. The ecological dangers from mercury pollution stem from the chemical transformations that can take place once mercury has been released into the environment. Methylation of mercury greatly increases its toxicity and microbe growth can be prevented by methyl mercuric chloride at only one-twentieth of the concentration that is needed by mercuric chloride, itself a very toxic compound. The methylation of mercury can take place enzymatically within the carcasses of rotting fish as well as anaerobically in lake sediments by the action of *Clostridium* spp. Aerobic methylation can also be accomplished by methanogenic fungi, for example *Neurospora crassa* and by varying species of *Pseudomonas* (Higgins and Burns, 1975).

Under alkaline conditions, methyl mercury can be further converted non-enzymatically to dimethyl mercury. Even the insoluble mercuric sulphide can be converted to the soluble sulphate and then methylated biologically (Fig. 10.30). The effect of mercury on higher plants depends greatly on it solubility, for this controls both its uptake and mobility once absorbed. It is in its highly toxic methyl form that mercury is most soluble and therefore poses the greatest danger to plant life. Fortunately plants do not readily absorb large quantities of mercury and even in crops that are sprayed with mercuric pesticides, the eventual tissue concentrations are rarely more than 0.1 ppm (Smart, 1968). When absorbed it acts in a manner similar to lead, causing the inhibition of mitosis in roots. These mitotic disturbances can be lethal and the only known remedial treatment under experimental conditions is to treat the plant with an excess of cysteine or glutathione.

The dangers of mercury poisoning, like those of lead discussed above, come from its mobility through the food chain. The most classic case in recent years has been the 'Minimata Incident'. At Minimata, a small industrial town on Japan's west coast where an 8-year epidemic (1953–61) of mercury poisoning, eventually leading to 65 deaths, was finally traced to the discharge of mercuric sulphate catalyst from an acetaldehyde-producing plant. This mercury was methylated in the bottom of the bay, concentrated by the molluscs and fish feeding in these waters and then transferred to the local human population. It took 5 years from the outbreak of the poisoning for the source of the mercury to be traced. It should be remembered, however, that this incident took place before the biological transformation of inorganic mercury to its lethal organic form was known. The dramatic effects of the 'Minimata Incident' should not make us think of mercury poisoning as a chance occurrence of unfortunate circumstances which might from time to time lead to disaster. Industrial activity is constantly adding to the level of heavy metal pollutants in our soil and water systems and we need all the ecological and biochemical informa-

tion we can gather on the cycling and sequestration mechanisms of these potentially dangerous pollutants.

10.3.5 EVOLUTION OF METAL TOLERANCE

Due to the localized nature of metal pollution either around smelters, mine spoil heaps or motorways it is possible to investigate the physiology and ecology of plants that can survive in these sites as compared with those that grow in unpolluted areas. Frequently the date of the onset of pollution is known and as this presents an opportunity to observe the processes of evolution on a micro-scale such situations have attracted much research (Bradshaw, 1984; Antonovics *et al.*, 1971). The frequency of metal-tolerant individuals on unpolluted sites is always low and when areas are first exposed to metal pollutants the lack of sufficient resistance results in extensive damage with the production of chlorotic or necrotic leaves. However, when subjected to pollution species vary in relation to the facility with which tolerant plants can be found. In *Agrostis tenuis* and *Dactylis glomerata* it is possible to obtain copper tolerant individuals in one cycle of selection starting with a population of 10 000 seeds (Gartside and McNeilly, 1974). In this same experiment other species such as *Anthoxanthum odoratum* and *Plantago lanceolata* showed no copper-tolerant individuals even although it was possible to find zinc and lead-tolerant genotypes. From such studies it can be seen that tolerance to one metal does not necessarily lead to tolerance of others. Plants growing on copper-rich sites are tolerant only of copper but not of zinc (Woolhouse, 1983). Long-lived woody plants such as trees and shrubs are unable to evolve metal-tolerant ecotypes and the most successful invaders of toxic sites are grasses and some dicotyledonous herbs.

Zinc is less toxic than copper and this may explain why zinc-tolerant ecotypes are more readily obtained than those tolerant of copper. It is even possible to find in the area of water drip from a galvanized fence, localized populations of zinc-tolerant grasses. The extreme localization of tolerant ecotypes not only reflects the facility with which they can evolve but also the disadvantages of the tolerant type in unpolluted areas. Metal-tolerant ecotypes invariably have growth rates that are 20–50 per cent lower than that of the intolerant plants.

Despite much research no evidence has been found to suggest that metal-tolerant ecotypes avoid toxicity by limiting uptake of the toxic ions. Where ion exchange capacity of roots is reduced there is a reduction in the absorption rate of copper and zinc, but selection for cation exchange capacity is genetically independent of heavy metal tolerance (Ernst, 1976). Instead, tolerance appears to depend on the toxic ions being accumulated in tissues where they do not interfere with the metabolic activity of the cytoplasm. *In vitro* studies of enzymes from tolerant and intolerant plants, like those from glycophytes and halophytes have failed to demonstrate any evolution of specifically adapted enzymes in relation to their tolerance of toxic ions. The sequestration of the toxic ions in higher plants appears to be in the cell wall. In zinc-tolerant ecotypes of *Agrostis tenuis* there is a very close correlation between zinc tolerance and zinc accumulation in the cell wall (Turner and Marshall, 1972) (Fig. 10.31). When the cell wall is digested with trypsin no significant release of zinc is obtained. However digestion with cellulase released 66 per cent of the bound zinc, which suggests that it is the carbohydrate molecules that are responsible for the sequestration of the ion (Wyn Jones *et al.*, 1971). When zinc-tolerant and zinc-intolerant ecotypes are grown in culture solution with 10^{-4} M Zn^{2+} it was found that both susceptible and non-susceptible plants accumulated similar quantities of zinc in their leaves. From this it was concluded that in water culture the cell-binding capacity of the cell wall accounted for only 1 per cent of the absorbed zinc (Wainwright and Woolhouse, 1978). Consequently the differential ability of the ecotypes to grow roots in a zinc-rich water culture

could not be explained solely in terms of the properties of the cell wall compartmentation.

Analyses of the tricarboxylic acids of a number of higher plants (Ernst, 1976) has shown that zinc-tolerant species have higher concentrations of malic acid than zinc-sensitive species. It is possible that malic acid may act in the transfer of zinc from the cytoplasm to the vacuole. Copper can form even more stable complexes with malic acid than zinc without reducing the copper sensitivity of the plants. Therefore malic acid complexes alone cannot explain zinc-tolerant plants to tolerate this ion. Differences have also been found in the amount of oxalate ions in zinc-tolerant and zinc-intolerant species and a scheme has been suggested (Mathys, 1977) in which malate tranports zinc through the cytoplasm and oxalate then sequesters it in the vacuole possible with the aid of other more stable ligands.

The mechanism by which plants are able to evolve ecotypes tolerant of toxic metals such as copper, zinc, and lead still remains obscure. Not enough is known of the reactions that metals can make with potential ligands at the surface of the plasmamembranes (Woolhouse, 1983). When this is better understood it is likely to shed light on the problem of differential metal toxicity in relation to pollutants

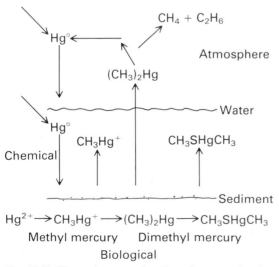

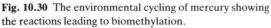

Fig. 10.30 The environmental cycling of mercury showing the reactions leading to biomethylation.

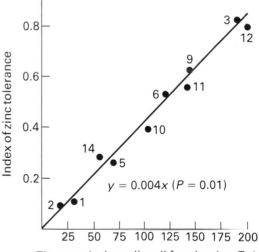

Fig. 10.31 Relationship between zinc accumulation by the cell wall and the index of zinc tolerance of 16 clones of *Agrostis tenuis*. The tolerance index was based on the mean length of the longest root in zinc solution. (Redrawn from Turner and Marshall 1972.)

such as zinc and copper as well as similar mysteries in relation to the differential response of plants to calcium, aluminium, and soil pH.

10.4 Plant survival after a nuclear war

Until recently the fear of nuclear war has centred on the devastating effects of blast, fire, and lethal radiation. Ecologically however its probable effects on climate are now being suggested as much more serious for the survival of plant and animal life. Observation of the effects of a dust storm on Mars in 1971 by the US spacecraft Mariner 9 showed that as the dust rose into the upper atmosphere and absorbed the incoming sunlight, the temperature at ground level dropped and returned to normal levels only after several months. This observation prompted a group headed by Carl Sagan to examine the effects of atmospheric dust on global temperatures. The dust level in the earth's upper atmosphere rises during periods of volcanic activity and the group

were able to relate the quantity of dust found in ice-cores with changes in temperature. Using this basic quantitative guide, predictions were then made from model systems as to the quantity of dust and temperature changes that might arise from a major nuclear exchange (Turco *et al.*, 1983). These predictions suggest that a 5000 megaton exchange would create enough dust to cause the surface temperature over land in the northern hemisphere to drop by 10–15°C. Depending on the time of year this temperature drop could last for several months or even up to a year and would be combined with a possible reduction in sunlight to less than 2 per cent of normal. This nuclear winter, especially if it were out of phase with the normal seasonal progression, would annihilate many plant and animal communities.

Ecologically it is of interest to speculate which plant communities would be the most likely to survive. Forests would appear most vulnerable as prolonged frost and fire would almost certainly be lethal. All plants with perennating organs at the surface of the soil would also be likely to suffer from

Fig. 10.32 The Indian water lotus *Nuphar nelumbo* a possible contender for a species able to survive a nuclear winter. The rhizomes would be protected by burial in sub-aquatic mud and the seeds of this species have been reported to germinate after burial in anaerobic mud for 250 years.

the prolonged cold and danger of fire. Seed banks would be drastically reduced and germinating seedlings would be likely to find environmental conditions too harsh for survival. The most likely plants to survive would be those that are deeply buried in soil or in the mud at the bottom of lakes. These species usually have large carbohydrate reserves. They also have the ability to remain dormant for prolonged periods and some can even survive months without oxygen (*see* p.116). Seeds buried in lake muds and wet peaty soils have the greatest chance of survival. Seeds of these habitats are known to survive burial longer than seeds in dry soils possibly due to the lack of oxidative damage to membranes in anaerobic environments. Seeds of the Indian water lotus have been reported to have survived burial in anaerobic mud for up to 250 years (Turrill, 1957). A post-nuclear war temperate landscape might therefore reveal blasted heath and dead burnt remains of forest but with lakes fringed with a green margin where the common reed (*Phragmites australis*), bullrush (*Schoenoplectus lacutris*), and the yellow flag iris (*Iris pseudacorus*) were able to emerge from the anaerobic mud after a period of enforced dormancy under the ice. The burial of the rhizomes of these species would also reduce their exposure to the high radiation levels that are likely to persist for many months after a nuclear war. Ice movement would bring to the surface the buried seeds of aquatic plants which due to their long viability would be more likely to germinate successfully than those from shallow dry soils (Fig. 10.32). It is likely therefore that apart from the oceans the habitats with the greatest powers of homeostasis after a nuclear winter would be the wetlands and freshwater lakes. Arctic vegetation may also be able to survive as many species of the polar floras can survive an entire growing season without making any net carbon gain and are able to survive frost at any season of the year. In northern Europe therefore a return to post-ice-age vegetation would appear a strong possibility with wetlands showing the first recovery then followed by grassland species. Depending on the devastation of the seed bank it would probably take many decades before there was an active forest regeneration.

References

Abeles F.B. (1973) *Ethylene in Plant Biology*. Academic Press, New York.

Ahmad I., Larher F. & Stewart G.R. (1979) Sorbitol a compatible osmotic solute in *Plantago maritima*. *New Phytol.* **82**, 671–678.

Ahmad I., Larher F. & Stewart G.R. (1981) The accumulation of Δ'-acetylornithine and other solutes in the salt marsh grass *Puccinellia maritima*. *Phytochemistry* **20**, 1501–1504.

Al-Ani A., Leblanc J.M., Raymond P. & Pradet A. (1982) Effet de la pression partielle d'oxygène sur la vitesse de germination des semences à réserves lipidiques et amylacées: rôle du metabolisme fermentaire. *C.R. Acad. Sci. Paris* **295**, 271–274.

Al-Ani A., Bruzau F., Raymond P., Saint-Ges V., Leblanc J.M. & Pradet A. (1985) Germination, respiration and adenylate energy charge of seeds at various oxygen partial pressures. *Plant Physiol.* **79**, 885–890.

Ambler J.E., Brown J.C. & Gauch H.G. (1971) Sites of iron reduction in soybean plants. *Agron. J.* **63**, 95–97.

Anderson M.C. (1964) Studies of the woodland light climate. II. Seasonal variation in the light climate. *Biol. Rev.* **39**, 425–486.

Antonovics J. & Primack R.B. (1982) Experimental ecological genetics in *Plantago*. VI. The demography of seedling transplants of *P. lanceolata*. *J. Ecol.* **70**, 55–75.

Antonovics J., Bradshaw A.D. & Turner R.G. (1971) Heavy metal tolerance in plants. *Adv. Ecol. Res.* **7**, 1–85.

Armstrong W. (1979) Aeration in higher plants. In *Advances in Botanical Research* (Ed. Woolhouse H.W.), Vol. 7, pp. 226–332.

Armstrong W. & Gaynard T.J. (1976) The critical oxygen pressures for respiration in intact plants. *Plant Physiol.* **37**, 200–206.

Arnold G.W. & Hill J.L. (1972) Chemical factors affecting selection of food plants by ruminants. In *Phytochemical Ecology* (Ed. Harborne J.B.), pp. 72–101. Academic Press, New York.

Ashmore M.R., Bell J.N..B. & Reily C.L. (1978) A survey of ozone levels in the British Isles using indicator plants. *Nature* **276**, 813–815.

Aspinall D. & Paleg L.G. (1981) Proline accumulation: physiological aspects. In *Drought Resistance in Plants* (Eds. Aspinall D. & Paleg L.G.), pp. 206–241. Academic Press, New York.

Auclair A.N. & Cottam G. (1971) Dynamics of black cherry (*Prunus serotina*) in Southern Wisconsin oak-forests. *Ecol. Monogr.* **41**, 53–77.

Bannister P. (1964) The water relations of certain heath plants with reference to their ecological amplitude. II. Field studies. *J. Ecol.* **52**, 481–497.

Bannister P. (1981) Carbohydrate concentration of heath plants of different origins. *J. Ecol.* **69**, 769–780.

Barber J., Horler D.N.H. & Chapman D.J. (1981) Photosynthetic pigments and efficiency in relation to the spectral quality of absorbed light. In *Plants and the Daylight Spectrum* (Ed. Smith H.), pp. 341–354. Academic Press, London.

Barclay A.M. & Crawford R.M.M. (1981) Temperature and anoxic injury in pea seedlings. *J. exp. Bot.* **32**, 943–949.

Barclay A.M. & Crawford R.M.M. (1982a) Winter desiccation and resting bud viability in relation to high altitude survival in *Sorbus aucuparia* L. *Flora* **172**, 21–34.

Barclay A.M. & Crawford R.M.M. (1982b) Plant growth and survival under strict anaerobiosis. *J. exp. Bot.* **33**, 541–549.

Barclay A.M. & Crawford R.M.M. (1983) The effect of anaerobiosis on carbohydrate levels in storage tissues of wetland plants. *Ann. Bot.* **51**, 255–259.

Barclay A.M. & Crawford R.M.M. (1984) Seedling emergence in the rowan *Sorbus aucuparia* from an altitudinal gradient. *J. Ecol.* **72**, 627–636.

Barnes B.V. (1966) The clonal growth of the American aspen. *Ecology* **47**, 439–447.

Battarbee R.W., Flower R.J., Stevenson A.C. & Rippey B. (1985) Lake acidification in Galloway: a palaeological test of competing hypotheses. *Nature* **314**, 350–352.

Batzli G.O., White R.G., Maclean Jr S.F., Pitelka F.A. & Collier B.D. (1980) The herbivore-based trophic system. In *An Arctic Ecosystem* (Eds. Brown J., Miller P.C., Tieszen L.L. & Bunnell F.L.). Dowden, Hutchinson & Ross, Stroudsburg, Pennsylvania.

Bawa K.S. (1980) Evolution of dioecy in flowering plants. *Ann. Rev. Ecol. Syst.* **11**, 15–39.

Bazzaz F.A. & Carlson R.W. (1984) The response of plants to elevated CO_2. I. Competition among an assemblage of annuals at two levels of soil moisture. *Oecologia* **62**, 196–198.

Begon M. & Mortimer M. (1981) *Population Ecology*. Blackwell Scientific Publications, Oxford.

Begon M., Harper J.L. & Townsend C.R. (1986) *Ecology: Individuals Populations and Communities*. Blackwell Scientific Publications, Oxford.

Bell E.A. (1971) Comparative biochemistry of non-protein

amino acids. In *Chemotaxanomy of the Leguminosae* (Eds. Harborne J.B., Boulter D. & Turner B.L.), pp. 179–204. Academic Press, New York.

Bell E.A. (1972) Toxic amino acids in the leguminosae. In *Phytochemical Ecology* (Ed. Harborne J.B.), pp. 163–177. Academic Press, New York.

Bell E.A. & Janzen D.H. (1971) Medical and ecological considerations of L-Dopa and 5-HTP in seeds. *Nature* **229**, 136–137.

Bell J.N.B. & Clough W.S. (1973) Depression in yield of ryegrass exposed to sulphur dioxide. *Nature* **241**, 47–48.

Bender M.M. (1968) Mass spectrometric studies of carbon-13 variations in corn and other grasses. *Radiocarbon* **10**, 468–472.

Benecke U. (1972) Wachstum CO_2-Gaswechsel und Pigmentgehalt einiger Baumarten nach Ausbringung in verschiedene Hoehenlagern. *Angew. Bot.* **46**, 117–135.

Bennert H.W. & Schmidt B. (1983) Untersuchungen zur Salzabscheidung bei *Atriplex hymenelytra* (Torr.) Wats. (Chenopodiaceae). *Flora* **174**, 341–355.

Bennet A.B. & Spanswick R.M. (1984) H^+-ATPase activity from storage tissue of *Beta vulgaris*. *Plant Physiol.* **74**, 545–548.

Bentley S. & Whittaker J.B. (1979) Effects of grazing by a chrysomelid beetle, *Gastrophysa viridula*, on competition between *Rumex obtusifolius* and *Rumex crispus*. *J. Ecol.* **67**, 79–90.

Benzing D.H. (1976). Bromeliad trichomes: structure function and ecological significance. *Selbyana* **1**, 330–348.

Berry J.A. & Raison J.K. (1981) Responses of macrophytes to temperature. In *Encyclopedia of Plant Physiology*, Vol. 12A, pp. 277–338. Springer Verlag, Berlin.

Bertani A. & Brambilla I. (1982a) Effect of decreasing oxygen concentration on some aspects of protein and amino-acid metabolism in rice roots. *Z. Pflanzenphysiol.* **107**, 193–200.

Bertani A. & Brambilla I. (1982b) Effect of decreasing oxygen concentration on wheat roots: growth and induction of anaerobic metabolism. *Z. Pflanzenphysiol.* **108**, 283–288.

Bewley J.D. (1972) The conservation of polyribosomes in the moss *Tortula ruraliformis* during total desiccation. *J. exp. Bot.* **23**, 692–698.

Bewley J.D. (1974) Protein synthesis and polyribosome stability upon desiccation of the aquatic moss *Hygrohypnum luridum*. *Can. J. Bot.* **52**, 423–427.

Bewley J.D. & Krochko J.E. (1982) Desiccation tolerance. In *Encyclopedia of Plant Physiology*, Vol. 12B, pp. 325–378. Springer Verlag, Berlin.

Bewley J.D. & Pacey J. (1978) Desiccation-induced ultrastructural changes in drought-sensitive and drought-tolerant plants. In *Dry Biological Systems* (Eds. Crowe J.H. & Clegg J.S.), pp. 52–73. Academic Press, New York.

Bewley J.D., Tucker E.B. & Gwozda E.A. (1974) The effect of stress on the metabolism of *Tortula ruralis*. In *Mechanisms of Plant Growth* (Eds. Bielski R.L., Ferguson A.R. & Creswell M.M.). *Roy Soc. New Zealand Bull.* **12**, 395–402.

Biebl R. (1964) Austrocknungsresistanz tropischen Urwaldmoose auf Puerto Rico. *Protoplasma* **59**, 277–299.

Billings W.D. (1974a) Adaptation and origins of alpine plants. *Arct. Alp. Res.* **6**, 129–142.

Billings W.D. (1974b) Arctic and alpine vegetation: plant adaptations to cold climates. In *Arctic and Alpine Environments* (Eds. Ives J.D. & Barry R.G.), pp. 403–443. Methuen & Co., Ltd, London.

Billings W.D., Godfrey P.J., Chabot B.F. & Bourque D.P. (1971) Metabolic acclimations to temperature in arctic and alpine ecotypes of *Oxyria digyna*. *Arct. Alp. Res.* **3**, 277–290.

Björkman O. (1968) Further studies on differentiation of photosynthetic properties in sun and shade ecotypes of *Solidago virgaurea* L. *Physiol. Plant.* **21**, 84–89.

Björkman O. (1981) Responses to different quantum flux densities. In *Encyclopedia of Plant Physiology*, Vol. 12A, pp. 57–107. Springer Verlag, Berlin.

Björkman O. & Holmgren P. (1961) Studies of climatic ecotypes of higher plants. Leaf respiration in different populations of *Solidago virgaurea*. *Kung. Landbrukshog. ann.* **27**, 279–304.

Björkman O., Nobs M., Mooney H., Troughton J., Berry J., Nicholson F. & Ward W. (1974) Growth responses of plants from habitats with contrasting thermal environments: transplant studies in Death Valley and the Bodega Head experimental garden. *Carnegie Inst. Wash. Yr Book.* **73**, 748–757.

Black C.R. & Black C.J. (1979) The effect of low concentrations of sulphur dioxide on stomatal conductance and epidermal cell survival in field bean (*Vicia faba* L.). *J. exp. Bot.* **30**, 291–298.

Blackman G.E. & Rutter A.J. (1946) Physiological and ecological studies of plant environment. II. The light factor and the distribution of the blue bell (*Scilla nonscripta*). *Ann. Bot. NS.* **10**, 361–390.

Blank L.W. (1985) A new type of forest decline in Germany. *Nature* **314**, 311–314.

Blenkinsop P.G. & Dale J.E. (1974) The effects of shade treatment and light intensity on ribulose 1, 5 phosphate carboxylase activity and fraction I protein level in the first leaf of barley. *J. exp. Bot.* **25**, 899–914.

Boardman N.K., Björkman O., Anderson J.M. & Goodchild D.J. (1975) Photosynthetic adaption of higher plants to light intensity: Relationship between chloroplast structure, composition of the photosystems and photosynthetic rates. In *Proceedings of 3rd International Congress on Photosynthesis* (Ed. Avron A.), pp. 809–827. Elsevier, Amsterdam.

Böcher T.W. (1972) Evolutionary problems in the Arctic flora. In *Taxonomy, Phytogeography and Evolution* (Ed. Valentine D.H.), pp. 101–113. Academic Press, London.

Bounds D.G. & Pope G.S. (1960) Light absorption and chemical properties of miroestrol the oestrogenic substance of *Pueraria mirifica. J. Chem. Soc.* 3696–3705.

Bowers W.S. (1982) Endocrine strategies for insect control. *Entomol. Exp. Appl.* **31**, 3–14.

Bowers W.S. (1984) Insect-plant interactions: endocrine defenses. In *Origins and Development of Adaptation*, Ciba Foundation Symposium 102, pp. 119–137. Pitman, London.

Bowers W.S., Fales H.M., Thompson M.J. & Uebel E.C. (1966) Juvenile hormone: identification of an active compound from balsam fir. *Science* **154**, 1020–1022.

Bradbury R.B. & White D.E. (1954) Oestrogens and related substances in plants. *Vit. Horm.* **12**, 207–223.

Bradford K.J. & Hsiao T.C. (1982) Physiological responses to moderate water stress. In *Encyclopedia of Plant Physiology*, Vol. 12B. pp. 223–234. Springer Verlag, Berlin.

Bradford K.J. & Young S.F. (1980) Xylem transport of 1-amino cyclopropane-1-carboxylic acid, an ethylene precursor in waterlogged tomato plants. *Plant Physiol.* **65**, 322–326.

Bradshaw A.D. (1984) Adaptation of plants to soils containing toxic metals — a test for conceit. In *Origins and Development of Adaptations*, Ciba Foundation Symposium 102, pp. 4–19. Pitman, London.

Brower L.P. (1969) Ecological chemistry. *Sci. Amer.* **220** (2), 22–29.

Brown S.A. (1981) Coumarins. In *The Biochemistry of Plants* (Eds. Stumpf P.K. & Conn E.E.), Vol. 7, pp. 269–300. Academic Press, New York.

Bryson R.A. (1966) Air masses, streamlines and the boreal forest. *Geog. Bull.* **8**, 228–269.

Bullard E.R., Shearer H.D.H., Day J.D. & Crawford R.M.M. (1987) Survival and flowering of *Primula scotica* (Hook). *J. Ecol.* **75**, 589–602.

Butler W.L. (1977) Chlorophyll fluorescence. A probe for electron transfer and energy transfer. In *Encyclopedia of Plant Physiology*, Vol. 5, pp. 149–167. Springer Verlag, Berlin.

Callow J.A. (1975) Plant lectins. *Current Adv. Plant Sci.* **7**, 181–193.

Carlson R.W. & Bazzaz F.A. (1982) Photosynthetic and growth response to fumigation with SO_2 at elevated CO_2 for C_3 and C_4 plants. *Oecologia* **54**, 50–54.

Chabot B.F. (1979) Metabolic and enzymatic adaptations to low temperature. In *Comparative Mechanisms of Cold Adaptation* (Eds. Underwood L.S., Tieszen L.L., Callahan A.B. & Folk G.E.), pp. 283–301. Academic Press, New York.

Chabot B.F., Chabot J.F. & Billings W.D. (1972) Ribulose 1, 5 diphosphate carboxylase activity in arctic and alpine populations of *Oxyria digyna. Photosynthetica* **6**, 364–369.

Chapin III F.S., Miller P.C., Billings W.D. & Coyne P.I. (1980) Carbon and nutrient budgets and their control in coastal tundra in Alaska. In *An Arctic Ecosystem* (Eds. Brown J., Miller P.C., Tieszen L.L. & Bunnell F.L.), pp. 458–482. Dowden, Hutchinson & Ross, Stroudsberg, Pennsylvania.

Chapman H.M. & Crawford R.M.M. (1981) Growth and regeneration in Britain's most northerly natural woodland. *Trans. Bot. Soc. Edinb.* **43**, 327–335.

Charlson R.J. & Rodhe H. (1982) Factors controlling the acidity of natural rainwater. *Nature* **295**, 683–685.

Chen D., Sarid S. & Katchalski E. (1968). The role of water stress in the inactivation of messenger RNA of germinating wheat embryos. *Proc. Nat. Acad. Sci.* **61**, 1378–1383.

Clark D.L. (1982) Origin, nature and world climate effect of Arctic Ocean ice-cover. *Nature* **300**, 321–325.

Clarke B.C. (1979) The evolution of genetic diversity. *Proc. R. Soc. Lond.* B **205**, 453–474.

Clarkson D.T., Hall K.C. & Roberts J.K.M. (1980) Phospholipid composition and fatty acid desaturation in the roots of rye during acclimatisation to low temperatures. Positional analysis of fatty acids. *Planta* **149**, 464–471.

Clutton-Brock T.H. & Harvey P.H. (1979) Comparison and adaptation. *Proc. R. Soc. Lond.* B **205**, 547–565.

Coley P.D., Bryant J.P. & Chapin III F.S. (1985) Resource availability and plant antiherbivore defense. *Science* **230**, 895–899.

Cook S.A. & Johnson M.P. (1968) Adaptation to heterogeneous environments. I. Variation in heterophylly in *Ranunculus flammula* L. *Evolution* **22**, 496–516.

Coombe D.E. (1966) The seasonal light climate and plant growth in a Cambridgeshire wood. In *Light as an Ecological Factor* (Eds. Bainbridge R., Evans G.C. & Rackham O.), 6th Symposium of the British Ecological Society, pp. 148–166. Blackwell Scientific Publications, Oxford.

Cooper A. (1982) The effects of salinity and waterlogging on the growth and cation uptake of salt-marsh plants. *New Phytol.* **90**, 263–275.

Cooper-Driver G.A. & Swain T. (1976) Cyanogenic polymorphism in bracken in relation to herbivore predation. *Nature* **260**, 604.

Corner E.J.H. (1964) *The Life of Plants*. Weidenfield & Nicolson, London.

Coutts M. & Philipson J.J. (1978) The tolerance of tree roots to waterlogging. II. Adaptation of sitka spruce and lodgepole pine to waterlogged soil. *New Phytol.* **80**, 71–77.

Cowan I.R. (1977) Stomatal behaviour and environment. *Adv. Bot. Res.* **4**, 117–228.

Cowan I.R. (1982) Regulation of water use in relation to carbon gain in higher plants. In *Encyclopedia of Plant Physiology*, Vol. 12B, pp. 589–613. Springer Verlag, Berlin.

Cowan I.R. & Milthorpe F.L. (1968) Plant factors affecting the state of plant tissues. In *Water Deficits and Plant Growth* (Ed. Kozlowski T.T.), Vol. 1, pp. 137–193. Academic Press, New York.

Crawford R.M.M. (1982a) Habitat specialisation in plants of cold climates. *Trans. Bot. Soc. Edinb.* **44**, 1–12.

Crawford R.M.M. (1982b) Physiological responses to flooding. In *Encyclopedia of Plant Physiology*, Vol. 12B, pp. 453–477. Springer Verlag, Berlin.

Crawford R.M.M. (1987) (Ed.) *Plant Life in Aquatic and Amphibious Habitats*. British Ecological Society Special Publication No. 5. Blackwell Scientific Publications, Oxford.

Crawford R.M.M. & Balfour J. (1983) Female predominant sex ratios and physiological differentiation in arctic willows. *J. Ecol.* **71**, 149–160.

Crawford R.M.M. & Palin M.A. (1981) Root respiration and temperature limits to the north south distribution of four perennial maritime plants. *Flora* **171**, 338–354.

Crawford R.M.M. & Wishart D. (1966) A multivariate analysis of the development of dune slack vegetation in relation to coastal accretion at Tentsmuir, Fife. *J. Ecol.* **54**, 729–743.

Crawford R.M.M. & Zochowski Z.M. (1984) Tolerance of anoxia and ethanol toxicity in chickpea seedlings. *J. exp. Bot.* **35**, 1472–1480.

Crawford-Sidebotham T.J. (1972) The role of slugs and snails in the maintenance of cyanogenesis polymorphisms of *Lotus corniculatus* and *Trifolium repens*. *Heredity* **28**, 405–411.

Crawley M.J. (1983) *Herbivory: the Dynamics of Animal-Plant Interactions*. Blackwell Scientific Publications, Oxford.

Cresswell E.G. & Grime J.P. (1981) Induction of light requirement during seed development and its ecological consequences. *Nature* **291**, 583–585.

Croizat L.C., Nelson G.J. & Rosen D.E. (1974) Centres of origin and related concepts. *Syst. Zool.* **23**, 265–287.

Cronquist A. (1977) On the taxonomic significance of secondary metabolites in plants. *Evol., Suppl.* **1**, 179–189.

Dacey J.W.H. (1980) Internal winds in water lilies: an adaptation for life in anaerobic sediments. *Science* **210**, 1017–1019.

Dacey J.W.H. & Klug M.J. (1982) Ventilation by floating leaves in *Nuphar*. *Amer. J. Bot.* **69**, 999–1003.

Dahl E. (1951) On the relation between summer temperatures and the distribution of alpine plants in the lowlands of Fennoscandinavia. *Oikos* **3**, 22–52.

Danell K., Elmqvist T., Ericson L. & Salomonson A. (1985) Sexuality in willows and preference by bark-eating voles: a defence or not? *Oikos* **44**, 82–90.

Darwin C. (1859) *The Origin of Species*. Harvard facsimile, 1st edn. (1964).

Daubenmire R. (1954) Alpine timberlines in the Americas and their interpretation. *Butler Univ. Stud.* **11**, 119–136.

Davies J.M. & Svensgaard D.J. (1987) Lead and child development. *Nature* **329**, 297–300.

Davies M.S. & Snaydon R.W. (1976) Rapid population differentiation in a mosaic environment. III. Measures of selection pressure. *Heredity* **36**, 59–66.

Davitaja F.F. & Melnik J.J. (1962) Radiation heating of the plant's active surface and the latitudinal and altitudinal limits of forest. (In Russian). *Meteord. Gidrol. Moskva* 3–9.

Delf E.M. (1912) Transpiration in succulent plants. *Ann. Bot.* **26**, 411–443.

Deverall B.J. (1977) *Defence Mechanisms of Plants*. Cambridge University Press, Cambridge.

Dickinson R.E. & Cicerone R.J. (1986) Future global warming from atmospheric trace gases. *Nature* **319**, 109–115.

Dirzo R. & Harper J.L. (1982) Experimental studies on slug-plant interactions. *J. Ecol.* **70**, 101–117; 119–138.

Drew M.C., Jackson M.B., Giffard S.C. & Campbell R. (1981) Inhibition by silver ions of gas space (aerenchyma) formation in adventitious roots of *Zea Mays* L. subjected to exogenous ethylene or to gaseous oxygen deficiency. *Planta* **153**, 217–224.

Duncalf W.G. (1976) *The Guinness Book of Plant Facts and Feats*. Guinness Superlatives, London.

Edmondson W.T. (1970) Phosphorus, nitrogen and algae in Lake Washington after diversion of sewage. *Science* **169**, 690–691.

Egle K. (1960) Menge und Verhältnis der Pigmente. In *Encyclopedia of Plant Physiology*, Vol. 5, pp. 492–496. Springer Verlag, Berlin.

Ehleringer J. & Björkman O. (1977) Quantum yields for CO_2 uptake in C_3 and C_4 plants. *Plant Physiol.* **59**, 86–90.

Ehleringer J. & Björkman O. (1978) A comparison of photosynthetic characteristics of *Encelia* species possessing glabrous and pubescent leaves. *Plant Physiol.* **62**, 185–190.

Eickmeier W.G. (1979) Photosynthetic recovery in the resurrection plant *Selaginella lepidophylla* after wetting. *Oecologia* **39**, 93–106.

Ellenberg H. (1958) Wald oder Steppe? Die natürliche Pflanzendecke der Anden Perus. I and II. *Umschau* **58**, 645–648, 679–681.

Ernst W. (1976) Physiological and biochemical aspects of metal tolerance. In *Effects of Air Pollutants on Plants* (Ed. Mansfield T.A.), Society for Experimental Biology Seminar Series, Vol 1, pp. 115–133. Cambridge University Press, Cambridge.

Etherington J.R. & Thomas O.M. (1986) Response to

waterlogging and differential divalent iron and manganese in clones of *Dactylis glomerata* L. derived from well-drained and poorly-drained sites. *Ann. Bot.* **58**, 109–119.

Evans L.S. (1979) Foliar responses that may determine plant injury by simulated acid rain. In *Polluted Rain* (Eds. Toribara T.Y., Miller M.W. & Morrow P.E.), pp. 239–257. Plenum Press, New York.

Evans L.S. & Ting I.P. (1973) Ozone-induced membrane permeability changes. *Amer. J. Bot.* **60**, 155–162.

Feeny P.P. (1975) Biochemical coevolution between plants and their insect herbivores. In *Coevolution of Plants and Animals* (Eds. Gilbert L.E. & Raven P.H.), pp. 3–19. University Texas, Austin.

Feeny P.P. (1976) Plant apparency and chemical defense. In *Recent Advances in Plant Biochemistry*, Vol. X. *Biochemical Interactions between Plants and Insects* (Ed. Wallace J.), pp. 1–40. Plenum Press, New York.

Fiala K. (1978) Underground organs in *Typha angustifolia* and *Typha latifolia*, their growth and production. *Acta Sc. Nat. Brno.* **12**, 1–43.

Firbas F. & Losert H. (1949) Untersuchung über die Entstehung der heutigen Waldstufen in der Sudeten. *Planta* **36**, 478–506.

Fisher R.A. (1930) *The Genetical Theory of Natural Selection.* Oxford University Press, Oxford.

Flower R.J., Battarbee R.W. & Appleby P.G. (1987) The recent palaeolimnology of acid lakes in Galloway, South-West Scotland: diatom analysis, pH trends, and the role of afforestation. *J. Ecol.* **75**, 797–824.

Flower-Ellis J.G.K. (1973) Growth and morphology in the evergreen dwarf shrubs *Empetrum hermaphroditum* and *Andromeda polifolia* at Stockholm. In *Primary Production and Production Processes, Tundra Biome. I. B. P.* (Eds. Bliss L.C. & Wielogolaski F.E.) Tundra Biome Steering Committee, Edmunton, Oslo.

Flowers T.J. (1972) The effect of sodium chloride on enzyme activities from four halophytic species of Chenopodiaceae. *Phytochemistry* **11**, 1881–1886.

Flowers T.J. & Hall J.L. (1978) Salt tolerance in the halophyte *Suaeda maritima* (L.) Dum: The influence of salinity of the culture solution on the content of various organic compounds. *Ann. Bot.* **42**, 1057–1063.

Flowers T.J., Hall. J.L. & Ward M.E. (1978) Salt tolerance in the halophyte *Suaeda maritima* (L.) Dum: Properties of malic enzyme and PEP carboxylase. *Ann. Bot.* **42**, 1065–1074.

Forbes J.C. & Kenworthy J.B. (1973) Distribution of two species of birch forming stands in Deeside, Aberdeenshire. *Trans. Bot. Soc. Edinb.* **42**, 101–110.

Fowler D., Cape J.N., Leith D. & Paterson I.S. (1986) Acid deposition in Cumbria. In *Pollution in Cumbria* (Ed. Ineson P.). Institute of Terrestrial Ecology Symposium No. 16, pp. 7–10.

Foyer C.H. & Hall D.O. (1980) Oxygen metabolism in the active chloroplast. *Trends Biochem. Sci.* **1980**, 188–191.

Frank H. & Frank W. (1985) Chlorophyll-bleaching by atmospheric pollutants. *Naturwissenschaften* **72**, 139–141.

Frankland B. (1981) Germination in shade. In *Plants and the Daylight Spectrum* (Ed. Smith H.), pp. 187–204. Academic Press, London.

Fridovich I. (1974) Superoxide dismutases. *Adv. Enzymol.* **41**, 35–97.

Fuhrer J. (1985) Formation of secondary air pollutants and their occurrence in Europe. *Experimentia* **41**, 286–301.

Gaff D.F. (1977) Desiccation tolerant plants of South Africa. *Oecologia* **31**, 95–109.

Gale J. & Poljakoff-Mayber A. (1970) Interrelations between growth and photosynthesis of salt bush (*Atriplex halimus* L.) grown in saline media. *Aust. J. Biol. Sci.* **23**, 937–945.

Gallagher J.L., Wolf P.L. & Pfeiffer W.J. (1984) Rhizome and root growth rates and cycles in protein and carbohydrate concentrations in Georgia *Spartina alterniflora* Loisel. plants. *Amer. J. Bot.* **71**, 165–169.

Garcilaso de la Vega (1608) *The Royal Commentaries of the Incas.* English ed. (1869) by the Hakluyt Society, London.

Gartside D.W. & McNeilly T. (1974) The potential for evolution of heavy metal tolerance in plants. II. Copper tolerance in normal populations of different plant species. *Heredity* **32**, 335–348.

Gates D.M. (1968) Transpiration and leaf temperature. *Ann. Rev. Plant Physiol.* **19**, 211–238.

Gates D.M. (1980) *Biophysical Ecology.* Springer Verlag, Berlin.

Gates D.M., Strain B.R. & Weber J.A. (1983) Ecological effects of changing atmospheric CO_2 concentration. In *Encyclopedia of Plant Physiology*, Vol. 12D, pp. 503–526. Springer Verlag, Berlin.

George W. (1964) *Biological Philosopher: a Study of the Life and Writing of Alfred Russell Wallace.* Abelard-Schuman, London.

Gessner F. (1968) Zür ökologischen Problematik der Überschwemmungswälder des Amazonas. *Int. Rev. Gesamten. Hydrobiol.* **53**, 525–547.

Gibbs R.D. (1974) *Chemotaxonomy of Flowering Plants.* McGill Queens University Press, London.

Gill C.J. (1970) The flooding tolerance of woody species — a review. *For. Abstr.* **31**, 671–688.

Gimingham C.H. (1964) Maritime and sub-maritime communities. In *The Vegetation of Scotland* (Ed. Burnett J.H.), pp. 67–142. Oliver & Boyd, Edinburgh.

Gleason M.L. & Zieman J.C. (1981) Influence of tidal inundation on internal oxygen supply of *Spartina alterniflora* and *Spartina patens*. *Est. Coast. Shelf Sci.* **13**, 45–57.

Glueck N. (1959) *Rivers in the Desert.* Weidenfeld & Nicolson, London.

Godzik S. & Linskens H.F. (1974) Concentration changes of free amino acids in primary bean leaves after continuous and interrupted SO₂ fumigation and recovery. *Environ. Pollut.* **7**, 25–38.

Goebel K. (1889–93) *Pflanzenbiologische Schilderungen*, Vols. 1 and 2. Marburg.

Goldsmith F.B. (1973) The vegetation of exposed sea cliffs at South Stack, Anglesey. II. Experimental studies. *J. Ecol.* **61**, 819–829.

Goodchild D.J., Björkman O. & Pyliotes N.A. (1972) Chloroplast ultrastructure, leaf anatomy and content of chlorophyll and soluble proteins in rainforest species. *Carnegie Inst. Wash. Yr Book* **71**, 102–107.

Gouyon P.H., Fort P. & Caraux G. (1983) Selection of seedlings of *Thymus vulgaris* by grazing slugs. *J. Ecol.* **71**, 299–306.

Grace J. (1977) *Plant Responses to Wind*. Academic Press, London.

Grace J. (1987) Climatic tolerance and the distribution of plants. *New Phytol.* **106** (Suppl.) 113–130.

Graham D. & Patterson B.D. (1982) Responses of plants to nonfreezing temperatures: proteins, metabolism and acclimation. *Ann. Rev. Plant Physiol.* **33**, 347–372.

Grant V. (1981) *Plant Speciation*, 2nd edn. Columbia University Press, New York.

Greenway H. (1962) Plant response to saline substrates. I. Growth and ion uptake of several varieties of *Hordeum* during and after sodium chloride treatment. *Aust. J. Biol. Sci.* **15**, 16–38.

Greenway H. & Osmond C.B. (1972) Salt response of enzymes from species differing in salt tolerance. *Plant Physiol.* **49**, 256–259.

Grenot C.J. (1974) Physical and vegetational aspects of the Sahara desert. In *Desert Biology* (Ed. Brown G.W.), pp. 103–164. Academic Press, New York.

Grime J.P. (1966) Shade avoidance and tolerance in flowering plants. In *Light as an Ecological Factor*. (Eds. Bainbridge R., Evans G.C. & Rackham O.), 6th Symposium of the British Ecological Society, pp. 281–301. Blackwell Scientific Publications, Oxford.

Grime J.P. (1979) *Plant Strategies and Vegetation Processes*. John Wiley, London.

Grime J.P. & Lloyd P.S. (1973) *An Ecological Atlas of Grassland plants*. Edward Arnold, London.

Grisebach A. (1872) *Die Vegetation der Erde nach ihrer klimatischen Anordnung. Ein Abriss der vergleichenden Geographie der Pflanzen*. Engelman, Leipzig.

Gusta L.V. & Fowler D.B. (1979) Cold resistance and injury in winter cereals. In *Stress Physiology in Crop Plants* (Eds. Mussell H. & Staples R.), pp. 159–178. Wiley, New York.

Gustafsson (1947) Apomixis in higher plants. *Lunds universitets Arsskrift* **42–43**, 1–370.

Guye M. (1986) *Adaptation to chilling: membrane composition, polamines and ethylene*. Unpublished PhD Thesis, University College of North Wales.

Hadac E. (1960) The history of the flora of Spitsbergen and Bear Island and the age of some arctic species. *Preslia* **32**, 225–53.

Hadley E.B. & Bliss L.C. (1964) Energy relationships of alpine plants on Mt. Washington, New Hampshire. *Ecol. Mon.* **34**, 331–357.

Haldemann C. & Braendle R. (1986) Jahrzeitliche Unterschiede im Reservstoffgehalt und von gärungsprozessen in Rhizomene von Sumpf— und Röhrichpflanzen aus dem Freiland. *Flora* **178**, 307–313.

Hall O. (1972) Oxygen requirements of root meristems in diploid and tetraploid rye. *Hereditas* **70**, 69–74.

Halliwell B. (1984) Toxic effects of oxygen on plant tissues. In *Chloroplast Metabolism. The Structure and Function of Chloroplasts in Green Leaf Cells*. pp. 180–206. Clarendon Press, Oxford.

Hamp R. & Ziegler H. (1974) Der Einfluss von Bleiionen auf Enzyme der Chlorophyllbiosynthese. *Z. Naturforsch.* **29c**, 552–558.

Harborne J.B. (1982) *Introduction to Ecological Biochemistry*. Academic Press, New York.

Harper J.L. (1977) *Population Biology of Plants*. Academic Press, London.

Harper J.L. (1982) After description. In *The Plant Community as a Working Mechanism* (Ed. Newman E.I.), pp. 11–25. Blackwell Scientific Publications, Oxford.

Harper J.L. & Sagar G.R. (1953) Some aspects of the ecology of buttercups in permanent grassland. *Proc. Br. Weed Control Conf.* 256–265.

Harrington J.F. (1973) Problems of seed storage. In *Seed Ecology* (Ed. Heydecker W.), Proc. 19 Easter School, Univ. of Nottingham.

Harrison D.E.F. (1976) *The Oxygen Metabolism of Microorganisms*. Meadowfield Press, Durham.

Hart R. (1977) Why are biennials so few? *Amer. Nat.* **111**, 792–799.

Harvey D.M.R., Hall J.L., Flowers T.J. & Kent B. (1981) Quantitative ion localisation within *Suaeda maritima* leaf mesophyll cells. *Planta* **151**, 550–560.

Haskell G. (1952) Polyploidy ecology and the British flora. *J. Ecol.* **40**, 265–282.

Heath R.L. (1975) Ozone. In *Responses of Plants to Air Pollution* (Eds. Mudd J.B. & Kozlowski T.T.), pp. 23–55. Academic Press, New York.

Heber U. & Santarius K.A. (1964) Loss of adenosine triphosphate synthesis caused by freezing and its relationship to frost hardiness problems. *Plant Physiol.* **39**, 712–719.

Heber U. & Santarius K.A. (1973) Cell death by cold and heat and resistance to extreme temperatures. Mechanisms of hardening and dehardening. In *Temperature and Life* (Eds. Precht H., Christorphersen J., Hensel H. & Larcher W.), pp. 232–263. Springer Verlag, Berlin.

Hedberg O. (1964) Features of Afro-alpine ecology. *Acta Phytogeogr. Suec.* **49**, 1–114.

Henckel P.A. & Pronina N.D. (1973) The euxerophytic affiliation of *Haberlea rhodopensis* (Family Gesneriaceae). *Sov. Plant Physiol.* **20**, 690−692.

Hendry G.A.F. & Brocklebank K.J. (1985) Iron-induced oxygen radical metabolism in water-logged plants. *New Phytol.* **101**, 199−206.

Hetherington A.M., Hunter M.I.S. & Crawford R.M.M. (1984a) Evidence contrary to the existence of storage lipid in leaves of plants inhabiting cold climates. *Plant Cell Environ.* **7**, 223−227.

Hetherington A.M., Hunter M.I.S. & Crawford R.M.M. (1984b) Survival of *Iris* species under anoxic conditions. *Ann. Bot.* **51**, 255−259.

Hewitt E.J., & Nicholas D.J.D. (1963) Cations and anions: Inhibitors and interactions in metabolism and in enzyme activity. In *Metabolic inhibitors* (Eds. Hochster R.M. & Quastel J.H.), Vol. II, pp. 311−436. Academic Press, New York.

Higgins I.J. & Burns R.G. (1975) *The Chemistry and Microbiology of Pollution.* Academic Press, New York.

Hiron R.W. & Wright S.T.C. (1973) The role of endogenous abscisic acid in the response of plants to stress. *J. exp. Bot.* **24**, 769−781.

Hitier H. (1925) Condensateurs des vapeurs atmospheriques dans l'antiquité. *C.R. Acad. Agric. Fr.* **11**, 679−683.

Hochachka P.W. & Somero G.N. (1973) *Strategies of Biochemical Adaptation*, pp. 249−253. W.B. Saunders, Philadelphia.

Holmen K. (1975) The vascular plants of Peary Land, North East Greenland. *Medd. Groenl.* **124**, 1−149.

Holtmeier F.K. (1971) Waldgrenzstudien in nordlichen Finnisch-Lappland und angrenzenden Norwegen. *Rep. Kevo Subarctic Res. Stn.* **8**, 53−62.

Hopf H., Gruber G., Zinn A. & Kandler O. (1984) Physiology and biosynthesis of lychnose in *Cerastium arvense*. *Planta* **162**, 283−288.

Howeler R.H. (1973) Iron induced oranging disease in rice in relation to physiochemical changes in a flooded oxisol. *Soil Sci. Soc. Amer. Proc.* **37**, 898−903.

Howells G.D. (1986) Acid rain: A CEGB view. In *Pollution in Cumbria* (Ed. Ineson P.), Institute of Terrestrial Ecology Symposium No. 16, pp. 11−16. N.E.R.C., Merlewood, Grange-over-Sands.

Huck M.G. (1970) Variation in taproot elongation as influenced by composition of the soil air. *Agron. J.* **62**, 815−818.

Hueck K. (1966) *Die Wälder Sudamerikas.* Gustav Fischer Verlag, Stuttgart.

Hughes A.P. (1966) The importance of light compared with other factors affecting growth. In *Light as an Ecological Factor* (Eds. Bainbridge R., Evans G.C. & Rackham O.), 6th Symposium of the British Ecological Society, pp. 121−146. Blackwell Scientific Publications, Oxford.

Hulten E. (1959) Studies in the genus *Dryas. Svensk Bot. Tidskrift* **53**, 507−542.

Hulten E. (1971) *Atlas over Vaxternas Utbredning i Norden.* A B Kartografiska Institutet, Stockholm.

Hunt R. (1978) *Plant Growth Analysis.* Arnold Studies in Biology, No. 96. Edward Arnold, London.

Hunter M.I.S., Hetherington A.M. & Crawford R.M.M. (1983) Lipid peroxidation — a factor in anoxia tolerance in *Iris* species. *Phytochemistry* **22**, 1145−1147.

Hutchinson T.C. (1984) Adaptation of plants to atmospheric pollution. In *Origins and Development of Adaptation*, Ciba Foundation Symposium 102, pp. 52−72. Pitman, London.

Hyvärinen H. (1972) Pollen analytic evidence for flandrian climatic change in Svalbard. In *Climatic Changes in Arctic Areas during the Last Ten-thousand Years.* (Eds. Vasari Y., Hyvarinen H. & Hicks S.), Acta Univ. Ouluensis A 3. Geol. Vol. 1, pp. 225−237.

Iljin W.S. (1957) Drought resistance in plants and physiological processes. *Ann. Rev. Plant Physiol.* **8**, 257−274.

Iversen J. (1944) *Viscum, Hedera* and *Ilex* as climatic indicators. *Geol. Foren. Stockh. Forh.* **66**, 463.

Ives J.D. (1974) Biological refugia and the nunatak hypothesis. In *Arctic and Alpine Environments* (Eds. Ives J.D. & Barry R.G.), pp. 605−636. Methuen & Co., Ltd, London.

Jackson M.B. (1982) Ethylene as a growth promoting hormone under flooded conditions. In *Plant Growth Substances* (Ed. Warcing P.F.), pp. 291−301. Academic Press, London.

Jackson M.B. & Campbell D.J. (1979) Effects of benzyladenine and gibberellic acid on the responses of tomato plants to anaerobic root environments and to ethylene. *New Phytol.* **82**, 331−340.

Jackson M.B. & Drew M.C. (1984) Effects of flooding on the growth and metabolism of herbaceous plants. In *Flooding and Plant Growth* (Ed. Kozlowski T.T.), pp. 47−128. Academic Press, New York.

Jackson M.B., Gales K., & Campbell D.J. (1978) Effects of waterlogged soil conditions on the production of ethylene and on water relationships in tomato plants. *J. exp. Bot.* **29**, 183−193.

Jackson M.B., Herman B. & Goodenough A. (1982) An examination of the importance of ethanol in causing injury of flooded plants. *Plant Cell Environ.* **5**, 163−172.

Jackson M.B., Fenning T.M., Drew M.C. & Saker L.R. (1985a) Stimulation of ethylene production and gas-space (aerenchyma) formation in adventitious roots of *Zea mays* L. by small partial pressures of oxygen. *Planta* **165**, 486−492.

Jackson M.B., Fenning T.M. & Jenkins W. (1985b) Aerenchyma (gas-space) formation in adventitious roots of rice (*Oryza sativa*). *J. exp. Bot.* **36**, 1566−1572.

Janzen D.H. (1969) Allelopathy by myrmecophytes: the ant *Azteca* as an allelopathic agent of *Cercropia*. *Ecology* **50**, 147−153.

Janzen D.H. (1975) *Ecology of Plants in the Tropics.*

Edward Arnold, London.

Janzen D.H. (1979) How to be a fig. *Ann. Rev. Ecol. Syst.* **10**, 13–51.

Janzen D.H. (1980) Specificity of seed-attacking beetles in a Costa Rica deciduous forest. *J. Ecol.* **68**, 929–952.

Janzen D.H. (1983) Physiological ecology of fruits and seeds. In *Encyclopedia of Plant Physiology*, Vol. 12C, pp. 625–655. Springer Verlag, Berlin.

Janzen D.H., Juster H.B. & Leiner I.E. (1976) Insecticidal action of the phytohemagglutinin in black beans on a bruchid beetle. *Science* **192**, 795–796.

Jeffries R.L., Davy A.J. & Rudmik T. (1979) In *Ecological Processes in Coastal Environments* (Eds. Jeffries R.L. & Davy A.J.), pp. 243–268. Blackwell Scientific Publications, Oxford.

Jena (1967) *World Atlas of Climatic Diagrams.*

Johnson A.W. & Packer J.G. (1965) Polyploidy and environment in arctic Alaska. *Science* **148**, 237–239.

Joly C.A. & Crawford R.M.M. (1982) Variation in tolerance and metabolic responses to flooding in some tropical trees. *J. exp. Bot.* **33**, 799–809.

Jones D.A. (1972) Cyanogenic glycosides and their function. In *Phytochemical Ecology* (Ed. Harborne J.B.), pp. 103–124. Academic Press, New York.

Jones H.E. & Etherington J.R. (1970) Comparative studies of plant growth and distribution in relation to waterlogging. I. The survival of *Erica cinerea* L. and *E. tetralix* L. and its apparent relationship to iron and manganese uptake in waterlogged soil. *J. Ecol.* **58**, 487–496.

Jones M.M., Turner N.C. & Osmond C.B. (1981) Mechanisms of drought resistance. In *The Physiology and Biochemistry of Drought Resistance in Plants* (Eds. Paleg L.G. & Aspinall D.), pp. 15–37. Academic Press, New York.

Jones P.D., Wigley T.M.L. & Wright P.B. (1986a) Global temperature variations between 1861 and 1984. *Nature* **322**, 430–434.

Jones R.J. & Mansfield T.A. (1970) Suppression of stomatal opening in leaves with abscisic acid in leaves. *J. exp. Bot.* **21**, 714–719.

Jones V.J., Stevenson A.C. & Battarbee R. (1986b) Lake acidification and the land-use hypothesis: a mid-postglacial analogue. *Nature* **322**, 157–158.

Kandler O. (1983) Waldsterben: Emissions — oder Epidemie-Hyothese? *Naturwiss. Rund.* **36**, 488–490.

Kandler O. (1987) Lichen and conifer recolonization in Munich's cleaner air. In *Effects of Air Pollution on Terrestrial and Aquatic Ecosystems*, Symposium of the Commission of the European Communities, 18–22 May, pp. 1–7.

Kandler O. & Hopf H. (1984) Biosynthesis of oligosaccharides in vascular plants. In *Storage of Carbohydrates in Vascular Plants* (Ed. Lewis D.H.), pp. 115–131. Cambridge University Press, Cambridge.

Kandler O., Miller W. & Ostner R.(1987) Dynamik der 'akuten vergilbung' der Fichte. *Allg. Forst. Z.* **27–29**, 715–723.

Kappen L. (1969) Kälteverträglichkeit und Zuckergehalt von Salzpflanzen. *Ber. Dtsch. Bot. Ges.* **82**, 103–106.

Kappen L. & Ullrich W.R. (1970) Verteilung von Chlorid und Zuckern in Blattzellen halophiler Pflanzen bei verrschiedene höher Frostresistenz. *Ber. Dtsch. Bot. Ges.* **83**, 265–275.

Kareiva P. (1987) Habitat fragmentation and the stability of predator-prey interactions. *Nature* **326**, 388–390.

Kausch A.P., Seago J.L. & Marsh L.C. (1981) Changes in starch distribution in the overwintering organs of *Typha latifolia* (Typhaceae). *Amer. J. Bot.* **68**, 877–880.

Keatinge T.H. & Dickson J.H. (1979) Mid-Flandrian changes in vegetation in mainland Orkney. *New Phytol.* **82**, 585–612.

Keeling C.D. (1978) Atmospheric carbon dioxide in the 19th century. *Science* **202**, 1109.

Kerr J.A., Calvert J.G. & Demerijian D.L. (1972) The mechanism of photochemical smog formation. *Chem. Brit.* **8**, 252–257.

Kimball B.A. (1983) Carbon dioxide and agricultural yield: an assemblage of 430 prior observations. *Agron. J.* **75**, 779–788.

Kimmerer T. & Kozlowski T.T. (1982) Ethylene, ethane, acetaldehyde and ethanol production by plants under stress. *Plant Physiol.* **69**, 840–847.

King T.J. (1975) Inhibition of seed germination under leaf canopies in *Arenaria serpyllifolia*, *Veronica arvensis* and *Cerastium* [sic] *holosteoides*. *New Phytol.* **75**, 87–90.

Kingsbury J.M. (1964). *Poisonous Plants of the United States and Canada*. Prentice Hall, New Jersey.

Kinzel H. (1982) *Pflanzenoekologie und Mineralstoffwechsel*. Ulmer, Stuttgart.

Kircher H.W. & Heed W.B. (1970) Phytochemistry and host plant specificity in *Drosophila*. *Rec. Adv. Phytochem.* **3**, 191–208.

Klemow K.M. & Raynal D.J. (1981) Population ecology of *Melilotus alba*. *J. Ecol.* **69**, 33–44.

Klepper B., Taylor H.M., Huck M.G. & Fiscus E.L. (1973) Water relations and growth of cotton in drying soil. *Agron. J.* **65**, 307–310.

Kluge M. & Osmond C.B. (1972) Studies on phosphoeolpyruvate carboxylase and other enzymes of Crassulacean acid metabolism of *Bryophyllum tubiflorum* and *Sedum praealtum*. *Z. Pflanzenphysiol.* **66**, 97–105.

Kluge M. & Ting I.P. (1978) Crassulacean acid metabolism: analysis of an ecological adaptation. Ecol Stud. Vol. 30 In *Ecological Studies* (Ed.), Vol. 30, Springer Verlag, Berlin.

Kluge M., Lange O.L., Eichman M.V. & Schmid R. (1973) Diurnaler Sauererythmus bei *Tillandsia usneoides*: Untersuchungen sowie die Abhängigkeit des CO_2

Gaswchsels von Lichintensitat, Temperatur und Wassergehalt der Pflanze. *Planta* **112**, 357−372.

Koller D. (1969) The physiology of dormancy and survival of plants in desert environments. *Symp. Soc. Exp. Bot.* **23**, 449−469.

Koller D. (1970) Analysis of the dual action of white light on germination of *Atriplex dimorphostegia* (Chenopodiaceae). *Isr. J. Bot.* **19**, 499−516.

Koller D. & Hadas A. (1982) Water relations in the germination of seeds. In *Encyclopedia of Plant Physiology*, Vol. 12B, pp. 402−431. Springer Verlag, Berlin.

Krogg J. (1955) Notes on temperature measurements indicative of special organization in arctic and sub-arctic plants for utilization of radiation heat from the sun. *Physiol. Plant* **8**, 836−839.

Kubitzki K. & Gottlieb O.R. (1984) Micromolecular pattern and the evolution and major classification of angiosperms. *Taxon* **33**, 375−391.

Kullman L. (1979) Change and stability in the altitude of the birch tree-limit in the southern Swedish Scandes 1915−1975. *Acta Phytogeogr. Suec.* **65**, 1−121.

Kunert von K.J. & Hofer G. (1987) Geben Veränderungen des antioxidativen Systems von Pflanzen, Hinweise auf die Wirkung von Luftschadstoffen. *Allg. Forst Z.* **27−29**, 697−700.

Lamb H.H. (1967) Britain's changing climate. *Geogrl. J.* **133**, 445−468.

Lamb H.H. (1982) *Climate, History and the Modern World*. Methuen & Co., Ltd, London.

LaMer V.K., Healy T.W. & Aylmore L.A.G. (1964) The transport of water through monolayers of long chain in paraffinic alcohols. *J. Colloid Sci.* **19**, 673−684.

Lange O.L. & Schulze E.D. (1966) Untersuchungen über Dickentwicklung der kutikularen Zellwandschichten bei der Fichtennadel. *Fortwiss. Zentralbl.* **85**, 27−38.

Lange O.L. & Medina E. (1979) Stomata of the CAM plant *Tillandsia recurvata* respond directly to humidity. *Oecologia* **40**, 357−363.

Lange O.L., Lösch R., Schulze E.-D. & Kappen L. (1971) Responses of stomata to changes in humidity. *Planta* **100**, 76−86.

Larcher W. (1977) Ergebnisse des IBP-Projekts 'Zwergstrauchheide Patscherkofel'. *Sitz. Ber. Oesterr. Akad. Wiss. Math-naturw. Kl, Abt. I*, **186**, 301−371.

Larcher W., Schmidt L. & Tschager A. (1973) Starkefettspeicherung und höher Kaloriengehalt bei *Loiseleuria procumbens* (L.) Desu. *Oekolog. Plant.* **8**, 377−383.

Larcher W. & Bauer H. (1981) Ecological significance of resistance to low temperature. In *Encyclopedia of Plant Physiology*, Vol. 12A, pp. 403−437. Springer Verlag, Berlin.

Larcher W. & Eggarter H. (1960) Anvendung des Triphenyltetrazoliumchlorids zur Beurteilung von Frostschaden in verschiedenen Achsengeweben bei *Pinus*

Arten und Jahresgang der Resistenz. *Protoplasma* **51**, 595−691.

Larher F., Hammelin J. & Stewart G.R. (1977) L'acide dimethylsulphonium-3 propanoique de *Spartina anglica*. *Phytochemistry* **16**, 2019−2020.

Lawrence D.B., Schoenicke R.E., Quispel A. & Bond G. (1967) The role of *Dryas drummondii* vegetation development following recession at Glacier Bay, Alaska with special reference to nitrogen fixation by root nodules. *J. Ecol.* **55**, 793.

Le Saint A.M. (1969) Variations comparées des teneurs en proline libre et en glucide solubles en relation avec l'inegale sensibilité au sel des organes de plant de choux de Milan, cult. Pontoise. *Compt. Rend. Acad. Sci. Paris* **268**, 310−313.

Lee E.H. & Bennett J.H. (1982) Superoxide dismutase. A possible enzyme against ozone injury in snap beans (*Phaseolus vulgaris* L.). *Plant Physiol.* **69**, 1444−1449.

Lendzian K.J. & Unsworth M.H. (1983) Ecological effects of atmospheric pollutants. In *Encyclopedia of Plant Physiology*, Vol. 12D, pp. 465−502. Springer Verlag, Berlin.

Leopold A.S., Erwin M., Oh J. & Browning B. (1976) Phytoestrogens: Adverse effects on reproduction in Californian quail. *Science* **191**, 98−99.

Levin, D.A. (1973) The role of trichomes in plant defense. *W. Rev. Biol.* **48**, 3−15.

Levin D.A. (1976) Alkaloid-bearing plants: an ecogeographic perspective. *Amer. Nat.* **110**, 261−284.

Levitt J. (1980) *Responses of Plants to Environmental Stresses*, 2nd edn. Academic Press, New York.

Lewis H. (1966) Speciation in flowering plants. *Science* **152**, 167−172.

Lincoln R.J., Boxshall G.A. & Clark P.F. (1982) *A Dictionary of Ecology, Evolution and Systematics*. Cambridge University Press, Cambridge.

Lindsay J.H. (1971) Annual cycle of leaf water potential in *Picea engelmannii* and *Abies lasiocarpa* at timberline in Wyoming. *Arct. Alp. Res.* **3**, 131−138.

Linhart Y.B. & Baker G. (1973) Intra-population differentiation of physiological responses to flooding in a population of *Veronica peregrina* L. *Nature* **242**, 275.

Long S.P. & Woolhouse H.W. (1978) The response of net photosynthesis to vapour pressure deficit and CO_2 concentration in *Spartina townsendii* (sensu latu), a C_4 species from a cool temperatue climate. *J. exp. Bot.* **29**, 567−577.

Long S.P., Incoll L.D. & Woolhouse H.W. (1975) C_4 photosynthesis in plants from cool temperate regions with particular reference to *Spartina townsendii*. *Nature* **257**, 622−624.

Lotfield J.V.G. (1921) The behaviour of stomata. *Publ. Carnegie Inst. Wash.* **314**, 1−104.

Löve A. & Löve D. (1974) Origin and evolution of the arctic alpine floras In *Arctic and Alpine Environments*

(Eds. Ives J.D. & Berry R.G.), pp. 571–603. Methuen & Co., Ltd, London.

Löve D., McLellan C. & Gamow I. (1970) Coumarins and coumarin derivatives in various growth-types of Engelmann spruce *Svensk. Bot. Tidskr.* **64**, 284–296.

Lüttge U. & Ball E. (1979) Electrochemical investigation of active malic acid transport into the vacuoles of the CAM plant *Kalanchoe daigremontiana. J. membr. Biol.* **47**, 401–422.

Lüttge U. & Osmond C.B. (1970) Ion absorption in *Atriplex* leaf tissue. III. Site of metabolic control of light-dependent chloride secretion to epidermal bladders. *Aust. J. Biol. Sci.* **23**, 17–25.

Lynch J.M. (1977) Phytotoxicity of acetic acid produced in the anaerobic decomposition of wheat straw. *J. Appl. Bacteriol.* **42**, 81–87.

Lyons J.M. & Raison J.K. (1970) Oxidative study of mitochondria isolated from plant tissues sensitive and resistant to chilling injury. *Plant Physiol.* **45**, 386–389.

Lyttle R.W. & Hull R.J. (1980) Photoassimilate distribution in *Spartina alterniflora* Loisel. I. Vegetative and floral development. *Agron. J.* **72**, 933–946.

McCord J.M. (1985) The role of superoxide in postischemic injury. In *Superoxide Dismutase* (Ed. Oberley L.W.), Vol. 3, pp. 143–150. CRC Press, Boca Raton.

McCree K.J. (1981) Photosynthetically active radiation. In *Encyclopedia of Plant Physiology*, Vol. 12A, pp. 41–55.

McDiarmid R.W., Ricklefs R.E. & Foster M.S. (1977) Dispersal of *Stemmadenia donell-smithii* (Apocynaceae) by birds. *Biotropica* **9**, 9–25.

MacDougal D.T. & Brown J.G. (1928) Living cells two and a half centuries old. *Science* **67**, 447.

McGraw J.B. & Antonovics J. (1983) Experimental ecology of *Dryas octopetala* ecotypes. I. Ecotypic differentiation and life-cycle stages of selection. *J. Ecol.* **71**, 879–897.

MacLean S.F. & Jensen T.S. (1985) Food plant selection by insect herbivores in Alaskan arctic tundra: the role of plant life form. *Oikos* **44**, 211–221.

MacNaughton S.J. (1983) Physiological and ecological implications of herbivory. In *Encyclopedia of Plant Physiology*, Vol. 12C, pp. 657–677. Springer Verlag. Berlin.

Maier-Maercker U. (1979) 'Peristomal transpiration' and stomata movement: A controversial view. I. Additional proof of peristomal transpiration by hygrophotography and a comprehensive discussion in the light of recent results. *Z. Pflanzenphysiol.* **91**, 5–43.

Mair B. (1968) Frosthartgradienten entlang den Knospenfolge auf Eschentreiben. *Planta* **82**, 164–169.

Malta N. (1921) Versuch über die Wiederstandsfähigkeit der Gametophyten und Sporophyten der Laubmoose. *Latu. Augstsk. Rak.* **1**, 125–129.

Malthus T.R. (1798) *An Essay on the Principle of Population.* London.

Marshall J.K. (1965) *Corynephorus canescens* (L.) P. Beauv.

as a model for the *Ammophila* problem. *J. Ecol.* **53**, 447–463.

Martin M.H. (1968) Conditions affecting the distribution of *Mercurialis perennis* in certain Cambridgeshire woodlands. *J. Ecol.* **56**, 777–793.

Maruyama K. (1971) Effect of altitude on dry matter production of primeval Japanese beech forest communities in Naeba mountains. *Mem. Fac. Agric. Niigata Univ.* **9**, 87–171.

Mathys W. (1977) The role of malate, oxalate and mustrad oil glucosides in the evolution of zinc-resistance in herbage plants. *Plant Physiol.* **40**, 130–136.

Mattocks A.R. (1972) Toxicity and metabolism of *Senecio* alkaloids. In *Phytochemical Ecology* (Ed. Harborne J.B.), pp. 179–200. Academic Press, New York.

Mattson W.J. (1980) Herbivory in relation to plant nitrogen content. *Ann. Rev. Ecol. Syst.* **11**, 119–161.

Maximov N.A. (1929) *The Plant in Relation to Water.* Allen & Unwin, London.

Maynard Smith J. (1976) *The Evolution of Sex.* Cambridge University Press, Cambridge.

Mayr E. (1942) *Systematics and the Origin of Species.* Columbia University Press, New York.

Mayr E. (1970) *Population Species and Evolution.* Harvard University Press, Cambridge, Mass.

Mazur P. (1969) Freezing injury in plants. *Ann. Rev. Plant Physiol.* **20**, 419–448.

Medina E. & Klinge H. (1983) Productivity of tropical forests and tropical woodlands. In *Encyclopedia of Plant Physiology*, Vol. 12D, pp. 281–303. Springer Verlag, Berlin.

Mehlhorn H. & Wellburn A.R. (1987) Stress ethylene formation determines plant sensitivity to ozone. *Nature* **327**, 417–418.

Metraux J.P. & Kende H. (1983) The role of ethylene in the growth response of submerged deep water rice. *Plant Physiol.* **72**, 441–446.

Metraux J.P., de Zacks R. & Kende H. (1982) The response of deep-water rice to submergence and ethylene. In *MSU-DOE Plant Research Laboratory Annual Report for 1981*, pp. 80–81. Michigan State University, East Lansing.

Meusel H. (1965) *Vergleichende Chorologie der Zentraleuropäischen Flora.* Gustav Fischer Verlag, Jena.

Michaelis P. (1934) Ökologischer Studien an der alpinen Baumgrenze. V Osmotischer Wert und Wassergehalt während des Winters in den verschiedene Höhenlagen. *Jahrb. Wiss. Bot.* **80**, 337–362.

Millar A. (1965) The effect of temperature and day length on the height growth of the birch (*Betula pubescens*) at 1900 feet in the northern Penines. *J. appl. Ecol.* **2**, 17–29.

Millar G.R. & Cummins R.P. (1982) Regeneration of Scots pine (*Pinus sylvestris*) at a natural tree-line in the Cairngorm Mountains, Scotland. *Holarct. Ecol.* **5**, 27–34.

Miller P.C., Webber P.J., Oechel W.C. & Tieszen L.L. (1980) Biophysical processes and primary production. In *An Arctic Ecosystem.* (Eds. Brown J., Miller P.C., Tieszen L.L. & Bunnell F.L.), pp. 66–101. Dowden, Hutchinson & Ross, Stroudsberg, Pennsylvania.

Miller P.R. & Elderman (Eds.) (1977) *Photochemical Oxidant Air Pollutant Effects on a Mixed Conifer Forest System.* Environmental Protection Agency, Corvallis, Oregon.

Mizrahi Y., Blumenfeld A. & Richmond A.E. (1972) The role of abscisic acid and salination in the adaptive response of plants to reduced root aeration. *Plant Cell Physiol.* **13**, 15–21.

Monk L.S. & Braendle R. (1982) Adaptation of respiration and fermentation to changing levels of oxygen in rhizomes of *Schoenoplectus lacustris* (L.) palla and its significance to flooding tolerance. *Z. Pflanzenphysiol.* **105**, 369–374.

Monk L.S., Crawford R.M.M. & Braendle R. (1984) Fermentation rates and ethanol accumulation in relation to flooding tolerance in rhizomes of monocotyledonous species. *J. Exp. Bot.* **35**, 738–745.

Monsi M. (1960) Dry matter production in plants. I. Schemata of dry matter reproduction. *Bot. Mag.* **73**, 81–90.

Monteith J.L. (1963) Dew: facts and fallacies. In *The Water Relations of Plants* (Eds. Rutter A.J. & Whitehead F.H.), pp. 37–56. Blackwell Scientific Publications, Oxford.

Montgomery E.G. (1912) Competition in cereals. *Bull. Nebr. Agric. Stn.* **26**, 1–22.

Mooney H.A. & Billings W.D. (1965) Effect of altitude on carbohydrate content of mountain plants. *Ecology* **46**, 750–751.

Mooney H.A., Gulmon S.L., Ehleringer J.R. & Rundel P.W. (1980) Further observations on the water relations of *Prosois tamaringo* of the northern Atacama desert. *Oecologia* **44**, 177–180.

Moran V.C. (1983) The phytophagous insects and mites of cultivated plants in South Africa: patterns and pest status. *J. appl. Ecol.* **20**, 439–450.

Morgan D.C. (1981) Shadelight quality effects on plant growth. In *Plants and the Daylight Spectrum* (Ed. Smith H.), pp. 205–221. Academic Press, London.

Morgan D.C. & Smith H. (1981) Non-photosynthetic responses to light quality. In *Encyclopedia of Plant Physiology*, Vol. 12A, pp. 109–134. Springer Verlag, Berlin.

Morgan J.M. (1984) Osmoregulation and water stress in higher plants. *Ann. Rev. Plant Physiol.* **35**, 299–319.

Moser M. & Haselwandter K. (1983) Ecophysiology of mycorrhizal symbioses. In *Encyclopedia of Plant Physiology*, Vol. 12C, pp. 392–421. Springer Verlag, Berlin.

Moser W. (1973) Licht, Temperatur und Photosynthese an der Station 'Hoher Nebelkogel' (3184 m). In *Okosystem-forschung* (Ed. Ellenberg H.), pp. 203–223. Springer Verlag, Berlin.

Mudd J.B. (1982) Effects of oxidants on metabolic function. In *Effects of Air Pollution in Agriculture and Horticulture* (Eds. Unsworth M.H. & Omrod D.P.), pp. 189–203. Butterworths Scientific, London.

Muller C.H. (1966) The role of chemical inhibition (Allelopathy) in vegetational composition. *Bull. Torrey Bot. Club* **93**, 332–351.

Mulroy T.W. & Rundel P.W. (1977) Annual plants; adaptations to desert environments. *Bioscience* **27**, 109–114.

Munns R., Brady C.J. & Barlow E.W.R. (1979) Solute accumulation in the apex and leaves of wheat during water stress. *Aust. J. Plant Physiol.* **6**, 379–389.

Munns R., Greenway H. & Kirst G.O. (1983) Halotolerant eukaryotes. In *Encyclopedia of Plant Physiology*, Vol. 3, pp. 59–135. Springer Verlag, Berlin.

Myers J. (1971) Enhancement studies in photosynthesis. *Ann. Rev. Plant Physiol.* **22**, 289–312.

Myerscough P.J. & Whitehead F.H. (1966) Comparitive biology of *Tussilago farfara, Chamaenerion angustifolium* and *Epilobium adenocaulon.* I. General biology and germination. *New Phytol.* **65**, 192–210.

Nash T.H. (1973) The effects of air pollution on other plants, particularly vascular plants. In *Air Pollution and Lichens* (Eds. Ferry B.W., Baddeley M.S. & Hawksworth D.L.), pp. 192–223. Athlone Press, London.

Neales T.F. (1973) Effect of night temperature on the assimilation of carbon dioxide by mature pineapple plants. *Aust. J. Biol. Sci.* **26**, 539–546.

Neftel A., Moor E., Oeschger H. & Stauffer B. (1985) Evidence from polar ice cores for the increase in atmospheric CO_2 in the past two centuries. *Nature* **315**, 45–47.

Neger F.W. (1908) Das Tannensterben in den saechsischen und anderen deutschen Mittelgebirgen. *Tharanter Forstl. Jhb.* **58**, 201–255.

Nir I., Klein S.S. & Polkjakoff-Mayber A. (1969) Effect of moisture stress on submicroscopic structure of maize roots. *Aust. J. Biol. Sci.* **22**, 17–33.

Nobel P.S. (1978) Surface temperature of cacti and influence of environmental and morphological factors. *Ecology* **59**, 986–996.

Olson J.S. (1958) Lake Michigan dune development. II. Plant as agents and tools in geomorphology. *J. Geol.* **66**, 345–431.

Oosting H.J. (1954) Ecological processes and vegetation of the maritime strand in the United States. *Bot. Rev.* **20**, 226–262.

Osawa T. & Ikeda H. (1976) Heavy metal toxicities in vegetable crops. I. The effect of iron concentration in the nutrient solution on manganese toxicities on vegetable crops. *J. Jpn. Soc. Hortic. Sci.* **45**, 50–58.

Osborne D.J. (1984) Ethylene in plants of aquatic and semi-aquatic environments: a review. *Plant Growth Reg.*

2, 167–185.

Oshino N., Oshino R. & Chance B. (1973) The characteristics of the 'peroxidatic' reaction of catalase in ethanoloxidation. *Biochem. J.* **131**, 555–567.

Osmond C.B. (1978) Crassulacean acid metabolism: a curiosity in context. *Ann. Rev. Plant Physiol.* **29**, 379–414.

Osmond C.B. & Greenway H. (1972) Salt response of carboxylation enzymes from species differing in salt tolerance. *Plant Physiol.* **49**, 260–263.

Ott E. (1978) Uber die Abhängigkeit des Radialzuwaches und der Oberhöhen bei Fichte und Lärche von der Meershöhe und Exposition im Lötschertal. *Schweiz. Z. Fortwes.* **129**, 169–193.

Palta J.P. & Li P.H. (1979) Frost hardiness in relation to leaf anatomy and natural distribution of several *Solanum* species. *Crop Sci.* **19**, 665–671.

Parker J. (1968) Drought resistance mechanisms. In *Water Deficits and Plant Growth* (Ed. Kozlowski, T.T.), Vol. 1, pp. 195–234. Academic Press, New York.

Parkhurst D.F. & Loucks O.L. (1972) Optimal leaf size in relation to environment. *J. Ecol.* **60**, 505–537.

Parsons J.A. & Rothschild M. (1964) Rhodanase in the larva and pupa of the Common Blue Butterfly (*Polymmatus icarus* [Rott.]) (Lepidoptera). *Entomol. Gaz.* **15**, 58.

Patrick W.H. (1978) Critique of 'Measurement and prediction of anaerobiosis in soils'. In *Nitrogen in the Environment* (Eds. Nielsen D.R. & MacDonald J.G.), Vol. 1, pp. 449–457. Academic Press, New York.

Patterson B.D., Kenrick J.R. & Raison J.K. (1978). Lipids of chill-sensitive and -resistant *Passiflora* species: Fatty acid composition and temperature dependence of spin label motion. *Phytochemistry* **17**, 1989–1992.

Pearcy R.W. & Calkin H.C. (1983) Carbon dioxide exchange of C_3 and C_4 tree species in the understorey of a Hawaiian forest. *Oecologia* **58**, 19–25.

Pears N.V. (1967) Present tree-line of the Cairngorm Mountains, Scotland. *J. Ecol.* **55**, 815–29.

Pears N.V. (1968) The natural altitudinal limit of forest in the Scottish Grampians. *Oikos* **19**, 71–80.

Pennington W. (1984) *Rep. Freshwater Biol. Ass.* **52**, 28–46.

Perring F.H. & Farrell L. (1977) *British Red Data Book*. I. *Vascular Plants*. Society for protection of nature conservation, Lincoln.

Perring F.H. & Walters S.M. (1976) *Atlas of the British Flora*. Botanical Society of the British Isles.

Pielou E.C. (1979) *Biogeography*. John Wiley, New York.

Pollard A. & Wyn Jones R.G. (1979) Enzyme activities in concentrated solutions of glycinebetaine and other solutes. *Plant* **144**, 291–298.

Polunin O. & Huxley A. (1965) *Flowers of the Mediterranean*. Chatto & Windus, London.

Ponnamperuma F.M. (1972) The chemistry of submerged soils. *Adv. Agr.* **24**, 29–96.

Pourrat Y. & Hubac C. (1974) Comparison des mécanismes de la résistance a la sécheresse chez deux plantes désertiques *Artemisia alba* et *Carex pachystylis* (J. Gray) Asch. et Graebn. *Physiol. Veg.* **12**, 135–147.

Pradet A. & Bomsel J.L. (1978) Energy metabolism in plants under hypoxia and anoxia. In *Plant Life in Anaerobic Environments* (Eds. Hook D.D. & Crawford R.M.M.), pp. 89–118. Ann Arbor, Michigan.

Pradet A. & Raymond P. (1983) Adenine nucleotide ratios and adenylate energy charge in energy metabolism. *Ann. Rev. Plant Physiol.* **34**, 199–224.

Prescott W.H. (1847) *History of the Conquest of Peru with a Preliminary View of the Civilization of the Incas.* London.

Pridgeon A.M. (1981) Absorbing trichomes in the Pleurothallidinae (Orchidaceae). *Amer. J. Bot.* **68**, 64–71.

Puckett K.J., Nieboer E., Flora W.P. & Richardson D.H. S. (1973) Sulphur dioxide its effect on photosynthetic ^{14}C fixation in lichens and suggested mechanisms of toxicity. *New Phytol.* **72**, 141–154.

Queiroz O. (1974) Circadian rhythms and metabolic patterns. *Ann. Rev. Plant Physiol.* **25**, 115–134.

Rackham O. (1975) Temperatures of plant communities as measured by pyrometric and other methods. In *Light as an Ecological Factor*. II. (Eds. Evans G.C., Bainbridge R. & Rackham O.), pp. 423–449. Blackwell Scientific Publications, Oxford.

Raison J.K. (1973) Temperature-induced phase changes in membrane lipids and their influence on metabolic regulation. *Symp. Soc. Exp. Biol.* **27**, 485–512.

Raison J.K., Berry J.A., Armond P.A. & Pike C.S. (1980) Membrane properties in relation to the adaptation of plants to temperature stress. In *Adaptation of Plants to Temperature Stress*. (Eds. Turner N.C. & Kramer P.J.), pp. 261–273. Wiley – Interscience, New York.

Ranft H. & Dässler H.G. (1970) Rauchhärtigtest an Gehölzen in SO_2 Kabinenversuch. *Flora* **159**, 573–588.

Ranwell D.S. (1959) Newborough Warren, Anglesey. I. The dune system and dune slack habitat. *J. Ecol.* **47**, 571–601.

Ranwell D.S. (1972) *Ecology of Salt Marshes and Sand Dunes*. Chapman & Hall, London.

Raschke K. (1975) Stomatal action. *Ann. Rev. Plant Physiol.* **26**, 309–340.

Raskin I. & Kende H. (1983) How does deep water rice solve its aeration problem? *Plant Physiol.* **72**, 447–454.

Raunkiaer C. (1904) Biological types with reference to the adaptation of plants to survive the unfavourable season. In *History of Ecology, Life Form of Plants and Statistical Plant Geography* (Ed. Egerton F. N.). Arno Press, New York. (Reprint 1977.)

Raynaud D. & Barnola J.M. (1985) An Antarctic ice core reveals atmospheric CO_2 variations over the past few centuries. *Nature* **315**, 309−311.

Read D.J. (1978) The biology of mycorrhiza in heathland ecosystems with special reference to the nitrogen nutrition of the Ericaceae. In *Microbial Ecology* (Eds. Loutit M.W. & Miles J.A.R.), pp. 324−328. Springer Verlag, Berlin.

Renberg I. & Wallin J-E. (1985) The history of the acidification of lake Gardsjon as deduced from diatoms and *Sphagnum* leaves in the sediment. *Ecol. Bull.* **37**, 47−52.

Reuss J.O., Cosby B.J. & Wright R.F. (1987) Chemical processes governing soil and water acidification. *Nature* **329**, 27−32.

Richards A.J. (1973) The origin of *Taraxacum* agamospecies. *Bot. J. Linn. Soc.* **66**, 182−241.

Riley H.P. (1938) A character analysis of colonies of *Iris fulva, Iris hexagona* var *giganticaerulea* and natural hybrids. *Amer. J. Bot.* **25**, 727−738.

Roberts J.K.M., Callis J., Jardetzky O., Walbot V. & Freeling M. (1984) Cytoplasmic acidosis as a determinant of flooding intolerance in plants. *Proc. nat. Acad. Sci.* **81**, 6029−6033.

Robinson T. (1974) Metabolism and function of alkaloids in plants. *Science* **184**, 433−435.

Rodhe H. & Granat L. (1984) An evaluation of sulphate in European precipitation 1955−1982. *Atmos. Environ.* **18**, 2627−2639.

Rosenthal G.A, Janzen D.H. & Dahlman D.L. (1977) Degradation and detoxification of canavanine by a specialised seed predator. *Science* **196**, 256−258.

Rothschild M. (1972) Some observations on the relationship between plants, toxic insects and birds. In *Phytochemical Ecology* (Ed. Harborne J.B.), pp. 1−12. Academic Press, New York.

Rowe R.N. & Beardsell D.V. (1973) Waterlogging of fruit trees. *Hortic. Abstr.* **43**, 223−232.

Rumpho M.E. & Kennedy R.A. (1981) Anaerobic metabolism in germinating seeds of *Echinochloa crus-galli* (barnyard grass): metabolism and enzyme studies. *Plant Physiol.* **68**, 165−168.

Rundel P.W. (1980) The ecological distribution of C_4 and C_3 grasses in the Hawaiian Islands. *Oecoligia* **45**, 354−359.

Rundel P.W. (1982) Water uptake by organs other than roots. In *Encyclopedia of Plant Physiology*, Vol. 12B, pp. 111−134. Springer Verlag, Berlin.

Rutter A.J. (1968) Water consumption by forests. In *Water Deficits and Plant Growth* (Ed. Kozlowski T.T.), pp. 23−84. Academic Press, New York.

St. Omer L. & Horvath S.M. (1983) Elevated carbon dioxide concentrations and plant senescence. *Ecology* **64**, 1311−1314.

Sakai A. (1961) Effect of polyhydric alcohols on frost hardiness in plants. *Nature* **189**, 416−417.

Salisbury E.J. (1916) The oak-hornbeam woods of Hertfordshire. I and II. *J. Ecol.* **4**, 83−117.

Salisbury E.J. (1952) *Downs and Dunes*. Bell, London.

Santarius K. & Heber U. (1972) Physiological and biochemical aspects of frost damage and winter hardiness in higher plants. In *Proceedings Colloquium on Winter Hardiness of Cereals*, pp. 7−29. Martonvasar, Hungary.

Schaller G.B. (1963) *The Mountain Gorilla*. University of Chicago Press, Chicago.

Schimper A.F.W. (1898) *Pflanzengeographie auf physiologischen Grundlage*, Jena.

Schindler D.W., Mills K.H., Malley D.F., Findlay D.L, Shearer J.A., Davies I.J., Turner M.A., Linsey G.A. & Cruikshank D.R. (1985) Long-term ecosystem stress: the effects of years of experimental acidification on a small lake. *Science* **228**, 1395−1401.

Schobert B. & Tschesche H. (1978) Unusual solution properties of proline and its interaction with proteins. *Biochim. Biophys. Acta* **51**, 270−277.

Schofield C.L. & Trojnar J.R. (1979) Aluminium toxicity to fish in acidified water. In *Polluted Rain* (Eds. Toribara T.Y., Miler M.W. & Morrow P.E.), pp. 347−366. Plenum Press, New York.

Schönherr J. (1982) Resistance of plant surfaces to water loss: Transport properties of cutin, suberin and associated lipids. In *Encyclopedia of Plant Physiology*, Vol. 12B, pp. 153−179. Springer Verlag, Berlin.

Schröder J.V. & Reuss C. (1883) *Die Schädigung der Vegetation durch Rauch*. Paul Parey, Berlin.

Schultz A.M. (1969) A study of an ecosystem. The Arctic Tundra. In *The Ecosystem Concept in Natural Resource Management* (Ed. Van Dyne G.M.), pp. 75−93. Academic Press, New York.

Schultz J.E. (1974) Root development of wheat at the flowering stage under different cultural practices. *Agric. Res.* **1**, 12−17.

Schulze E.-D. (1970) Der CO_2 Gaswechsel der Buche (*Fagus sylvatica* L.) in Abhängigkeit von Klimafactoren in Freiland. *Flora* **159**, 177−232.

Schulze E.-D. (1982) Plant life forms and their carbon, water and nutrient relations. In *Encyclopedia of Plant Physiology*, Vol. 12B, pp. 616−676. Springer Verlag, Berlin.

Schulze E.-D., Lange O.L., Buschbom U., Kappen L. & Evenari M. (1972) Stomatal response to change in humidity in plants growing in the desert. *Planta* **108**, 259−270.

Schulze E.-D., Oren R.E. & Zimmerman R. (1987) Die Wirkung von Immissionen auf 30 jährige Fichten in mittleren Höhenlagen des Fichtelgebirges auf Phyllit. *Allg. Forst Z.* **27−29**, 725−730.

Scott Russell R. (1977) *Plant Root Systems*. McGraw Hill, London.

Scuderi L.A. (1987) Late holocene upper timberline variation in the southern Sierra Nevada. *Nature* **325**, 242−244.

Seddon B. (1971) *Introduction to Biogeography*. Duckworth, London.

Sekiya J., Wilson L.G. & Filner P. (1982). Resistance to injury by sulphur dioxide. Correlation with its reduction to, and emission of, hydrogen sulphide in Cucurbitaceae. *Plant Physiol.* **70**, 437−441.

Sellmair J. & Kandler O. (1970) Zur Physiologie von Hamamelose und Hamamelit in *Primula clusiana* Tausch. *Z. Pflanzenphysiol.* **63**, 65−83.

Silvertown J. (1980) Leaf canopy induced seed dormancy in a grassland flora. *New Phytol.* **85**, 109−918.

Silvertown J. (1984) Death of the elusive biennial. *Nature* **310**, 271.

Simon E.W. (1974) Phosopholipids and plant membrane permeability. *New Phytol.* **73**, 377−420.

Simpson G.C. (1929) Dew: does it rise or fall? *Nature* **124**, 578.

Sionit N., Hellmers H. & Strain B. (1980) Growth and yield of wheat under CO_2 enrichment and water stress conditions. *Crop Sci.* **20**, 687−690.

Slama K. & Williams C.M. (1965) Juvenile hormone activity for the bug *Pyrrhocoris apterus*. *Proc. Natl. Acad. Sci.* **54**, 411−414.

Slatyer R.O. (1962) Methodology of a water balance study conducted on a desert woodland (*Acacia aneura* F. Muell.) community in central Australia. *Arid Zone Res.* **16**, 15−26.

Smart N.A. (1968) Use and residues of mercury compounds in agriculture. *Residue reviews* **23**, 1−36.

Smirnoff N. & Crawford R.M.M. (1983) Variation in the structure and response to flooding of root aerenchyma in some wetland plants. *Ann. Bot.* **51**, 237−249.

Smirnoff N. & Stewart G.R. (1985) Stress metabolites and their role in coastal plants. *Vegetatio* **62**, 273−278.

Smith H. (1982) Light quality, photoperception and plant strategy. *Ann. Rev. Plant Physiol.* **33**, 481−518.

Smith H. (1984) Plants that track the sun. *Nature* **26**, 774.

Smith P. (1966) Effect of the plant alkaloid sparteine on the distribution of the aphid *Acyrthosiphon spartii* Koch. *Nature* **212**, 213−214.

Solbrig O.T. & Simpson B.B. (1974) Components of regulation of a population of dandelions in Michigan. *J. Ecol.* **62**, 473−486.

Sørensen T. (1941) Temperature relations and phenology of the Northeast Greenland flowering plants. *Meddr. Grønland.* **125**, 1−305.

Southwood T.R.E., Brown V.K. & Reader P.M. (1986) Leaf palatability and herbivore damage. *Oekologia* **70**, 544−548.

Sowell J.B. (1985) Winter water relations of trees at alpine timberlines. *Proc. 3rd IUFRO Workshop Eidg. Anst. forstl. Versuchswes. ber.* **270**, 71−77.

Sowell J.B. & Spomer G.C. (1986) Ecotype variation in root respiration rate among elevational populations of *Abies lasiocarpa* and *Picea engelmannii*. *Oecologia* **68**, 375−379.

Spence D.H.N. (1979) *Shetlands Living Landscape.* The Thule Press, Shetland.

Stace C.A. (1975) *Hybridization in the British Flora.* Academic Press, New York.

Stebbins G.L. (1971) *Chromosomal Evolution in Higher Plants.* Edward Arnold, London.

Steinmann F. & Braendle R. (1984a) Carbohydrate and protein metabolism in the rhizomes of the bullrush (*Schoenoplectus lacustris* L. Palla) in relation to natural development of the whole plant. *Aqua. Bot.* **19**, 53−63.

Steinmann F. & Braendle R. (1984b) Auswirkung von Halmverlusten auf der Kohlenhydratstoffwechsel ueberfluteter Seebinsenrhizome (*Schoenoplectus lacustris* L. Palla). *Flora* **175**, 295−299.

Stenseth N.C. (1978) Do grazers maximise individual plant fitness? *Oikos* **31**, 299−306.

Steponkus P.L. (1981) Responses to extreme temperatures. Cellular and sub-cellular bases. In *Encyclopedia of Plant Physiology*, Vol. 12A, pp. 371−402. Springer Verlag, Berlin.

Sternberg L.O., Deniro M.J. & Ting I.P. (1984) Carbon, hydrogen and oxygen isotope ratios of cellulose from plants having intermediary photosynthetic modes. *Plant Physiol.* **74**, 104−107.

Steudle E., Lüttge U. & Zimmerman U. (1975) Water relations of the epidermal bladder cells of the halophytic species *Mesembryanthemum crystallinum*: direct measurement of hydrostatic pressure and hydraulic conductivity. *Planta* **126**, 229−246.

Steudle E., Zimmerman U. & Lüttge U. (1977) Effect of turgor pressure and cell size on wall elasticity of plant cells. *Plant Physiol.* **59**, 285−289.

Stewart C.R. (1981) Proline accumulation: biochemical aspects. In *The Physiology and Biochemistry of Drought Resistance in Plants* (Eds. Paleg L.G. & Aspinall D.), pp. 243−259. Academic Press, New York.

Stewart C.R., Bogess S.F., Aspinall D. & Paleg L. (1977) Inhibition of proline oxidation by water stress. *Plant Physiol.* **59**, 930−932.

Stewart G.R. & Ahmad J. (1983) Adaptation to salinity in angiosperm halophytes. In *Phytochemistry Society Symposium of Metals and Biosystems*, pp. 33−50. Pitman Press, London.

Stewart G.R. & Lee J.A. (1974) The role of proline accumulation in halophytes. *Planta* **120**, 279−289.

Stewart W.S. & Bannister P. (1973) Seasonal changes in carbohydrate content of three *Vaccinium* species with particular reference to *V. uliginosum* and its distribution in the British Isles. *Flora* **162**, 134−155.

Stewart W.S. & Bannister P. (1974) Dark respiration rates in *Vaccinium* spp. in relation to altitude. *Flora* **163**, 134−155.

Stillmark H. (1888) Über Rizin, ein gifteges Ferment aus dem Samen von *Ricinus communis* L. und einigen anderen

Euphorbiacean. *Inaug. Dis. Dorpat.*

Storey R. & Wyn Jones R.G. (1975) Betaine and choline levels in plants and their relationship to NaCl stress. *Plant Sci. Lett.* **4**, 161.

Strain B.R. & Bazzaz F.A (1983) Terrestrial communities. In *CO₂ and Plants: The Response of Plants to Rising Levels of Carbon Dioxide* (Ed. Lemon E.), AAAS Selected Symposium **84**. Westview Press, Boulder, Colorado.

Stuiver M. (1978) Atmospheric carbon dioxide and carbon reservoir changes. *Science* **199**, 253−258.

Swain T. (1977) Secondary compounds as protective agents. *Ann. Rev. Plant Physiol.* **28**, 479−501.

Szarek S.S., Johnson H.B. & Ting I.P. (1973) Drought adaptation in *Opuntia basiliaris. Plant Physiol.* **52**, 539−541.

Tadano T. (1975) Devices of rice roots to tolerate high iron concentration in growth media. *Jpn. Agric. Res. Q.* **9**, 34−39.

Tal M. & Imber D. (1971) Abnormal stomatal behaviour and hormone imbalance in *Flacca* a wilty mutant of tomato. *Plant Physiol.* **46**, 373−376.

Taylor G.E. & Murdy W.H. (1975) Population differentiation of an annual species *Geranium carolinianum* L. in response to sulphur dioxide. *Bot. Gaz.* **136**, 362−368.

Taylorson R.B. & Borthwick H.A. (1969) Light filtration by foliar canopies: significance of light controlled weed seed germination. *Weed Sci.* **17**, 48−51.

Theophrastus *Enquiry into plants.* Trans. Sir Arthur Holt (1916).

Tieszen L.L., Miller P.C. & Oechel W.C. (1980) Photosynthesis. In *An Arctic Ecosystem* (Eds. Brown J., Miller P.C., Tieszen L.L. & Bunnell F.L.), pp. 335−410. Dowden, Hutchinson & Ross, Stroudsburg, Pennsylvania.

Ting I.P. & Rayder L. (1982) Regulation of C₃ to CAM shifts. In *Crassulacean Acid Metabolism* (Eds. Ting I.P & Gibbs M.) Proceedings 5th Annual Symposium, American Society of Plant Physiology, pp. 193−207.

Torres A.M. & Diedenhofen U. (1979) Baker sunflower populations revisited. *J. Hered.* **70**, 275−276.

Tranquillini W. (1979) *Physiological Ecology of the Alpine Timberline.* Springer Verlag, Berlin.

Tranquillini W. & Turner H. (1961) Untersuchungen über die Pflanzentemperaturen in der subalpinen Stufe mit besonderer Berücksichtigung der Nadeltemperaturen der Zirbe. *Mitt. Forstl. Bundesversuchanst. Mariabrunn* **59**, 127−151.

Treharne K.J. (1972) Biochemical limits to photosynthetic rates. In *Crop Processes in Controlled Environments* (Eds. Rees A.R., Cockshull K.E., Hand D.W. & Hurd R.C.), pp. 280−303. Academic Press, New York.

Treichel S.P., Kirst G.O. & von Willert D.J. (1974) Verüunderung der Aktivität des Phosphoenolpyruvat Carboxylase durch NaCl bei Halophyten verschiedener Biotope. *Z. Pflanzenphysiol.* **71**, 437−441.

Treshow M. (1970) *Environment and Plant Response.* McGraw-Hill, New York.

Tschermak L. (1950) *Waldbau und Pflanzengeographie Grundlage.* Springer Verlag, Vienna.

Turco R.D., Toon O.B., Ackerman T.P., Pollack J.B. & Sagan C. (1983) Nuclear winter: global consequences of multiple nuclear explosions. *Science* **222**, 1283−1300.

Turesson G. (1922) The genotypical responses of plant species to the habitat. *Hereditas* **3**, 211−350.

Turner R.G. & Marshall C. (1972) The accumulation of zinc by subcellular fractions of roots of *Agrostis tenuis* Sibth in relation to zinc tolerance. *New Phytol.* **71**, 671−676.

Turrill W.B. (1929) *The Plant-life of the Balkan Peninsula.* Clarendon Press, Oxford.

Turrill W.B. (1957). Germination of seeds. V. The vitality and longevity of seeds. *Gardn. Chron.* **142**, 37.

Tyler G. (1971) Hydrology and salinity of Baltic seashore meadows. *Oikos* **22**, 1−20.

Unsworth M. (1987) Adding ethylene to injury. *Nature* **327**, 364.

van Breemen N., Burrough P.A., Velthorst E.J., Dobben H.F., De Wit T., Ridder T.B. & Reijnders H.F.R. (1982) Soil acidification from atmospheric ammonium sulphate in forest canopy through fall. *Nature* **299**, 548−550.

van Emden H.H. (1972) Insects as phytochemists. In *Phytochemical Ecology* (Ed. Harborne J.B.), pp. 25−43. Academic Press, New York.

Vartapetian B.B., Snkhchian H.H. & Generozova I.P. (1987) Mitochondrial fine structure in imbibing seeds and seedlings of *Zea mays* L. under anoxia. In *Plant Life in Aquatic and Amphibious Habitats* (Ed. Crawford R.M.M.). British Ecological Society Special Publication No. 5, pp. 205−223. Blackwell Scientific Publications, Oxford.

Vasek F.C. (1980) Creosote bush: long-lived clones in the Mohave Desert. *Amer. J. Bot.* **67**, 246−255.

Vazquez-Yanes C. (1980) Light quality and seed germination in *Cercropia obtusifolia* and *Piper auritum* from a tropical rainforest in Mexico. *Phyton* **38**, 33−35.

Vermeulen A.J. (1979) The acidic precipitation phenomenon. In *Polluted Rain* (Eds. Toribara T.Y., Miller M.W. & Morrow P.E.), pp. 7−60. Plenum Press, New York.

Veron A., Lambert C.E., Isley A., Linet P. & Grousset F. (1987) Evidence of recent lead pollution in deep north-east Atlantic sediments. *Nature* **326**, 278−280.

Vieweg G.H. & Ziegler H. (1969) Zür Physiologie von *Myrothamnus flabellifolia. Ber. Dtsch. Bot. Ges.* **82**, 29−36.

Wagner E., Vollbrecht P., Janistyn B., Gross K. & Woerth J. (1987) Monoterpen-vermittelte Zerstörung des Photosyntheseapparates von Waldbäumen. *Allg. Forst Z.* **27−29**, 705−708.

Wainwright S.J. & Woolhouse H.W. (1978) Some physiological aspects of copper and zinc tolerance in *Agrostis stolonifera* Sibth. *J. exp. Bot.* **29**, 525–531.

Waisel Y. (1972) *Biology of Halophytes.* Academic Press, New York.

Walter H. (1939) Grassland, Savanna und Busch der ariden teile Afrikas in ihrer ökologischen Bedingheit. *Jahrb. Wiss. Bot.* **87**, 750–860.

Walter H. (1962) *Die Vegetation de Erde.* Gustav Fischer Verlag, Jena.

Walter H. (1971) *Ecology of Tropical and Sub-tropical Vegetation.* Van Nostrand, New York.

Walter H. (1973) *Vegetation of the Earth.* Springer Verlag, Berlin.

Walter H. (1981) Über Hochswerte der Produktion von natürlichen Pflanzenbestanden in NO Asien. *Vegetatio* **44**, 33–41.

Walter H. & Stadelman E. (1974) A new approach to the water relations of desert plants. In *Desert Biology.* II. (Ed. Brown G.W.). Academic Press, New York.

Walter H. & Straka H. (1970) *Arealkunde: Floristischhistorische Geobotanik.* Verlag Eugen Ulmer, Stuttgart.

Wardle P. (1968) Engelmann spruce (*Picea engelmannii*) at its upper limits on the front range Colorado. *Ecology* **49**, 483–495.

Wardle P. (1971) An explanation for alpine timberlines. *N. Z. J. Bot.* **9**, 371–402.

Wardle P. (1972) New Zealand timberlines. Tussock grassland and mountain land. *Inst. Rev.* **23**, 31–48.

Wardle P. (1974) Alpine timberlines. In *Arctic and Alpine Environment Research.* (Eds. Ives J.D. & Barry R.G.), pp. 371–402. Methuen & Co., Ltd, London.

Warming E. (1909) *Ecology of Plants. An Introduction to the Study of Plant Communities.* Clarendon Press, Oxford.

Webb S.J. (1965) *Bound Water in Biological Integrity.* Thomas Springford, Illinois.

Weber N.A. (1950) The role of lemmings at Point Barrow, Alaska. *Science* **111**, 552–553.

Wellburn A.R., Majernik O. & Wellburn F.A.M. (1972) Effects of SO_2 and NO_2 polluted air on the ultrastructure of chloroplasts. *Environ. Pollut.* **3**, 37–49.

Wellington P.S. & Durham V.M. (1961) Studies on the germination of cereals. III. The effect of the covering layer on the uptake of water by the embryos of the wheat grains. *Ann. Bot.* **25**, 185–196.

Went F.W. (1975) Water vapour absorption in *Prosopis.* In *Physiological Adaptations to the Environment* (Ed. Vernberg F.J.), pp. 67–75. Intext Educational Publications, New York.

Wesson G. & Wareing P.F. (1969) The role of light in the germination of naturally occurring populations of buried weed seeds. *J. exp. Bot.* **20**, 402–213.

White K.D. (1970) *Roman farming.* Thames & Hudson, London.

Whitehead F.H. (1973) The relationship between light intensity and reproductive capacity. In *Plant Response to Climatic Factors* (Ed. Slatyer R.O.), pp. 73–75. UNESCO, Paris.

Wild A. (1979) Physiologie der Photosynthese höherer Pflanzen. Die Anpassung an Lichtbedinungen. *Ber. Dtsch. Bot. Ges.* **92**, 341–364.

Williams W.T. & Barber D.A. (1961) The functional significance of aerenchyma in plants. *Symp. Soc. Exp. Biol.* **15**, 132–54.

Willis A.J. & Jeffries R.L. (1963) Investigation on the water relations of sand dune plants under natural conditions. In *The Water Relations of Plants* (Ed. Rutter A.J. & Whitehead F.H.), pp. 168–89. Blackwell Scientific Publications, Oxford.

Willis A.J., Folkes B.F., Hope-Simpson J.F. & Yemm E.W. (1959) Braunton Burrows: the dune system and its vegetation. I and II. *J. Ecol.* **47**, 1–24.

Wilson J.M. (1976) The mechanism of chill and drought-hardening of *Phaseolus vulgaris* leaves. *New Phytol.* **76**, 257–270.

Wilson J.M. (1978) Lipid changes in plants at temperatures inducing chill-hardiness. *Pestic. Sci.* **9**, 173–183.

Winter K. (1973) NaCl-induzierter Crassulaceensäurestoffwechsel bei einer weitern *Aizoacee: Carpobrotus edulis.* *Planta* **115**, 187–188.

Winter K. (1974) Evidence for the significance of Crassulacean acid metabolism as an adaptive mechanism to water stress. *Plant. Sci. Lett.* **3**, 279–281.

Winter K. & von Willert D.J. (1972) NaCl-induzierter Crassulacean-Saurestoffwechsel bei *Mesembryanthemum crystallinum.* *Z. Pflanzenphysiol.* **67**, 166–170.

Winter K., Lüttge U., Winter E. & Troughton J.H. (1978) Seasonal shift from C_3 photosynthesis to Crassulacean acid metabolism in *Mesembryanthemum crystallinum* growing in a natural environment. *Oecologia* **34**, 341–346.

Wolff E. & Peel D.A. (1985) The record of global pollution in polar snow and ice. *Nature* **313**, 535–540.

Woodell S.R.J., Mooney H.A. & Hill A.J. (1969) The behaviour of *Larrea divaricata* (creosote bush) in response to rainfall in California. *J. Ecol.* **75**, 37–44.

Woodmansee R.G., Dodd J.L., Bowman R.A., Clark F.E. & Dickinson C.E. (1978) Nitrogen budget of a shortgrass prairie ecosystem. *Oecologia* **34**, 363–376.

Woodwell G.M. & Bodkin D.B. (1970) Metabolism of terrestrial ecosystems by gas exchange techniques: the Brookhaven approach. In *Analysis of Temperate Forest Ecosystems* (Ed. Reichle D. E.), pp. 73–85. Springer Verlag, Berlin.

Woolhouse H.W. (1983) Toxicity and tolerance in the response of plants to metals. In *Encyclopedia of Plant Physiology,* Vol. 12C, pp. 245–300. Springer Verlag, Berlin.

Wright M. & Simon E.W. (1973) Chilling injury in cucumber

leaves. *J. exp. Bot.* **24**, 400−411.

Wright S. (1931) Evolution in Mendelian populations. *Genetics* **16**, 97−159.

Wyn Jones R.G. & Gorham J. (1983) Osmoregulation. In *Encyclopedia of Plant Physiology*, Vol. 12C, pp. 35−58. Springer Verlag, Berlin.

Wyn Jones R.G., Sutcliffe M. & Marshall C. (1971) Physiological and biochemical basis for heavy metal tolerance in clones of *Agrostis tenuis*. In *Recent Advances in Plant Nutrition* (Ed. Samish R.M.), Vol. II. Gordon and Breach Science Publishers Inc., New York.

Young D. (1969) *St Andrews*. Cassell, London.

Young S.F. & Hoffman N.E. (1984) Ethylene biosynthesis and its regulation in higher plants. *Ann. Rev. Plant Physiol.* **35**, 155−189.

Zeiger E., Field C. & Mooney H.A. (1981) Stomatal opening at dawn: Possible roles of the blue light response in nature. In *Plants and the Daylight Spectrum* (Ed. Smith H.), pp. 391−407. Academic Press, London.

Ziegler H. (1974) Chemische Anpassungen der Pflanzen an extreme standortbedingungen. *Biologisch Rundschau* **12**, 81−95.

Ziegler H. & Vieweg G.H. (1970) Poikilohydere Pteridophyta (Farngewächse). Poikilohydere Spermatophyta (Samenpflanzen). In *Die Hydration und Hydratur des Protoplasmas der Pflanzen und ihre ökologischen Bedeutung. Protoplasmalogia* Eds. Walter H. & Kreeb K.), Vol. 11, pp. 88−108. Springer Verlag, Berlin.

Ziegler H., Osmond C.B., Stichler W. & Trimborn P. (1976) Hydrogen isotope discrimination in higher plants: correlation with photosynthetic pathways and environment. *Planta* **128**, 85−92.

Ziegler I. & Libera W. (1975) The enhancement of CO_2 fixation in isolated chloroplasts by low sulfite concentrations and by ascorbate *Z. Naturforsch.* **30C**, 634−637.

Index